Undergraduate Topics in Computer Science

'Undergraduate Topics in Computer Science' (UTiCS) delivers high-quality instructional content for undergraduates studying in all areas of computing and information science. From core foundational and theoretical material to final-year topics and applications, UTiCS books take a fresh, concise, and modern approach and are ideal for self-study or for a one- or two-semester course. The texts are all authored by established experts in their fields, reviewed by an international advisory board, and contain numerous examples and problems, many of which include fully worked solutions.

The UTiCS concept relies on high-quality, concise books in softback format, and generally a maximum of 275–300 pages. For undergraduate textbooks that are likely to be longer, more expository, Springer continues to offer the highly regarded Texts in Computer Science series, to which we refer potential authors.

More information about this series at https://link.springer.com/bookseries/7592

Maurits Kaptein · Edwin van den Heuvel

Statistics for Data Scientists

An Introduction to Probability, Statistics,
and Data Analysis

 Springer

Maurits Kaptein
Department of Methodology and Statistics
Tilburg University
Tilburg, Noord-Brabant, The Netherlands

Data Analytics
Jheronimus Academy of Data Science
's-Hertogenbosch, Noord-Brabant
The Netherlands

Edwin van den Heuvel
Department of Mathematics and Computer Science
Eindhoven University of Technology
Eindhoven, Noord-Brabant
The Netherlands

Preventive Medicine & Epidemiology
Boston University School of Medicine
Boston, Massachusetts, USA

ISSN 1863-7310   ISSN 2197-1781 (electronic)
Undergraduate Topics in Computer Science
ISBN 978-3-030-10530-3   ISBN 978-3-030-10531-0 (eBook)
https://doi.org/10.1007/978-3-030-10531-0

This Springer imprint is published by the registered company Springer Nature Switzerland AG
The registered company address is: Gewerbestrasse 11, 6330 Cham, Switzerland

Preface

Data science is a rapidly growing field. While there is still a raging academic debate on the exact position of data science in between—or encapsulating—statistics, machine learning, and artificial intelligence, the number of data science programs offered at schools around the world is growing rapidly (Donoho 2017). It has to, since data scientists are among the most sought after people on the job market today and there is a strong need for people who are trained and well educated in this discipline. Whatever your view is on data science, it is hard to deny that an understanding of statistics is useful, if not necessary, for any successful data scientist. This book aims to provide an initial understanding of statistics meant for data scientists.

This book grew out of us teaching a bachelor level (undergraduate) course in statistics to students enrolled in the data science program at the Eindhoven University of Technology and the University of Tilburg, both in the Netherlands. The program was new and we believed that this course had to be developed from scratch. Both of us have extensive experience in teaching statistics to mathematics, social science, and computer science students, but we felt data science needed something else. Statistics for data scientists should focus on making sense out of data with a good understanding of analytical tools. It should be practical and hands on, as statistics is commonly thought of in the social sciences, but the data scientist should at least also know the basic mathematical theory underlying the procedures she or he carries out, although perhaps not to the level of a mathematician. The data scientist should also be versed in the latest statistical methods, not just those used in the field 30 years ago. Furthermore, we wanted to discuss tools that would make sense practically and not resort to theoretically convenient tools, as many introductory statistics books discuss. Finally, we expect the data scientist to use both analytical as well as computational tools to solve data problems. This is what this book offers.

For Whom is This Book for?

This book was designed to support a single semester course for undergraduate students of data science. We used the material in the first year of the data science program, placed in the curriculum just after basic calculus and a programming course. However, we have since used this material also for teaching other students: for computer science, artificial intelligence, and econometrics students. We have used this material during their undergraduate programs to introduce them to modern statistical methods and the relevant probability theory. For social science students (psychology, sociology), we have used this material even at the master's level; for students with a little mathematical background and no programming experience, the material in this book provides a challenging next step in their capacity for dealing with data. Next to these targeted students, we feel this book is relevant for anyone with limited knowledge of statistics, but some familiarity with basic mathematics and programming, who wants to not just gain procedural knowledge of statistical inference, but properly understand the basic principles and its modern applications.

What Makes This Book Different?

We are well aware that this is not the first nor the last undergraduate statistics text and we also present material that can be found in many other books on statistics. However, this book is different:

1. This book provides an accessible introduction to applied statistics for data scientists. The book uniquely combines hands-on exercises with mathematical theory and an introduction to probability theory.
2. This book uniquely contains modern statistical methods that are often deemed advanced material; we cover bootstrapping, Bayesian methods, equivalence testing, study designs with relevant association measures, and bivariate measures of association.
3. This book provides a unique introduction to sampling and uncertainty; where most textbooks either ignore the underlying mathematical theory or introduce solely the asymptotic theory, we introduce statistical inference in a natural way using finite samples and real data.
4. We try to avoid emphasizing oversimplified topics that are appealing for their mathematical ease but that are practically unrealistic (e.g., the z-test with known variance). The focus is on statistical principles that have direct practical relevance.

The balance between practical application, theoretical insight, and modern methods is unique.

Structure of the Book and Its Chapters

To get started, in Chap. 1, we immediately dive into the analysis of a real dataset. Using R, students become acquainted with basic statistics used to describe data and learn to visualize data. Immediately setting this book apart from many other applied statistics texts, and directly relevant to data science students, we also discuss methods of dealing with (extremely) large datasets. After this quick start, we take a step back to reflect on what we have been doing; in Chap. 2, we more thoroughly study how our data may have come about or should have been collected, and how it can be used to make inferences about a population. In this chapter, students learn about sampling methods, and they learn to evaluate the quality of different estimators (in terms of bias and mean squared error) based on finite populations. Note that this chapter has proven challenging for students, and in bachelor level courses, we have often skipped parts of the chapters relating to specific sampling plans or estimators.

Chapters 3 and 4 take us—seemingly—a step away from data and practical application. However, the finite population approach introduced in Chap. 2 that provides students with a strong intuition regarding sampling and estimation can only take us so far; to advance further, we need more theoretical tools to deal with uncertainty (or randomness). Chapter 3 introduces basic probability theory starting from the probabilities of events; students learn the probability axioms and learn how to calculate probabilities in simple settings. We also introduce 2×2 contingency tables and their common set of association measures. Chapter 4 takes us a step further; in this chapter, students are introduced to discrete and continuous random variables and their probability distributions.

We don't, however, want to linger on the theory too long; we want to cover only the parts we deem necessary for further application. In Chap. 5, we return to our data and examine how we can use our newly developed theory to advance beyond our understanding of sample characteristics as we did in Chap. 2. We discuss normal populations, the central limit theorem, confidence intervals, and maximum likelihood estimation.

In Chap. 6, we elaborately discuss dependencies between multiple random variables and statistical measures for quantifying these dependencies for numerical, categorical, and binary variables (e.g., correlation, agreement, and similarity). We make explicit the difference between population characteristics and estimators (i.e., calculations from data) of these characteristics. We also discuss confidence intervals for many of the dependency measures. We have included topics on how dependency can be constructed mathematically and how this is connected to measures of dependency. We kept these pieces separate to make this more complicated topic separate from what is normally discussed in introductory statistics monographs.

In Chap. 7, we take on what everyone expects in an (applied) statistics book: the topic of statistical testing. We discuss standard hypothesis testing, but in contrast to the many contemporary texts, we clearly relate hypothesis testing to estimation and discuss alternative ways (e.g., bootstrapping) to quantify estimation precision. Furthermore, we cover equivalence testing: a testing method that likely gains importance as datasets grow.

Finally, in Chap. 8, we introduce students to the Bayesian statistics; we detail both the alternative philosophy and its pros and cons. Here, we do not choose a side in the—in our view meaningless—Frequentist vs. Bayesians battle; rather, we emphasize relationships between the two camps.

Teaching Statistics to Data Scientists Using This Book

We have tried various approaches to teaching the material in this book and made changes to the content and organization based on our experiences. For several years, we have simply aligned the individual chapters of the book with the weeks that made up our teaching quartile: eight weeks of teaching of four lecture hours and four hours of practicals, one week for recaps, and a tenth week for a multiple choice exam. Often we would use the first week to make sure students had R installed, had sufficient mathematical background, and discuss the first chapter. The subsequent weeks would then match with the remaining seven chapters of the book. Often we would skip parts of the chapters that are perceived to be particularly challenging depending on the students' progress: especially parts of Chaps. 2, 5, 6, and 7 prove challenging. We also often prioritize the assignments for each chapter such that students know where to start and which assignments to possibly skip. The lectures would consist primarily (using an old-fashioned blackboard) of discussing the material per chapter, while students were encouraged to do the assignments, and during lab sessions, the assignments were discussed. The recap week of the course was usually devoted to review of topics that students wanted to see addressed again and for discussions. A discussion of a practice exam was often included in this week as well.

This relatively old style of teaching seemed to work well, but recently we have been experimenting with alternatives. One alternative has now been to allow students to work on the assignments in groups during lecture hours (with the possibility of asking questions to a teaching assistant), while prior to the lecture, students were encouraged to (a) study the material in the book and (b) watch a number of videos introducing the material in the different chapters. We have combined this setup with a number of discussion lectures in which prior to the lecture, students were able to post questions and discussion topics to an online forum, which were subsequently discussed in front of the class. This latter setup seemed to work well, but students seem to prefer the traditional style of teaching over the video lectures.

Whatever teaching style is preferred, we believe that the full content of our book is best taught in a semester to be able to go through all the content of the book. This would also help students obtain a deeper understanding of all the topics included in our book and have more time to get experienced in working with R.

Datasets Used Throughout This Book

In this book, we often use actual empirical data to illustrate the methods and techniques we are discussing. All the datasets used in this book, as well as all the relevant R code, can be found at http://www.nth-iteration.com/statistics-for-data-scientist.

Here, we provide a quick overview of the datasets used in this book:

- face-data.csv—The dataset face-data.csv contains a subset of the publicly available dataset face_clean_with_demo.tab which can be found at https://doi.org/10.7910/DVN/Q0LJVI. The original dataset describes the data collected for an experiment that is described in detail in the publication "Uncovering noisy social signals: Using optimization methods from experimental physics to study social phenomena" by Kaptein, van Emden, and Iannuzzi which itself can be found at https://doi.org/10.1371/journal.pone.0174182. In this paper, the authors examined the effects of changing a synthetically generated face along two different dimensions (the distance between the eyes, and the brow-nose-chin ratio). Next, they measured participants' subjective experience of attractiveness of the generated face, rated on a scale ranging from 1 (not attractive) to 100 (attractive). Each participant rated a single face which is drawn from the 100×100 faces that could be rendered based on varying the two dimensions. Examples of the faces involved are shown in Figure 1.2. We took a subset of the original data collected in the study that describes the responses and demographics of 3,628 participants. The resulting dataset contains the following seven variables:

1. Id—A unique identifier for each participant.
2. Dim1—The value of the first dimension of the face (distance between the eyes). As this value is randomly generated, it contains several decimal places. However, in the actual generation of the face, only 100 unique values could be rendered.
3. Dim2—The value of the second dimension of the face (brow-nose-chin ratio). Also for this variable, 100 unique values could be rendered. In total, the number of distinct faces was, thus, $100 \times 100 = 10,000$.
4. Rating—The attractiveness rating.
5. Gender—The gender of the participant.
6. Age—The age group of the participant.
7. Edu—The highest education completed by the participant.

We use this dataset extensively in Chap. 1 to illustrate the computation of several sample statistics. The 3,628 participants in this study were recruited on Amazon Mechanical Turk and participated in the study from their own homes. The set of participants contained in the current dataset was presented with a face generated using randomly (not uniform) selected values for dim1 and dim2; hence, each participant was confronted with a slightly different synthetic face. After seeing the face, participants could rate the attractiveness of the face by adjusting a slider.

- voting-demo.csv—The dataset voting-demo.csv is used in the assign-ments in a number of the chapters, starting from Chap. 1. This dataset contains synthetic data modeled based on German election data. The dataset contains data of 750 individuals and it contains the following variables:

1. ID—A unique identifier for each individual in the study.
2. Vote—Whether or not the individual voted.
3. Age—The age in years of the individual.
4. Church—A variable indicating whether or not the individual identifies as religious.
5. Choice—The political party the individual voted for or would have voted for had they cast their vote.
6. Educ—The education level of the individual on a five-point scale. Note that the labeling of this variable is inconsistent: an issue the student should find and resolve.
7. Agegr—A recoding of the age of the individual into two groups (young and old).

This dataset is used in Chaps. 1, 6, and 7.

- demographics-synthetic.csv—The dataset demographics-synthetic.csv contains simulated data and is used in the assignments in Chap. 1. It contains data regarding $n = 500$ units, describing the following variables:

1. Gender—The gender of the participant.
2. Age—The age of the participant.
3. Weight—The weight of the participant.
4. Height—The height of the participant.
5. Voting—The political party the participant voted for (numeric).

The code to generate this data is provided in the R script generate-data.R which is printed in full in Assignment 1.6.

- high-school.csv—The dataset high-school.csv describes a number of properties of over 50,000 first- and second-year high-school students in the Netherlands from more than 2,250 classes from almost 240 schools. The dataset was collected by the Central Bureau of Statistics (CBS) Netherlands to support the education of statistics at high schools in the Netherlands. The CBS used a questionnaire to collect all information. The dataset was used to help students get a better flavor of the field of statistics. The main question asked to the students was "Who is the average student?". We will use the dataset for different purposes throughout the book. The dataset contains different variables:

1. Number—A unique indicator of the student.
2. Gender—The gender of the student.
3. Age—The age of the student in years.

4. Class—The class level (1 or 2) of the student.
5. Height—The height of the student in centimeters.
6. Sport—The number of hours per week the student practices sport.
7. TV—The number of hours per week the student watches TV.
8. Computer—The number of hours per week the student uses the computer.
9. Topic—The favorite topic at high school of the student (e.g., mathematics, English, physical training, etc.).
10. Allowance—The allowance per week for the student.
11. Work—The amount of money per week earned by the student.
12. Breakfast—An indicator of whether the student ate breakfast on the day the information was collected.

This dataset is used in Chaps. 2, 5, 6, and 7.

- potatoes.csv—The dataset potatoes.csv contains six genetic profiles of different potatoes. Two readings of a Bintje potato, two genetically modified potatoes, one experimental potato, and one Maris Piper potato. Each profile contains more than 47,000 binary signals. They represent the presence or absence of an active gene. Comparing genetic profiles may be important for a good understanding of the biological characteristics. The data was provided by Wageningen University & Research.

1. X—A unique indicator of the gene.
2. Bintje1—The first reading of the genetic profile of a Bintje potato.
3. Bintje2—The second reading of the genetic profile of a Bintje potato.
4. Maris_Piper—A reading of the genetic profile of a Maris Piper potato.
5. Experimental—A reading of the genetic profile of an experimental potato.
6. GMO1—A reading of the genetic profile of a genetically modified potato.
7. GMO2—A reading of the genetic profile of a second genetically modified potato.

This dataset is used in Chaps. 6 and 7.

Assignments

Each chapter contains multiple assignments; we feel that much of the material can only be properly understood when applied. The assignments in each chapter do not introduce new results or material; rather, they are designed to allow the student to practice. The answer manual is available online at http://www.nth-iteration.com/statistics-for-data-scientist under the title "Statistics for

Data Scientist; Answers to the assignments" and was written by Florian Böing-Messing, Maurits Kaptein, and Edwin van den Heuvel.

Nijmegen/'s-Hertogenbosch, The Netherlands   Maurits Kaptein
Eindhoven, The Netherlands/Boston, USA   Edwin van den Heuvel
December 2021

Acknowledgements During the writing of this book, we have greatly benefited from students taking our course and providing feedback on earlier drafts of the book. We would like to explicitly mention the help of Florian Böing-Messing who, as a teaching assistant for the course, provided invaluable feedback regarding the structure of the book, caught numerous mistakes (the remaining ones are ours), and created a number of the assignments. We would also like to thank Abu Manju, Zhuozhao Zhan, Marta Regis, and Linda Sloot, who have all provided many valuable comments and helped improve the text. Furthermore, we are very grateful to the Central Bureau of Statistics Netherlands and the Wageningen University & Research for allowing to use their data. Clearly, we want to thank our family members Rosa, Rachel, Zoë, and Isabel for their support. Isabel, who followed our course as Data Science student, was one of the students who provided feedback. Finally, we want to thank Helen Desmond of Springer for her support and her patience with us in finishing the book.

Reference

D. Donoho, 50 years of data science. J. Comput. Graph. Stat. **26**(4), 745–766 (2017).

The original version of the book was revised: The frontmatter have been updated with changes. The correction to the book is available at https://doi.org/10.1007/978-3-030-10531-0_9

Contents

Notation and Code Conventions

Here, we clarify and illustrate the mathematical notation and coding conventions we use throughout the book. We try to do this as much as possible in order of appearance in the book, but sometimes we reorder things to highlight relationships between different concepts. We start with the mathematical notation:

- **Observations**: $x, x_i, \mathbf{x}_i, y, y_i, \mathbf{y}_i, \ldots$

  In this book, we often consider observed data: measurements obtained from individual units. By convention, we use lower-case Roman letters to denote observations. The reader will encounter these immediately in Chap. 1 and in most of the following chapters. We use the following notation:

  x It denotes a single scalar observation. Note that we also use x predominantly as arguments to mathematical functions ($f(x)$) when these are discussed in isolation.

  x_i We use the subscript i to index observations from units (see also below). Hence, x_1 refers to the scalar observation on variable x from unit 1.

  $\vec{x}_i$ When we have multiple observations on a single variable or observations on multiple variables from unit i, we use $\mathbf{x}_i, \mathbf{y}_i, \ldots$ explicitly, but we sometimes use vector notation $\vec{x}_i$. Here, $\vec{x}_i$ is a row vector containing all p observations of a single unit.

  $x_{(k)}$ In a limited number of places, we use bracketed subscripts for the kth order statistics. In this case, $x_{(1)}$ denotes the minimum of the n observations, and $x_{(n)}$ denotes the maximum.

  R^{x_i} We occasionally use R^{x_i} to denote the rank score of x_i. If $x_1 = 5, x_2 = 1$, $x_3 = 7, x_4 = 2$, and $x_5 = 4$, then $R^{x_1} = 4, R^{x_2} = 1, R^{x_3} = 5, R^{x_4} = 2$, and $R^{x_5} = 3$.

- **Indices, subscripts, and superscripts**: $x_i, \theta_k, \ldots$

  We have tried to use consistent indexing throughout the book. The following indices are used:

  i We use $i = 1, \ldots, i = n$ to index units in a sample. n is used for the size of the sample.

k Primarily in Chap. 2, we use $k = 1, \ldots, k = K$ to index possible samples originating from a population. Thus, $\hat{\mu}_k$ is the estimator (by virtue of the hat) used to estimate the population parameters μ (see below) based on sample k. Note that k is used in Chap. 3 to index the values of a discrete random variable. The index k is also used in Chaps. 4, 5, 7, and 8 to indicate that we may have multiple parameters.

Finally, note that we sometimes use subscripts h, i, j, or k to index different strata, probabilities, events, or order statistics; this should be clear from the context.

- **Statistics**: $\bar{x}, s^2, \ldots$
  In Chap. 1, we introduce *statistics*; these are simply functions of data. Where available, we use commonly used symbols for statistics that students will often find in statistical texts. In the book, we introduce the following descriptive statistics for numerical variables:

$\bar{x}$ The sample mean: $\bar{x} = \sum_{i=1}^{n} x_i / n$.

s^2 The sample variance: $s^2 = \frac{\sum_{i=1}^{n}(x_i - \bar{x})^2}{n-1}$. Note that by convention, this is the *unbiased* sample variance; we simply use MSD in Chap. 1 to refer to the biased version: $MSD = \frac{\sum_{i=1}^{n}(x_i - \bar{x})^2}{n}$.

s The sample standard deviation: $s = \sqrt{s^2}$. The notation VAR is also used in other chapters to indicate the variance of an estimator (see below).

g_1 The sample skewness: $g_1 = \frac{1}{n}\sum_{i=1}^{n}(x_i - \bar{x})^3 / s^3$. It is a measure of "asymmetry" of the data.

g_2 The sample kurtosis: $g_2 = \frac{1}{n}\sum_{i=1}^{n}(x_i - \bar{x})^4 / s^4$. It is a measure of how "peaked" the data is.

S_{XY} The sample covariance for variables x and y.

r_P The sample correlation—Pearson's correlation—between variables x and y, sometimes also denoted as r_{xy} or r_{XY}, is given by $r_P = \frac{\sum x_i y_i - n\bar{x}\bar{y}}{(n-1)s_x s_y}$, where s_x is the sample standard deviation of data collected on variable x.

r_S Another often used measure of association, next to the correlation detailed above, is Spearman's correlation; this is given by simply computing Pearson's correlation above on rank scores R^{x_i} and R^{y_i}.

r_P The third measure of correlation is Kendall's tau: $\frac{1}{n(n-1)}\sum_{i=1}^{n}\sum_{j=1}^{n}\text{sgn}(x_j - x_i)\text{sgn}(y_j - y_i)$, with $\text{sgn}()$ the sign function. Kendall's tau correlation coefficient is a measure of concordance.

We also introduce many statistics for quantifying the dependency between two categorical variables x and y (including two nominal, ordinal, and binary variables). For two categorical variables, the data are often summarized in a contingency table, where the frequencies N_{xy} for all the combinations of levels of x and y are reported. Based on these frequencies, many statistics can be defined. In Chap. 3, we introduce the risk difference, relative risk, and odds ratio for binary variables (which are evaluated again in Chap. 6). In Chap. 6, we introduce Cohen's Kappa κ_C statistic, Pearson's χ^2 statistic, Pearson's ϕ coefficient, Cramer's V coefficient, Goodman and Kruskal's γ statistic, Yule's Q coefficient of association, and several families of similarity measures S_θ, T_θ, and the $\mathscr{L}$-family S.

In Chap. 7, we introduce several test statistics for testing the hypothesis on population parameters. The following statistics are introduced:

$(\bar{y} - \mu_0)/(s/\sqrt{n})$

The one-sample t-test for testing the hypothesis that the population mean μ is equal to the known value μ_0, with $\bar{y}$ the sample average and s the sample standard deviation.

$(\bar{y}_1 - \bar{y}_2)/\left(s_p\sqrt{n_1^{-1} + n_2^{-1}}\right)$

The two samples t-test for testing the hypothesis that the means μ_1 and μ_2 of two populations are equal, with $\bar{y}_k$ the sample average for population k and $s_p^2 = [(n_1 - 1)s_1^2 + (n_2 - 1)s_2^2]/[n_1 + n_2 - 2]$ the pooled variance of the two sample variances of the two populations. There is also a version of the t-test where another variance than the pooled variance is used. We also discuss equivalence testing for testing the hypothesis that the difference in population means is not close or not equivalent.

s_1^2/s_2^2

The F-test for testing the hypothesis that the variances σ_1^2 and σ_2^2 from two independent populations are equal, with s_k the sample standard deviation of population k. A non-parametric version of the F-test, called Levene's test, is also introduced.

$\bar{d}/s_d$

The paired t-test for testing the null hypothesis that two means μ_1 and μ_2 from paired samples are equal, where $\bar{d}$ is the average of the difference in the observations of the pairs and s_d is the standard deviation of these differences.

U

The Mann-Whitney U test for testing the hypothesis that the values of one population are not stochastically larger than the values of another population (two independent samples).

$\sum_{i=1}^{n} 1_{(y_{1,i} > y_{2,i})}$

The sign test for paired samples: testing the null hypothesis that values from the first dimension are not stochastically larger than the values of the second dimension.

W^+

The Wilcoxon signed rank test for testing the hypothesis that differences from paired samples are on average equal for positive and negative differences.

$r_P\sqrt{n - 2}/\sqrt{1 - r_P^2}$

The correlation test statistic for testing the hypothesis of independence between two numerical variables x and y, with r_P Pearson's correlation coefficient. The test statistic also plays a role in quantifying

	the dependence between the continuous variables in Chap. 6.
χ_P^2	Pearson's chi-square test for testing the hypothesis of independence between two categorical variables x and y.
W	The Shapiro-Wilk test for testing the hypothesis that the population distribution $F(\cdot)$ is equal to the normal distribution $\Phi((\cdot - \mu)/\sigma)$.
G	The Grubbs test for testing the hypothesis that there is not an outlier in a single sample of data. We also discuss John Tukey's approach of identifying outliers, without doing a formal test. An approach that is implemented in the box-plot of a variable.

If we introduce a statistic that is not directly used as an estimator for a population parameter (see below) and that is less common (and hence does not have its own standard symbol), we simply use acronyms: for example, in Chap. 1, we use *MAD* for the mean absolute deviation. Note that when discussing general statistics, we often use T_n; we use F_{T_n} from Chap. 5 onwards to describe the distribution (CDF) of statistics over repeated random sampling (see also "Distributions" below).

- **Population parameters**: $\theta, \mu, \ldots$
  Parameters of a population are, by convention, referred to using lower-case Greek letters. For a general population parameter, we use θ, and for a vector of population parameters, we use $\boldsymbol{\theta}$. Next, we use the following standards for population parameters of numerical variables:

μ	We use μ to refer to the population mean. We also use μ_x or μ_X to denote the mean of variable X when needed.
σ^2	We use σ^2 to refer to the population variance.
σ	We use σ to refer to the population standard deviation.
σ_{xy}	We use σ_{xy} or σ_{XY} to refer to the population covariance between X and Y.
ρ	We use ρ, ρ_{xy}, or ρ_{XY} to refer to the population correlation between X and Y.

- **Estimators**: $\widehat{\mu}, \widehat{\theta}, \ldots$
  We often use sample statistics as an *estimator* for population parameters. Simply put, every estimator is a statistic, but only some statistics are useful estimators. When we use statistics as estimators, we identify them as such by placing a hat on top. Hence, $\widehat{\mu}$ is an estimator for μ (which is commonly the statistic $\bar{x}$ discussed above), and $\widehat{\boldsymbol{\theta}}$ is a vector estimator of general population parameters $\boldsymbol{\theta}$. If needed, we use $\widehat{\mu}_x$ to denote that we are concerned with variable x.

- **Events and their probability**: $\Pr(D), \Pr(D|E), \Pr(D \cap E), \ldots$
  Primarily in Chap. 3, but also later in the book, we are concerned with events. Throughout this chapter, we denote events using capital Roman letters A, B, C, etc. These should not be confused with the later use of capital roman letters for

random variables (see below) as used in Chap. 4. This, however, is convention, and we stick to it. We use the notation Pr for probability. We use the following:

$\Pr(A)$ to denote the probability of event A happening,

$\Pr(A^c)$ to denote the probability of the complement of A happening (or "not A"),

$\Pr(A|B)$ to denote the probability of event A happening given that B has happened,

$\Pr(A \cup B)$ to denote the probability that event A or B happens, and

$\Pr(A \cap B)$ to denote the probability that event A and B both happen.

Relationships between $\Pr(A)$, $\Pr(A|B)$ and $\Pr(A \cap B)$ are given in Chap. 3

- **Probability distributions and random variables**: Starting from Chap. 4, we move beyond the probabilities of individual events to discuss distributions and random variables. Here, we list our main conventions:

X We use capital Roman letters to denote random variables. When we talk about multiple random variables, we either use different letters (X, Y, Z) as we do in Chap. 6 or we use subscripts $X_{h,i}$. We use lower-case Roman letters (e.g., $x_{h,i}$) to denote the realization of a random variable.

$f(x)$ We use lower-case $f()$ to denote both probability mass functions—which we refer to as PMFs in the text—and probability density functions (PDFs) for discrete or continuous random variables, respectively. Note that the following properties hold:

Discrete: For discrete random variables, $f(x) = \Pr(X = x)$. We sometimes use the latter expression explicitly. Note that we also use the shorthand $p_k = \Pr(X = k)$ which allows us to easily enumerate all possible values of X which for discrete random variables are natural numbers. Using set notation, this is written as $x \in \mathbb{N}$.

Continuous: For continuous random variables, $\int_a^b f(x)dx = \Pr(a \le X \le b)$, where $x \in \mathbb{R}$.

$F(x)$ We use capital Roman $F()$ to denote cumulative distribution functions or CDFs. Here, $F(x) = \Pr(X \le x)$ which in the discrete case is given by $\sum_{k=0}^{x} \Pr(X = k) = \sum_{k=0}^{x} p_k$ and in the continuous case by $\int_{-\infty}^{x} f_X(t)dt$. Note here the use of the subscript x to denote the fact that f is the PDF of random variable X, and not of T. When the targeted distribution is clear—which is often the case when we focus on a single random variable—we omit the subscripts.

$f_{XY}(x, y)$ In Chap. 6, we explicitly discuss properties of multiple random variables. Here, the subscripts mentioned above become important. $f_{XY}(x, y)$ denotes the joint PMF or PDF of X and Y. This should not be confused with $f_{X|Y}(x, y)$, which is also a function of two variables (x and y), but now $f_{X|Y}$ refers to the PMF or PDF of X conditional on Y.

Further relationships between $f(x)$, $F(x)$, $f_{XY}(x, y)$ and $f_{X|Y}(x, y)$ are discussed in Chap. 6

- **Expectation, variance, covariance**: While technically operators (see below), we discuss separately our notation for expectations, variances, and covariances when we are doing analysis with random variables. Given random variables X and Y, we use the following:

  $\mathbb{E}(X)$ for the expectation of X which in the discrete case is given by $\sum_{k=0}^{\infty} k p_k$ (using the shorthand $p_k = \Pr(X = k)$ introduced above) and by $\int_{\mathbb{R}} x f(x) dx$ in the continuous case,

  $\mathsf{VAR}(X)$ for the variance of X which is given by $\mathbb{E}[(X - \mathbb{E}(X))^2]$, and

  $\mathsf{COV}(X, Y)$ for the covariance of X and Y which is given by $\mathbb{E}[(X - \mathbb{E}(X))(Y - \mathbb{E}(Y))]$.

- **Operators**: We expect familiarity with standard mathematical notation ($+$, $-$, $\times$, and, $f'(x)$, $\int f(x) dx$, etc.). However, we explicitly detail the following operators:

  $\sum_{i=1}^{n} x_i$ The summation operator: $\sum_{i=1}^{n} x_i = x_1 + x_2 + \cdots + x_n$.

  $\prod_{i=1}^{n} x_i$ The multiplication operator: $\prod_{i=1}^{n} x_i = x_1 \times x_2 \times \cdots \times x_n$.

Next to the mathematical symbols used in the text, we also present the R code. Code blocks, and their respective output, are presented as

```
> x<-c(1:10)
> x
[1]  1  2  3  4  5  6  7  8  9  10
```

while inline code is presented as `x <- c(1:10)`. Note that for the code blocks, we use > to start lines with code that is actually interpreted by R; lines that do not start with > are output. Lines starting with ># are comments and are not executed but are used to make the code more readable.

Finally, we assume some basic familiarity with the concepts of classes, functions, and objects. Since we are working with R, it is important for the reader to know that everything in R is an object. Hence, x defined in the example code above is an object. Objects are instances of a class, which itself can be thought of as the blueprint for an object. The object x above is a so-called instantiation of the "integer" class (you can see this by executing `class(x)`). Abstractly, an object is a data structure that has attributes and methods which can act on these attributes. Methods for the class integer include the standard mathematical operators * and +.

Chapter 1
A First Look at Data

1.1 Overview and Learning Goals

For data scientists, the most important use of statistics will be in making sense of data. Therefore, in this first chapter we immediately start by examining, describing, and visualizing data. We will use a dataset called `face-data.csv` throughout this chapter; this dataset, as well as all the other datasets we use throughout this book, is described in more detail in the preface. The dataset can be downloaded at http://www.nth-iteration.com/statistics-for-data-scientist.

In this first chapter we will discuss techniques that help visualize and describe available data. We will use and introduce R, a free and publicly available statistical software package that we will use to handle our calculations and graphics. You can download R at https://www.r-project.org.

In this first chapter we will cover several topics:

- We will learn how to open datasets and inspect them using R. Note that a more extensive overview of how to install R can be found in Additional Material I at the end of this chapter.[1]
- We will run through some useful basic R commands to get you started (although you should definitely experiment yourself!)
- We will explain the different types of variables and how they relate to different descriptive measures and plots for summarizing data.
- We will discuss basic methods of summarizing data using measures of frequency, central tendency (mean, mode, median), spread (mean absolute deviation, variance, standard deviation), and skewness and kurtosis.

---

[1] A number of the chapters in this book contain additional materials that are positioned directly after the assignments. These materials are not essential to understand the material, but they provide additional background.

© Springer Nature Switzerland AG 2022
M. Kaptein and E. van den Heuvel, *Statistics for Data Scientists*, Undergraduate Topics in Computer Science, https://doi.org/10.1007/978-3-030-10531-0_1

- We will discuss the basic plotting functionality of R to create line graphs, bar charts, scatter plots, box plots, and multi-panel density plots, and we will discuss how to interpret these plots.
- You will learn to reason about which plots are most informative for which variables. We will discuss how the measurement levels relate to the types of plots we could or should make.
- We will look briefly at using R for plotting mathematical functions.

1.2 Getting Started with R

R is a programming language: you can write code and have R execute that code. R is very well suited for analyzing data, and has many statistical operations build-in, but in the end it can be used to built all kinds of things. However, in this book we will mainly use it for analyzing data.

There are many different ways in which you can use R. One way is to use what is called the R console, which is shown in Fig. 1.1. The console comes with any default installation of R that you can find at https://www.r-project.org. You can use this console to type in R commands, and execute them line by line. The figure shows the execution of the line `print("hello world")` which prints the string "hello world" to the screen. Everything you do within a session (thus, without closing the console) will be remembered; if you close the console you will lose the work you have not saved explicitly.

The console is, however, not the easiest way of working with R. There are two often used alternative ways of using R:

1. Using a code editor: You can use any text/code editor, such as TextMate or Sublime text to write and store the R (analysis) code that you end up writing. Good code editors will allow you to run the code directly from the editor by sending it to the R console. If you have already programmed in some other language using a code editor that supports R this might be your best option.
2. Using a graphical user interface: You can also use a point and click solution such as RStudio. For downloads see https://www.rstudio.com. RStudio is preferred by many of our students, and hence we explain installing and using RStudio in more detail in the additional materials at the end of this Chapter.

RStudio is very popular these days, but this book is not tied to using RStudio. Find something you are comfortable with and get used to it. In the end, it's all a matter of preference.

1.2.1 Opening a Dataset: *face-data.csv*

We begin our studies by simply opening the dataset called `face-data.csv`, which contains the data we will be using in this first chapter in .csv format. The dataset contains data (or records) from $n = 3,628$ participants in a scientific study (Kaptein

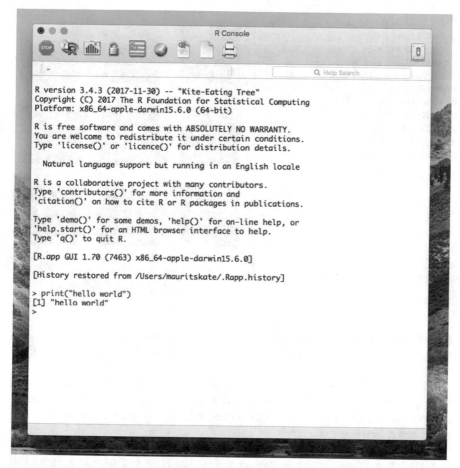

Fig. 1.1 A screenshot of the R console with the execution of the command `print("hello world")`

et al. 2016). In the study participants were asked to rate the attractiveness of a synthetically generated face. For each participant the generated face differed slightly along one of two dimensions: either the distance between the eyes, or the brow-nose-chin ratio was different. Figure 1.2 provides an example of nine synthetically generated faces.

After seeing a face, participants rated its attractiveness on a scale from 1 to 100. Next to the rating, the dataset also contains some demographic information about the participants (their age, gender, education level). For a full description of the dataset and the setup of the experiment, see the preface.

Opening the data in R can be done using some simple R commands:

```
> path <- "data-files/"
> file <- "face-data.csv"
> face_data <- read.csv(paste(path, file, sep=""),
    stringsAsFactors = TRUE)
```

Fig. 1.2 Nine examples of different faces rated by participants in the experiment differing along two dimensions. In the actual experiment faces were generated randomly with different values for the distance between the eyes and the brow-nose-chin ratio for each participant

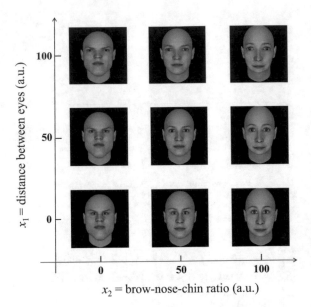

x_1 = distance between eyes (a.u.)

x_2 = brow-nose-chin ratio (a.u.)

Note that in the above code snippet the commands are preceded by a ">", all other printouts will be responses from R. We will stick to this convention throughout the book.

The core of the code above is the call to the *function* `read.csv` which allows us to open the datafile. In programming, a function is a named section of a program that performs a specific task. Functions often take one or more *arguments* as their input, and return some *value*.[2]

Since this is the first R code we cover in this book, we go through it line by line. However, whenever you encounter code in the book and you are unsure of its functionality, make sure to run the code yourself and inspect the output.

The first line of the code creates (or instantiates) a new variable called `path` with the value `data-files/`. The quotes surrounding the value indicate that the variable is of type *string* (a set of characters) as opposed to *numeric* (a number). We discuss this in more detail below. Similarly, the second line defines a string variable called `file` with the value `face-data.csv`. Jointly these indicate the folder of the datafile (relative to R's working directory which you can find by executing the function `getwd()`) and the name of the datafile. The working directory can be changed by running `setwd(dir)`, where `dir` is a character string specifying the desired working directory (e.g., `setwd("/Users/username")` on a Mac). Note that the path needs to be specified using the forward slash / as the separator.

The third line is more complex as it combines two function calls: first, inside the brackets, a call to `paste(path, file, sep="")`, and second, a call to

---

[2] In R you can always type `?function_name` to get to the help page of a function. You should replace `function_name` by the name of the function you want to see more information about.

`read.csv` to open the file. The call to `paste` combines the strings `path` and `file`. This can be seen by inspecting the output of this function call:

```
> paste(path, file, sep="")
[1] "data-files/face-data.csv"
```

Hence, in this code `paste` ensures that the path to the actual datafile, relative to R's current working directory, is supplied to the `read.csv` function. Note that the third argument supplied to the `paste` function, named `sep`, specifies how the two strings `path` and `file` should be combined.

The `read.csv` function opens the csv data and returns a so-called `data.frame` object containing the data in the file. We will discuss the `data.frame` object in more detail in Sect. 1.2.3. The function `read.csv` is one of many functions to read data into R; there are separate functions for opening different types of files (such as .sav[3] data files or .txt files). Thus, if you encounter other types of files make sure to find the right function to open them: in such cases, Google (or any other search engine of your choice) is your friend. Finally, note the `stringsAsFactors = TRUE` argument: this specifies that any variable containing strings will be considered a factor (see below). This option has been R's default for a long time, but since R version 4.0.0, we have to supply it manually.[4]

The function `read.csv` takes a number of additional arguments. We could for example make the following call:

```
> face_data <- read.csv("data-files/face-data.csv", sep=",",
    header=TRUE, dec=".", stringsAsFactors = TRUE)
```

where we make explicit that the values in the file are separated by comma's, that the first line contains variable names (headers), and that we use a point (".") to indicate the starting of decimals in numbers.

After running this code we have, in the current session, an *object* called `face_data` which contains the data as listed in the file in an R `data.frame`; the data frame is just one of the many ways in which R can store data, but it is the one we will be using primarily in this book. In this example, we have opened a ".csv" file, which contains data stored in rows and columns. The values are separated by commas (it is a *comma-separated value* file), and the different rows present data of different people. This is really nice and structured. However, during actual work as a data scientist you will encounter data of all sorts and guises. Likely, very often the data will not all be neatly coded like this, and it will not be at all easy to read data into R or any other software package. We will not worry about these issues too much in this book, but Crawley (2012) provides an excellent introduction to ways of opening, re-ordering, re-shuffling, and cleaning data.[5]

---

[3] .sav files are data files from the statistical package SPSS.

[4] As is true for many programming languages, R is continuously updated. For the interested reader here is a discussion regarding the change of the default `stringsAsFactors` argument: https://developer.r-project.org/Blog/public/2020/02/16/stringsasfactors/.

[5] Note that the `read.csv` function loads all the data into RAM; be aware that this might not be feasible for large datasets.

1.2.2   Some Useful Commands for Exploring a Dataset

So, now we have a `data.frame` object named `face_data` that contains our data.
We can inspect the contents of this object by simply typing the name of the object
into the R console:

```
> face_data
  id dim1     dim2     rating gender      age           edu
1 1 13.02427 13.54329 5      Male    25 to 34 years High school
2 2 24.16519 22.42226 59     Female  25 to 34 years Some college
3 3 19.71192 22.54675 5      Female  25 to 34 years 4 year college
4 4 16.33721 13.46684 10     Female  25 to 34 years 2 year college
5 5 26.67575 27.99893 25     Male    25 to 34 years Some college
6 6 12.02075 13.62148 75     Female  35 to 44 years Some college
### AND ON AND ON ####
```

This shows a very long list consisting of all of the records in our dataset. Clearly, if
you have data on millions of people this will quickly become an impenetrable mess.[6]
It is therefore usually much easier to inspect a dataset using the functions `head`,
`dim`, and `summary` which we will explain here.

The `head` function prints the first few lines of a data frame. If you do not specify
a number of lines, then it wil print 6 lines:

```
> head(face_data)
  id dim1     dim2     rating gender      age           edu
1 1 13.02427 13.54329 5      Male    25 to 34 years High school
2 2 24.16519 22.42226 59     Female  25 to 34 years Some college
3 3 19.71192 22.54675 5      Female  25 to 34 years 4 year college
4 4 16.33721 13.46684 10     Female  25 to 34 years 2 year college
5 5 26.67575 27.99893 25     Male    25 to 34 years Some college
6 6 12.02075 13.62148 75     Female  35 to 44 years Some college
```

You could also try out the function `tail(face_data,10L)` to get the last 10
lines.

The `summary` function provides a description of the data using a number of
summary statistics; we will cover these in detail in Scct. 1.4.

```
> summary(face_data)
id                    dim1              dim2
Min.   :   1.0     Min.   :11.12     Min.   :12.08
1st Qu.: 907.8     1st Qu.:25.55     1st Qu.:40.81
Median :1814.5     Median :35.53     Median :52.54
Mean   :1814.5     Mean   :35.97     Mean   :49.65
3rd Qu.:2721.2     3rd Qu.:46.74     3rd Qu.:60.61
Max.   :3628.0     Max.   :63.09     Max.   :70.28

rating             gender
Min.   :  1.00       :  14
1st Qu.: 41.00     Female:1832
```

---

[6] To prevent a messy output the R console will stop printing at some point, but still, the output will
be largely uninformative.

```
Median : 63.00      Male :1782
Mean   : 58.33
3rd Qu.: 78.00
Max.   :100.00

age                              edu
  : 14                 4 year college:1354
18 to 24 years : 617   Some college : 969
25 to 34 years :1554    2 year college: 395
35 to 44 years : 720   High school  : 394
45 to 54 years : 413   Masters degree: 384
55 to 64 years : 237   Doctoral     :  95
Age 65 or older:  73   (Other)      :  37
```

In the output above, note the difference between the way in which gender and rating are presented: for gender we see counts (or frequencies) of each possible value, while for rating we see a minimum value, a maximum value, and a number of other descriptive statistics. This is because R distinguishes numerical values such as rating from *strings* which contain text, such as gender in our case. We will look at this in more detail in Sect. 1.3.1 and we will see that there are apparently 14 cases for which no gender and age are observed.

Alternatively, gender could have also been stored as a number (for example using the values 0 and 1 for females and males respectively); if that were the case R would have reported its minimum and maximum values by default. If we want R to interpret numeric values not as numbers, but as distinct categories, we can force R to do so by using the as.factor function. The code below prints a summary of the variable rating, and subsequently prints a summary of the factor rating; the difference is pretty striking: in the first case R reports a number of so-called descriptive statistics which we will discuss in more detail in Sect. 1.4, while in the second case R prints how often each unique rating occurs in the dataset (the value 1 occurs 84 times apparently). More information on the type of variables can be found in Sect. 1.2.3.

```
> summary(face_data$rating)
   Min. 1st Qu.  Median    Mean 3rd Qu.    Max.
   1.00   41.00   63.00   58.33   78.00  100.00
> summary(as.factor(face_data$rating))
  1   2   3   4   5   6   7   8   9  10  11  12  13  14  15  16  17  18  19  20  21
 84  12   6  14  12   8  12  14  15  23  17  17  11  11  22  21  12  24  16  31   8
 22  23  24  25  26  27  28  29  30  31  32  33  34  35  36  37  38  39  40  41  42
 16   6  25  50  16  15  13  14  65  13  13  29  20  54  13  19  27  19  79  21  28
 43  44  45  46  47  48  49  50  51  52  53  54  55  56  57  58  59  60  61  62  63
 12  21  50  30  12  18  34 133  35  38  48  20  69  35  25  16  40 139  35  42  37
 64  65  66  67  68  69  70  71  72  73  74  75  76  77  78  79  80  81  82  83  84
 33  71  48  34  46  27 162  40  36  37  41 199  41  41  44  41 162  62  39  40  38
 85  86  87  88  89  90  91  92  93  94  95  96  97  98  99 100
122  46  25  29  34  89  16  25  16  11  39   7   8   3   1  41
```

Note that in the code-snippet above the "$" is used to address the variable rating in the dataset by its (column) name.

Finally, the function `dim` specifies the number of rows and columns in the
`data.frame`:

```
> dim(face_data)
[1] 3628    7
```

This shows that we have 3,628 rows of data and 7 columns. Often the rows are
individual units (but not always), and the columns are distinct variables.

1.2.3   Scalars, Vectors, Matrices, Data.frames, Objects

The `data.frame` is just one of the many *objects* that R supports.[7] We can easily
create other types of objects. For example, if we run:

```
> id <- 10
```

we create the object called `id`, which is a variable containing the value 10. The object
`id` lives outside or next to our dataset. Thus object `id` should not be confused with
the column `id` in our dataset. Just as with our dataset (the `face_data` object), we
can easily inspect our new object by just typing its name:

```
> id
[1] 10
```

To see the column `id` we should have used the R code

```
> face_data$id
```

indicating that the column `id` lives in the data frame `face_data`.

To gain some more understanding regarding R objects and their structure, we will
dig a bit deeper into the `face_data` object. The `face_data` object is of type
`data.frame`, which itself can be thought of as an extension of another type of
object called a `matrix`.[8] A matrix is a collection of numbers ordered by rows and
columns. To illustrate, the code below creates a matrix called M consisting of three
rows and three columns using the `matrix()` function. We populate this matrix
using the values $1, 2, \ldots, 9$ which we generate with the `c(1:9)` command.[9]

---

[7] In this book we do not provide a comprehensive overview of R; we provide what you need to know
to follow the book. A short introduction can be found in Ippel (2016), while for a more thorough
overview we recommend Crawley (2012).

[8] While it is convenient to think of a `data.frame` as a generalization of a matrix object, it
technically isn't. The `data.frame` is "a list of factors, vectors, and matrices with all of these
having the same length (equal number of rows in matrices). Additionally, a data frame also has
names attributes for labelling of variables and also row name attributes for the labelling of cases.".

[9] In R actually the command `1:9` would suffice to create the vector; however, we stick to using the
function `c()` explicitly when creating vectors.

```
> M <- matrix(c(1:9), nrow=3)
> M
     [,1] [,2] [,3]
[1,]   1    4    7
[2,]   2    5    8
[3,]   3    6    9
```

The `data.frame` object is similar to the `matrix` object, but it can contain different types of data (both numbers and strings), and it can contain column names: we often call these variable names.

We can access different elements of a matrix in multiple ways. This is called *indexing*. Here are some examples:

```
> M[2, 3]
[1] 8
> M[3, 1]
[1] 3
> M[1, ]
[1] 1 4 7
> M[, 2]
[1] 4 5 6
```

Both a row or a column of numbers is called a *vector*, and a single numerical entry of a vector is called a *scalar*. Hence, the object id that we defined above was a scalar (a single number) while the command c(1:9) generates a vector. Note that "under the hood" R always works using vectors, which explains the [1] in front of the value of id when we printed it: R is actually printing the first element of the vector id.

We can add more elements to the vector id and access them using their index:

```
> id
[1] 10
> id <- c(id, 11)
> id
[1] 10 11
> id[2]
[1] 11
```

In the code above, the function c() *concatenates* the arguments that are passed to this function into a single vector.

Just like a matrix, we can also index a data frame using row and column numbers.

```
> face_data[3, 5]
[1] Female
Levels:  Female Male
> face_data[3, ]
  id    dim1     dim2 rating gender        age              edu
3  3 19.71192 22.54675      5 Female 25 to 34 years 4 year college
```

If you use `face_data[1, ]` you would obtain the first row of data and not the variable names.

The data frame allows us to index and retrieve data in many different ways that the matrix does not allow. Here are some examples:

```
> face_data$gender[3]
[1] Female
Levels:  Female Male
> face_data$gender[1:5]
[1] Male   Female Female Female Male
Levels:  Female Male
> face_data$gender[face_data$rating>95]
 [1] Male   Male   Female Female Female Male Female Male Female Female Female
[12] Female Female Female Female Female Female Female Male Male Female Female
[23] Male   Female Male   Male   Female Female Male Female Female Male Female
[34] Female Female Female Female Female Male Female Male Male Male  Female
[45] Male   Male   Female Female Female Female Male Female Female Female Female
[56] Male   Male   Female Female Female
Levels:  Female Male
```

The first command returns the third value of the variable (column) called gender, and the second command returns the first five values of this variable. The R syntax in the last line selects from the dataset called face_data the gender variable, but only for those values in which the rating variable is larger than 95; thus, this shows the gender of all the participants that provided a very high rating.

1.3   Measurement Levels

We have already seen a difference between gender and rating in their "type" of data. A bit more formally, often four different measurement levels are distinguished (Norman 2010):

1. **Nominal:** Nominal data makes a distinction between groups (or sometimes even individuals), but there is no logical order to these groups. Voting is an example: there is a difference between those who vote, for instance, Democrat, or Republican, but it's hard to say which is better or worse, or how much better or worse. Nominal data is often encoded as a factor in R. When a variable is encoded as a factor in R, which can be forced by using the as.factor() function, the values that the variable can take are stored separately by R and are referred to as the different *levels* of the factor.
2. **Ordinal:** Ordinal data also distinguishes groups or individuals, but now imposes an order. An example is the medals won at a sports event: Gold is better than Silver, but it's unclear how much better.
3. **Interval:** The interval scale distinguishes groups (or actually, often individuals), imposes an order, and provides a magnitude of the differences in some unit. For example, we can say that the Gold winner has a score of 0 s, the Silver winner 10 s (being 10 s slower), and the Bronze winner 12 s.
4. **Ratio:** This contains all of the above, but now also imposes a clear reference point or "0". The interval scale level does not really allow one to say whether the Gold winner was "twice as fast" as the Silver winner; we know she was 10 s faster, but we don't know how long the total race took. If we measure the speed from the

start of the race, we have a fixed "0", and we can meaningfully state things like: "the winner was twice as fast as the Bronze medalist."

Note that each consecutive measurement level contains as much "information"—in a fairly loose sense of the word—as the previous one *and more*. As a consequence of this, note that if you have (e.g.) ratio data, you could summarize it into nominal data, but not the other way around. We can see this in our dataset for the variable `age`: while `age` could have been recorded as a ratio variable (in years), our dataset contains `age` as an ordinal variable: it only specifies the age group a specific person belongs to. Operations that are meaningful on ratio data (such as addition and multiplication) are often nonsensical on nominal or ordinal data.

Nominal and ordinal data are often called *categorical data*, while interval and ratio data are referred to as *numerical data*. We also make a distinction between *continuous* and *discrete* numerical data. Theoretically, continuous variables can assume any value. This means that the continuous variable can attain any value between two different values, no matter how close the two values are. For discrete variables this would be untrue.[10] Examples of continuous variables are temperature, weight, and age, while discrete data is often related to counts, like the number of text messages, accidents, microorganisms, students, etc. We will return to measurement levels later in this chapter when we describe which summaries and visualizations are well-suited for measurements of a specific level.

1.3.1 Outliers and Unrealistic Values

If you paid close attention before, you might have noticed that for 14 units the value of `gender` in the dataset we opened was empty. Thus, the factor gender contains three levels: "Female", "Male", and "". It is unclear what this last, empty, level tries to encode. You will find this a lot in real datasets: often a datafile contains entries that are clearly erroneous, non-sensical, or otherwise unclear. Quite a big part of data science is actually knowing what to do with such data. For now, however, we will just recognize these units and remove them from our analysis. Again, the book by Crawley (2012) provides more pointers regarding these issues.

The rows containing the level "" can be removed using the following commands:

```
> face_data <- face_data[face_data$gender == "Female" | face_data
    $gender == "Male", ]
> nrow(face_data)
[1] 3614
```

The code above deletes all the rows for which the value of gender is empty by selecting all the rows that contain the value Female (`=="Female"`) or (`|`) Male (`== "Male"`) and subsequently prints the number of rows in the dataset (which

---

[10] In practice, continuous data does not exist, since we record data always with a finite number of digits and hence the property that there would be a value in between any two values is lost. Thus data is essentially always discrete.

is now 3,614 as opposed to 3,628). Note that just deleting the rows that contain the empty values does not force R to delete the empty value as a possible level of the factor gender; we can achieve this by dropping this level explicitly:

```
> summary(face_data$gender)
       Female   Male
    0    1832   1782
> face_data$gender <- droplevels(face_data$gender)
> summary(face_data$gender)
Female   Male
  1832   1782
```

where in the first command it is clear from the summary that 0 rows have the value "", while in the last command the summary shows that the level "" is no longer considered a possible value of the factor gender.

We will make sure that we also drop the superfluous levels of the variable age by running

```
> face_data$age <- droplevels(face_data$age)
```

Besides unclear values we also often encounter very extreme values; for example, we might find that the age of a participant in a study is 622. Such extreme values are often called "outliers" (a more formal definition exists, but we won't get into that now; we will, however, discuss some statistical methods of finding outliers in Chap. 7). There are essentially two types of outliers: implausible and plausible outliers. An age of 622 years old would be implausible (so far), but a weight of a person of 500 kg, would be very extreme, but not impossible. How to properly deal with outliers is a topic in its own right, but by and large the options are (a) ignoring them, (b) removing them or (c) trying to substitute them, using statistical methods, with a more plausible alternative.

Removing implausible values is common practice, but removing plausible outliers should never be common practice. Similarly, many datasets contain "missing data"; for example when someone refused to fill out their gender; this is likely what happened in the face_data dataset. Missing values are often either ignored (the units removed as we did above), or they are filled in using statistical prediction models: this filling-in is called *imputation*.[11] Researchers also like to use special values to indicate missing data, like 88, 888, 99, and 999. These values should be used only when they represent implausible values for the variable. Thus 99 for a missing age is not recommended, while 999 would be proper value for a missing age.

---

[11] Data imputation is a field in its own right (see, e.g., Vidotto et al. 2015); we will not discuss it in this topic further in this book. However a very decent introduction is provided by Baguley and Andrews (2016).

1.4 Describing Data

We now turn to *describing* the data that we are looking at. We have seen some examples already, for example when we were using the `summary` function, but here we will discuss the R output and a number of popular data descriptions in a bit more detail. Note that some *descriptive statistics* (or just *descriptives*) that we introduce are often used for data of a certain measurement level; we will indicate the most appropriate measurement levels for each descriptive statistic discussed below.

1.4.1 Frequency

Nominal and ordinal data are often described using frequency tables; let's do this for the variable `age`:

```
> t1 <- table(face_data$age)
> t2 <- transform(t1, cumulative = cumsum(Freq), relative = prop.
    table(Freq))
> t2
            Var1 Freq cumulative relative
1  18 to 24 years 617        617 0.17072496
2  25 to 34 years 1554      2171 0.42999447
3  35 to 44 years 720       2891 0.19922524
4  45 to 54 years 413       3304 0.11427781
5  55 to 64 years 237       3541 0.06557831
6 Age 65 or older 73        3614 0.02019923
```

The code above presents the frequency of occurrence of each value of the variable, the so-called *cumulative* frequency, and the *relative* frequency.[12] The frequency is just the number of times a value occurs in the dataset. We may generally denote the values by a set of numbers $\{x_1, x_2, \ldots, x_m\}$, with m the number of levels, and the frequency for x_j given by f_j, $j = 1, 2, \ldots, m$. In the example above this means we may use

$$x_1 = 1 \text{ for 18 to 24 years,}$$
$$x_2 = 2 \text{ for 25 to 34 years,}$$
$$\ldots,$$
$$x_7 = 7 \text{ for 65 years and older,}$$

which have frequencies

---

[12] The cumulative frequency makes more sense for ordinal data than for nominal data, since ordinal data can be ordered in size, which is not possible for nominal data.

$$f_1 = 617,$$
$$f_2 = 1554,$$
$$\dots,$$
$$f_7 = 73.$$

Using this notation the cumulative frequency for a specific value x_j is given by $\sum_{k=1}^{j} f_k$, the relative frequency is given by $f_j / \sum_{k=1}^{m} f_k$, and the *cumulative relative frequency* is given by $\sum_{k=1}^{j} f_k / \sum_{k=1}^{m} f_k$.[13]

The code for computing the frequencies requires some explanation: in the first line we create a new object called `t1` by using the `table` function; this line creates a table that contains only the frequency counts but not yet the cumulative sums and relative frequencies. This is done in the second line using the `transform` function: this function creates a new table, called `t2`, which transforms `t1` into a new table with additional columns.

Frequencies are often uninformative for interval or ratio variables: if there are lots and lots of different possible values, all of them will have a count of just one. This is often tackled by discretizing (or "binning") the variable (which, note, effectively "throws away" some of the information in the data). Here is the code to discretize the variable `rating` into five bins and create a frequency table:

```
> bins <- 5
> rating_binned <- factor(cut(face_data$rating, breaks=bins))
> t3 <- table(rating_binned)
> t4 <- transform(t3, cumulative = cumsum(Freq), relative = prop.
    table(Freq))
> t4
  rating_binned Freq cumulative relative
1  (0.901,20.8]  381        381 0.1054234
2   (20.8,40.6]  513        894 0.1419480
3   (40.6,60.4]  821       1715 0.2271721
4   (60.4,80.2] 1214       2929 0.3359159
5    (80.2,100]  685       3614 0.1895407
```

Here the `cut()` function is used to effectively cut up the continuous rating into five distinct categories. Note that the interval length is 19.8, except for the first interval: here the default choice made by R is a bit mysterious.

1.4.2 Central Tendency

When we work with numerical data, we often want to know something about the "central value" or "middle value" of the variable, also referred to as the *location* of the data. Here we have several measures that are often used:

---

[13] Not all readers will be familiar with summation notation; let $x_1, x_2, \dots, x_j$ be a set of numbers, than $\sum_{k=1}^{j} x_k = x_1 + x_2 + \dots + x_j$.

1. **Arithmetic mean:** The arithmetic mean of a set of numbers, which is often denoted $\bar{x}$ when we are referring to the sample mean of a variable x, is given by:

$$\bar{x} = \frac{1}{n} \sum_{i=1}^{n} x_i \tag{1.1}$$

where n is the total number of observed units, and x_i the score on variable x by unit i. Note that all data points weigh equally in computing the mean, and that it is affected quite a bit by extreme values or outliers. In R we can use the build-in R function mean() to compute the mean of a variable.

2. **Mode:** The mode is merely the most frequently occurring value. And yes, there might be multiple modes. R does not have a built in function to compute it, so let's write our own:

```
> get_mode <- function(v){
+    uniqv <- unique(v)
+    uniqv[which.max(tabulate(match(v, uniqv)))]
+    }
> get_mode(face_data$rating)
[1] 75
```

The code above introduces a number of new concepts. Here, in the first line, we create a new function called getmode, which takes the argument v. In the second line the function unique is used to generate a vector containing all the unique elements of the vector v. The next line, which is a bit involved, creates a table of the counts of each of the unique elements of v, and subsequently selects the highest value with the highest count.[14] Thus, the function eventually returns the value of the mode, not how often that specific value occurred.[15]

3. **Median:** The median is a value that divides the ordered data from small to large (or large to small) into two equal parts: 50% of the data is below the median and 50% is above. The median is not necessarily a value that is present in the data. Practically, we sort the data and choose the middle-most value when n is odd, or the average of the two middle values when n is even. Hence, the median of the data $2, 5, 6, 4$ (which, when ordered is $2, 4, 5, 6$) is 4.5. In R you can use the function median().

4. **Quartiles, deciles, percentiles, quantiles:** Instead of using 50% for the median, we may use any cut-off value. The function quantile() provides as standard (its default) the cut-off values 0%, 25%, 50%, 75%, and 100%.

```
> quantile(face_data$rating)
  0%  25%  50%  75% 100%
   1   41   63   78  100
```

---

[14] It is good practice to execute parts of a complicated line of code like this separately so that you understand each command (in this case match, tabulate, and which.max).

[15] Finally, note that the smallest most frequently occurring value is reported; it is an interesting exercise to change this function such that it returns all the modes if multiple modes exist.

A cut-off of 0% is used to indicate the minimum. Thus 1 is the smallest rating in the dataset. The cut-off 100% is used for the maximum rating. The rating 41 years belongs to a cut-off of 25%, which means that 25% of the participants contained in the data provided a rating below or equal to 41.

More theoretically, a *quantile* x_q is a value that splits the ordered data of a variable x into two parts: $q \cdot 100\%$ of the data is below the value x_q and $(1 - q) \cdot 100\%$ of the data is above. The parameter q can take any value in the interval $[0, 1]$. When $q = 0.25$, $q = 0.50$, and $q = 0.75$ the quantiles are referred to as the first, second,[16] and third *quartiles*, respectively. We call quantiles *deciles* when q is restricted to the set $\{0.1, 0.2, \ldots, 0.9\}$ and percentiles when the q is restricted to $\{0.01, 0.02, 0.03, \ldots, 0.99\}$. Thus quartiles and deciles are also percentiles. In R we can compute different quantiles by passing the desired value q to the quantile function as a second argument:

```
> quantile(face_data$rating, c(.2))
20%
 35
```

Quantiles can be calculated in different ways, depending on the way we "interpolate" between two values. To illustrate this let us calculate the first quartile of the data $\{2, 5, 6, 4\}$ that we used to illustrate the calculation of the median (which was 4.5). If we order the data ($\{2, 4, 5, 6\}$), the first quartile can be seen as the median of the data $\{2, 4\}$ on the left of the median, which would give a value of 3. Alternatively and more generically, we could map the ordered values equally spaced on the interval $(0, 1)$, where the ith ordered value of the data is positioned at the level $q = i/(n + 1)$ in the interval $(0, 1)$, with n being the number of data points. Thus in our example we would put the value 2 at the level $q = 1/5$, the value 4 at $q = 2/5$, the value 5 at $q = 3/5$, and the value 6 at $q = 4/5$ (see Fig. 1.3). Now the first quartile is the value that should be positioned at level $q = 0.25$. In our example, the value 3 would be located at 0.3, since it is exactly in the middle of the levels 0.2 and 0.4 and 3 is exactly in the middle of the values 2 and 4. Thus the value 2.5 would be positioned at the level $q = 0.25$. The first quartile is 2.5 instead of 3. Note that the median of 4.5 is precisely in the middle of 4 and 5 which are positioned at level 0.4 and 0.6, respectively.

This procedure is available in many software packages, often as default setting, but R uses a different default. Instead of putting the data $\{2, 4, 5, 6\}$ at the quantiles $\{0.2, 0.4, 0.6, 0.8\}$ as seen in Fig. 1.3, the function quantile() in R uses the quantiles $(i - 1)/(n - 1)$ for the ordered data. Thus the data $\{2, 4, 5, 6\}$ is positioned at quantiles $\{0, 1/3, 2/3, 1\}$ and then an interpolation is applied. The following R code shows the results:

```
> x <- c(2, 4, 5, 6)
> quantile(x)
  0%  25%  50%  75% 100%
2.00 3.50 4.50 5.25 6.00
```

---

[16] Yes, the second quartile is equal to the median.

Fig. 1.3 Mapping the ordered data to the values $i/(n+1)$, with i the ordered number of the value

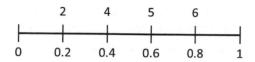

With the option `type=6` in the function `quantile()`, we obtain the results described earlier, i.e.

```
> quantile(x, type=6)
  0%  25%  50%  75% 100%
2.00 2.50 4.50 5.75 6.00
```

Note that quantiles are very useful to get some idea of the so-called "distribution" of a variable.

1.4.3 Dispersion, Skewness, and Kurtosis

Knowing the "central tendency" or locations of data points might not be sufficient to really understand the data that you are looking at. Dispersion measures help us understand how far apart the data are away from the center or from each other, while skewness and kurtosis are measures that describe features of the shape of the frequency plot of the data.

1. **Range and interquartile range:** Range is the difference between the maximum and minimum. It quantifies the maximum distance between any two data points. The range is easy to calculate in R:

```
> max(face_data$rating) - min(face_data$rating)
[1] 99
```

Clearly, the range is sensitive to outliers. Instead of using the minimum and maximum, we could use the difference between two quantiles to circumvent the problem of outliers. The interquartile range (IQR) calculates the difference between the third quartile and the first quartile. It quantifies a range for which 50% of the data falls within.

```
> quantile(face_data$rating, c(0.75)) - quantile(face_data$
    rating, c(0.25))
 37
```

Thus 50% of the `rating` data lies within a range of 37. The interquartile range is visualized in the boxplot, which we discuss later in this chapter.

2. **Mean absolute deviation:** We can also compute the average distance that data values are away from the mean:

$$MAD = \frac{1}{n} \sum_{i=1}^{n} |x_i - \bar{x}| \tag{1.2}$$

where $|\cdot|$ denotes the absolute value. R does not have a built-in function for this, but *MAD* can easily be computed in R:

```
> sum(abs(x-mean(x)))/length(x)
```

3. **Mean squared deviation, variance, and standard deviation:** Much more common than the mean absolute difference is the mean squared deviation about the mean:

$$MSD = \frac{1}{n} \sum_{i=1}^{n} (x_i - \bar{x})^2 \qquad (1.3)$$

It does the same as *MAD*, but now it uses squared distances with respect to the mean. The **variance** is almost identical to the mean squared deviation, since it is given by $s^2 = \sum_{i=1}^{n}(x_i - \bar{x})^2/(n-1) = n \cdot MSD/(n-1)$. For small sample sizes the *MSD* and variance are not the same, but for large sample sizes they are obviously very similar. The variance is often preferred over the *MSD* for reasons that we will explain in more detail in Chap. 2 when we talk about the bias of an estimator. The sample **standard deviation** is $s = \sqrt{s^2}$. The standard deviation is on the same scale as the original variable, instead of a squared scale for the variance.

4. **Skewness and kurtosis:** Both measures, skewness and kurtosis, are computed using so-called *standardized* values $z_i = (x_i - \bar{x})/s$ of x_i which are also called z-values. Standardized values have no unit and the mean and variance of the standardized values are equal to 0 and 1, respectively.[17]
Skewness is used to measure the asymmetry in data and kurtosis is used to measure the "peakedness" of data. Data is considered skewed or asymmetric when the variation on one side of the middle of the data is larger than the variation on the other side. The most commonly used measure for skewness is

$$g_1 = \frac{1}{n} \sum_{i=1}^{n} \left(\frac{x_i - \bar{x}}{s} \right)^3 \qquad (1.4)$$

When g_1 is positive, the data is called skewed to the right. The values on the right side of the mean are further away from each other than the values on the left side of the mean. In other words, the "tail" on the right is longer than the "tail" on the left. For negative values of g_1 the data is called skewed to the left and the tail on the left is longer than the tail on the right. When g_1 is zero, the data is considered symmetric around its mean. In practice, researchers sometimes compare the mean with the median to get an impression of the skewness, since the median and mean are identical under symmetric data, but this measure is more difficult to interpret than g_1. Data with skewness values of $|g_1| \leq 0.3$ are considered close to symmetry,

---

[17] Note that the average of the standardized values is equal to $\frac{1}{n} \sum_{i=1}^{n} z_i = \frac{1}{n} \sum_{i=1}^{n} x_i/s - \bar{x}/s = 0$ and that the sample variance is equal to $\frac{1}{n-1} \sum_{i=1}^{n} (z_i - 0)^2 = \frac{1}{n-1} \sum_{i=1}^{n} (x_i - \bar{x})^2/s^2 = s^2/s^2 = 1$.

since it is difficult to demonstrate that data is skewed when the value for g_1 is close to zero.

The most commonly used measure for kurtosis is

$$g_2 = \frac{1}{n} \sum_{i=1}^{n} \left(\frac{x_i - \bar{x}}{s} \right)^4 - 3 \tag{1.5}$$

When g_2 is positive, the data is called *leptokurtic* and the data has long heavy tails and is severely peaked in the middle, while a negative value for g_2 is referred to as *platykurtic* and the tails of the data are shorter with a flat peak in the middle. When g_2 is zero the data is called *mesokurtic*. Similar to g_1, it is difficult to demonstrate that data is different from mesokurtic data when g_2 is close to zero, since it requires large sample sizes. Values of g_2 in the asymmetric interval of $[-0.5, 1.5]$ indicate near-mesokurtic data.

Note that the measures g_1 and g_2 are unchanged when all values $x_1, x_2, \ldots, x_n$ are shifted by a fixed number or when they are multiplied with a fixed number. This means that shifting the data and/or multiplying the data with a fixed number does not change the "shape" of the data.

1.4.4 A Note on Aggregated Data

In practice we might sometimes encounter aggregated data: i.e., data that you receive are already summarized. For instance, income data is often collected in intervals or groups: $[0, 20, 000)$ euro, $[20, 000, 40, 000)$ euro, $[40, 000, 60, 000)$ euro, etc., with a frequency f_j for each group j. In the dataset face_data age was recorded in seven different age groups. Measures of central tendency and spread can then still be computed (approximately) based on such grouped data. For each group j we need to determine or set the value x_j as a value that belongs to the group, before we can compute these measures. For the example of age in the dataset face_data, the middle value in each interval may be used, e.g., $x_1 = 21.5$, $x_2 = 30$, etc. For the age group "65 years and older", such a midpoint is more difficult to set, but 70 years may be a reasonable choice (assuming that we did not obtain (many) people older than 75 years old). The mean and variance for grouped data are then calculated by

$$\bar{x} = \frac{\sum_{k=1}^{m} x_k f_k}{\sum_{k=1}^{m} f_k}, \qquad s^2 = \frac{\sum_{k=1}^{m} (x_k - \bar{x})^2 f_k}{\sum_{k=1}^{m} f_k - 1} \tag{1.6}$$

with m the number of groups. Similarly, many of the other descriptive statistics that we mentioned above can also be computed using aggregated data. The average age and the standard deviation in age for the dataset face_data, using the aggregated data and the selected midpoints, are equal to 35.6 and 11.75 years, respectively.

1.5   Visualizing Data

Next to inspecting data by computing summaries, we often visualize our data. Visualization, when done well, can make large and even high-dimensional datasets (relatively) easy to interpret. As with many topics we introduce in this book, visualization is a topic in its own right. We refer the interested reader to Rahlf (2017) or Young and Wessnitzer (2016); here we merely provide some basics to get started making simple data visualizations with R.

Since we will be making plots, the easiest thing to do is to just call the function plot on the data object we have and see what happens:

```
> plot(face_data)
```

This code produces Fig. 1.4. Note that variable id is a variable that represents the order of participants entering the study.

Admittedly, blindly calling plot is not very esthetically appealing nor is it extremely informative. For example, we can see that gender only has two levels: this is seen in the fifth row of panels, where gender is on the y-axis and each dot—each of which represents an observation in the dataset—has a value of either 1 or 2. However, the panels displaying gender do not really help us understand the differences between males and females. On the other hand, you can actually see a few things in the other panels that are meaningful. We can see that there is a pretty clear positive relationship between id and dim1: apparently the value of dim1 was increased slowly as the participants arrived in the study (see the first panel in the second row). Evaluate Fig. 1.4 carefully so you know what's going on with the other variables as well.

Interestingly, R will change the "default" functionality of plot based on the objects that are passed[18] to it. For example, a call to

```
> plot(face_data$rating)
```

produces Fig. 1.5, which is quite different from the plot we saw when passing the full data.frame as an argument. Thus, based on the type of object passed to the plotting function—whether that is a data.frame, a numerical vector, or a factor—the behavior of the function plot will change.

Admittedly, the default behavior of R is not always the best choice: you should learn how to make the plots you want yourself without relying on the R defaults. We will look at some ways of controlling R plots below.

---

[18] Each argument provided to a function is of a certain type, for example a vector or a data.frame as we discussed before. R uses this type information to determine what type of plot to produce.

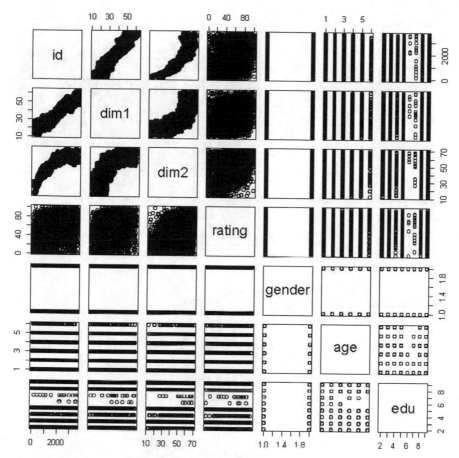

Fig. 1.4 Calling `plot` on a data frame. Note that on both the columns and the rows of the grid the variables in the dataset are listed. Each panel presents a visualization showing one variable in the dataset on the *y*-axis, and one on the *x*-axis

1.5.1 Describing Nominal/ordinal Variables

One way to start thinking about informative plotting is by considering the measurement levels of variables; just as frequencies are useful for describing nominal (and sometimes ordinal) variables, but less so for interval and ratio variables, certain types of plots are useful for nominal and ordinal variables, while others are less useful. A simple way of plotting the frequencies is a *bar chart*. The following code produces Fig. 1.6 and provides a simple example:

```
> counts <- table(face_data$age)
> barplot(counts)
```

Fig. 1.5 Calling `plot` on a
single continuous variable

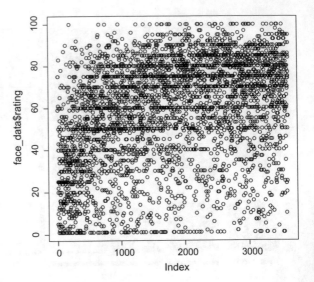

Fig. 1.6 Example of a bar
chart for displaying
frequencies

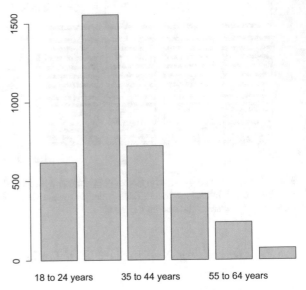

You can also make a *pie chart* using the same frequencies. The following code
produces Fig. 1.7.

```
> pie(counts)
```

Pie charts can be useful for easy comparisons of relative frequencies, while bar charts
are more meaningful for absolute numbers or frequencies. The bar chart allows you
to compare the heights of the bars more easily (and hence the size of the groups).
However, the pie chart more clearly visualizes the relative share of each value. Hence,

Fig. 1.7 Example of a pie chart for displaying frequencies

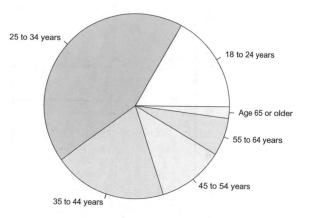

by using different visualizations you can emphasize different aspects of the data. For a first exploratory analysis of a dataset it is therefore often useful to look at multiple visualizations. Obviously, more ways of showing frequencies exist, but these are the most basic versions that you should know and should be able to construct.

1.5.2 Describing Interval/ratio Variables

We can now look at ways of visualizing *continuous* variables. Figure 1.8 shows a so-called *box and whiskers plot* (or *box plot*); these are useful for getting a feel of the spread, central tendency, and variability of continuous variables. Note that the middle bar denotes the median and the box denotes the middle 50% of the data (with Q_1 the first quartile at the bottom of the box and Q_3 the third quartile as the top of the box). Next, the whiskers show the smallest value that is larger or equal to $Q_1 - 1.5IQR$ and the largest value that is smaller than or equal to $Q_3 + 1.5IQR$. Finally, the dots denote the values that are outside the interval $[Q_1 - 1.5IQR, Q_3 + 1.5IQR]$, which are often identified or viewed as outliers. Box and whiskers plots can be very useful when comparing a continuous variable across subgroups of participants (e.g. males and females)—see Sect. 1.5.3. The figure was produced using the following code:

```
> boxplot(face_data$rating)
```

Next to box and whiskers plots, histograms (examples are shown in Fig. 1.9) are also often used to visualize continuous data. A histogram "bins" the data (discretizes it), and subsequently shows the frequency of occurrence in each bin. Therefore, it is the continuous variant of the bar chart. Note that the number of bins selected makes a big difference in the visualization: too few bins obscure the patterns in the data, but too many bins lead to counts of exactly one for each value. R "automagically" determines the number of bins for you if you pass it a continuous variable; however,

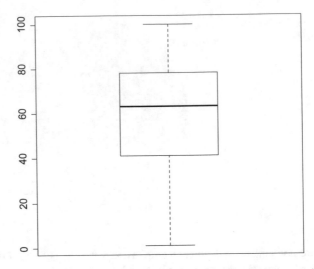

Fig. 1.8 Example of a box and whiskers plot useful for inspecting continuous variables

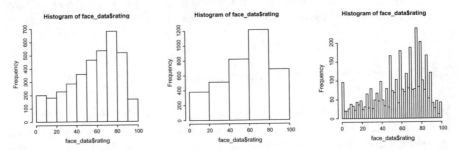

Fig. 1.9 Examples of histograms with different numbers of breaks (R's default on the left, 5 in the middle, 50 on the right). Determining the number of breaks to get a good overview of the distribution of values is an art in itself

you should always check what things look like with different settings. The following code produces three different histograms with different numbers of breaks (or bins).[19]

```
> hist(face_data$rating)
> hist(face_data$rating, breaks=5)
> hist(face_data$rating, breaks=50)
```

Finally, a *density* plot—at least in this setting—can be considered a "continuous approximation" of a histogram. It gives per range of values of the continuous variable the probability of observing a value within that range. We will examine densities in more detail in Chap. 4. For now, the interpretation is relatively simple: the higher the

---

[19] The bins don't always correspond to exactly the number you put in, because of the way R runs its algorithm to break up the data, but it gives you generally what you want. If you want more control over the exact breakpoints between bins, you can be more precise with the `breaks` option and give it a vector of breakpoints.

Fig. 1.10 Example of a density plot on the variable `rating`

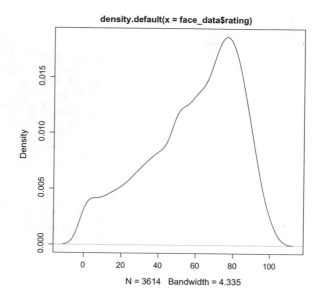

density.default(x = face_data$rating)

N = 3614   Bandwidth = 4.335

line, the more likely are the values to fall in that range.[20] The density plot shown in Fig. 1.10 is produced by executing the following code:

```
> plot(density(face_data$rating))
```

It is quite clear that values between 40 and 100 quite often occur, while values higher than 100 are rare. This could have been observed from the histogram as well.

1.5.3 Relations Between Variables

We often want to visualize multiple variables simultaneously instead of looking at single variables similar to Fig. 1.4. That way we can see relations between variables. The most often used visualization of a relationship between two continuous variables (and therefore R's default) is the *scatterplot* (see Fig. 1.11). Note that here each dot denotes a unique observation using an (x, y) pair. The following code produces such a plot:

```
> plot(face_data$dim1, face_data$rating)
```

This plot shows that there is a (kind of nonlinear) relationship between the first dimension of the face and the rating that participants give: for a very low value of the distance between the eyebrows people seem to provide low attractiveness ratings

---

[20] The interpretation of densities is different than the interpretation of histograms.

Fig. 1.11 A scatterplot
denoting the relationship
between dim1 and rating

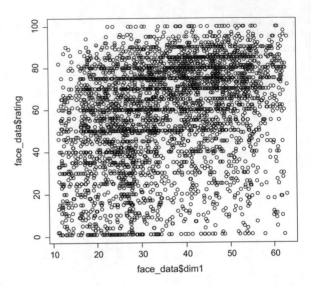

more frequently. However, the relationship is quite noisy: not everyone provides the
exact same rating for the same value of dim1.[21]

To see the relation between a categorical variable (nominal or ordinal) and a
continuous variable we can make box plots separately for each level of the categorical
variable. Figure 1.12 shows the boxplots for rating as a function of gender. This is
what R produces automatically using the plot function. The code to generate the
plot is:

```
> plot(face_data$gender, face_data$rating)
```

This default behavior of the plot function in this case is nice. However, we feel the
default is somewhat inconsistent: when passing the full data.frame to the plot
function (see Fig. 1.4) it would have been nice if the same boxplots were shown for
relationships between categorical and continuous variables (e.g., rows 5, 6, and 7).
However, this is regretfully not the case.

1.5.4 Multi-panel Plots

We often plot multiple panels in one single figure. This can be done using the
par(mfrow=c(y, x)) command in R. This basically sets up a *canvas* with
y rows and *x* columns for plotting. The following code produces Fig. 1.13 and thus
combines six panels into one figure.

```
> par(mfrow=c(3, 2))
```

---

[21] This makes sense: (a) people often do not respond in the exact same way, and (b) the value for
dim2 differs as well!.

Fig. 1.12 Box plots split for gender

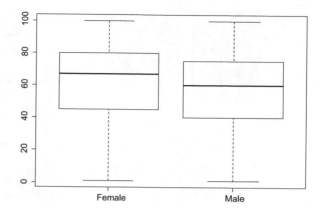

```
> plot(face_data$gender)
> plot(face_data$gender, face_data$rating)
> pie(counts)
> barplot(counts)
> hist(face_data$dim1)
> plot(density(face_data$dim2))
```

1.5.5 Plotting Mathematical Functions

Often we want to plot mathematical functions (such as $y = x^2$, or $z = sin(x) + cos(y)$). Plotting (2D) functions in R is simple and there are multiple ways. We first create the actual function in R and try it out to make sure you understand how this function actually works: we pass it the *argument* x, and it returns the *value* x^2. Here is a small example:

```
> square <- function(x) {x^2}
> square(3)
[1] 9
```

Next, we generate a sequence of numbers—using the built-in function `seq`. Try to type `?seq` into R to see the different uses of the function. Here we make a sequence from -2 to 2:

```
> x <- seq(-2, 2, by=.1)
> x
 [1] -2.0 -1.9 -1.8 -1.7 -1.6 -1.5 -1.4 -1.3 -1.2 -1.1 -1.0 -0.9
     -0.8 -0.7 -0.6
[16] -0.5 -0.4 -0.3 -0.2 -0.1 0.0 0.1 0.2 0.3 0.4 0.5 0.6 0.7
     0.8 0.9
[31] 1.0 1.1 1.2
> y <- square(x)
```

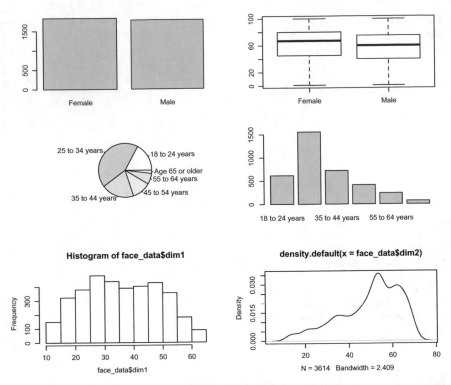

Fig. 1.13 Demonstrating a plot with multiple panels. Note that by now you should be able to name, and interpret, each of the panels

Make sure you understand the result that is generated when you run `print(y)`. We can then easily make the (x, y) plot using the following code (Fig. 1.14):

```
> par(mfrow=c(1, 2))
> plot(x, y)
> plot(x, y, type="l")
```

Note that Fig. 1.14 displays both the standard plot (on the left), and the plot using the additional argument `type="l"`; the latter makes an actual line as opposed to plotting the separate points. If you want to have both a line and points you can provide the option `"b"`.

Alternatively, we can also plot functions directly: the following code produces Fig. 1.15.

```
> curve(square, xlim=c(-2, 2))
```

This will work as long as the function you are plotting accepts an x argument. Note that you can pass additional arguments to `curve` to determine the range and domain of the function: in this case we specify the domain by adding `xlim=c(-2,2)`.

Fig. 1.14 Plotting a simple function by evaluating it at a sequence of points

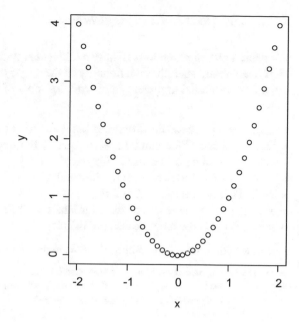

Fig. 1.15 Plotting a function directly using the `curve` function

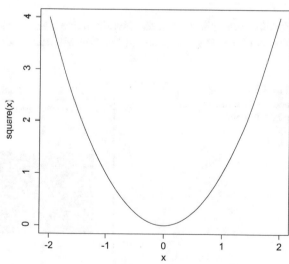

1.5.6   Frequently Used Arguments

R's default plotting functions all have additional arguments to style the created plots. We have already seen the functionality of the `type="1"` argument for plot. Here are a few additional arguments that work for most of the simple plotting functions in R:

- `type`: We just saw this difference between either plotting points, or a solid line.
- `xlim, ylim`: These can be used to specify the upper and lower limits of the x- and y-axes using `c(lower, upper)`.
- `lwd`: This can be used to set the line width.
- `col`: This can be used for colors.
- `xlab, ylab`: These can be used for labels on the axis by passing a string.
- `main`: This is the main heading of the plot

So, we can now do something like this to produce Fig. 1.16:

```
> plot(face_data$gender, face_data$rating,
+       col=c(1, 2), lwd=5, main="Comparing␣ratings",
+       xlab="Gender␣(males␣and␣females)",
+       ylim=c(1, 100)
+ )
```

Finally, note that we can use functions like `lines`, `points`, and `abline` to *add* lines and points to an existing plot. You will play with this in the assignments.

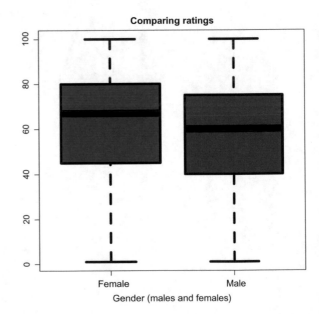

Fig. 1.16 Demonstrating additional arguments for the plot function

1.6 Other R Plotting Systems (And Installing Packages)

Up till now we have been using the "standard" plotting functions of R. However, R's functionality is greatly expanded by the fact that developers can easily contribute extension *packages* to R. A number of specific plotting packages have been added to the R language over the years. Here we briefly cover two of these, namely, `ggplot` and `Lattice`. Both provide more stylized plots than the standard functions. Here we briefly introduce each of these plotting systems—and in doing so show how you can include additional packages. However, both `ggplot` and `Lattice` require quite some time to really master; this is outside the scope of this book.

1.6.1 Lattice

`Lattice` is older than `ggplot`, but it is very useful for making multi-panel plots and for splitting plots according to some variables. Figure 1.17 is an example of a lattice plot that is generated using the following code:

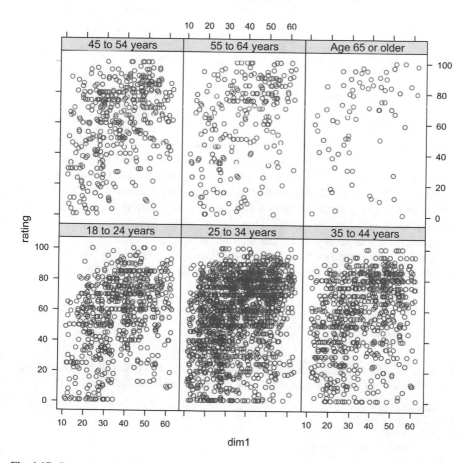

Fig. 1.17 Demonstrating a `lattice` plot and the ability to easily make multi-panel plots using formula's

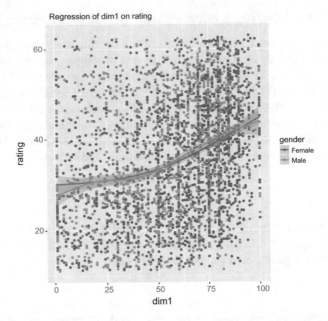

Fig. 1.18 Demonstrating ggplot2. Shown is a scatterplot with a *trend line* for each gender

```
> # install.packages("lattice")
> library(lattice)
> xyplot(rating ~ dim1 | age, groups=gender, data=face_data)
```

Here, the xyplot function generates the plot. The function takes as its first argument an R formula object: the syntax rating   dim1 | age indicates that we want to plot rating as a function of dim1 for each level of age. Next, we specify that we want to plot this relationship grouped by gender; this creates two different colors for the two genders in each panel. Finally, we point to the data object that contains these variables using the data=face_data argument.

In the above code-snippet, the line # install.packages("lattice") is a comment; the # ensures that the line is not executed. However, removing the # makes this code run, and tells R to automagically download the "lattice" package.

1.6.2   GGplot2

GGplot2 makes very nice looking plots, and has great ways of doing explorative data analysis. Just to give a brief example, the following code produces Fig. 1.18:

```
> library(ggplot2)
> qplot(rating,
+       dim1,
+       data=face_data,
```

```
+        geom=c("point", "smooth"),
+        color=gender,
+        main="Regression␣of␣rating␣on␣dim1",
+        xlab="dim1", ylab="rating")
```

See http://ggplot2.org/book/qplot.pdf for more info on GGplot2. Obviously, this code will only work after you have installed (using `install.packages()` and loaded (using `library()`) the `ggplot2` package.

Problems

1.1 This first assignment deals with the dataset `demographics-synthetic.csv`. Make sure to download the dataset from http://www.nth-iteration.com/statistics-for-data-scientist and carry out the following assignments:

1. Compute the mean, mode, and median of the variables `Age`, `Weight`, and `Voting`. Provide a short interpretation for each of these numbers.
2. In the dataset, find any values that you believe are erroneous. Hence, look for outliers or coding errors. Note which values you found (by specifying their row and column index). Next, remove the rows that contain these "errors" (yes, you will have to search the web a bit on how to delete rows from a data frame; this was not explicitly covered in the chapter). How many rows are left in the resulting dataset?
3. Compute the mean, mode, and median of the variable `Age` again. How did these values change by removing the outliers?
4. Compare the (sample) variance of the variables `Height` and `Weight`. How do they differ? What does this mean?
5. Use the `quantile` function to compute the 18th percentile of `Age`. What does this score mean?
6. Create a scatterplot relating the `Weight` (x-axis) and `Age` (y-axis) of participants. Do you see a relation?
7. Redo the same plot, but now color the points and add meaningful labels to the axis. Also, provide a nice title for the plot.
8. Next, add a horizontal line to the plot at `Age` = 30 and add a vertical line to the plot at `Weight` = 90.
9. Create a box plot comparing the distribution of `Age` for males and females.
10. Create a figure with two panels, one with the scatterplot you just created and one with the box plot you just created.
11. Create a histogram of the variable `Weight`. What do you think is a good number of breaks? Why?

1.2 This second assignment concerns the dataset `voting-demo.csv`.

1. Inspect the dataset: How many observations (rows) does it contain? And how many variables (columns)? What do the variables mean?

2. Write down the measurement level of each of the variables.
3. Compute all of the descriptive statistics described in this chapter for each of the variables. Try to interpret each of them. Do they all make sense? If not, why not?
4. Do you see any clear differences between this real—and hence not simulated—dataset you are looking at right now and the previous simulated dataset that we looked at in Problem 1.1? Think of a method by which you could tell the difference between simulated and real data in this case.

1.3 The next assignments should strengthen your R programming skills. To keep up with the book and to use R efficiently in the upcoming chapters, please make sure you can do the following:

1. Using the `for`, `sum`, and `length` functions (and whatever else you need), write your own function to create a frequency table (hence, the number of times each unique numerical value occurs in a vector that is passed to the function).
2. Create a function to compute the mean of a variable using only the `sum` and `length` functions.[22]
3. Create a function to compute the mean of a variable using only the `for` function.
4. Discuss how the above two functions for computing a mean differ. Does this difference change if you compute the mean of a variable with more and more observations? Use the `Sys.time` function to see how long each of the two functions takes to compute the mean as a function of the number of observations.
5. Run the command `x <- rnorm(100, mean=0, sd=1)` to create a new variable called `x`. What is the size of `x`?[23]
6. Compute descriptive statistics for `x` that you think are useful.
7. Visualize the data in `x`. What plot do you select and why?
8. Now try to examine in the same way, by computing descriptives and by plotting, the variable `x2 <- c(rnorm(1000, mean=0, sd=1), rnorm(1000, mean=4, sd=2))`.

1.4 Suppose we asked a group of 8 persons how old they are and recorded the following ages: 30, 23, 29, 36, 68, 32, 32, 23.

1. Use these data to calculate the following descriptive statistics by hand: arithmetic mean, median, mode, range, mean absolute deviation, variance, and standard deviation. Give a short interpretation of each statistic you calculated.
2. The age of 68 is quite an outlier in this dataset since it is considerably greater than the remaining ages. Remove the age of 68 from the dataset and calculate the above descriptive statistics again based on the remaining seven ages. What do you notice? Are some descriptive statistics more sensitive to the outlier than others?
3. Compute standardized (z) scores for each person in the 8 person dataset. Next, compute the mean and variance of the z-scores.

---

[22] Obviously, you can use *operators* such as + and *.

[23] We will discuss the command `rnorm` in more detail in Sect. 4.8.1.

1.5 Suppose you are given aggregated (or grouped) data describing the `Age` in years of a sample of n people. The dataset does not contain the raw n data points, but rather it contains K age values denoted by $v_1, \dots, v_K$. For example, if $v_1 = 12$, this means that the age of the first group of people is 12. Next, the data also contains frequencies f_k, which indicate how often a specific age v_k is present in the original (i.e., unaggregated) dataset. Using this setup, answer the following questions:

1. What is the result of the sum $\sum_{k=1}^{K} f_k$?
2. Give a formula for computing the variance of `Age` for the n people based on the aggregated data.
3. Can you give a formula for computing the median of `Age` based on the grouped data? If so, provide it. If not, then why is this not possible?
4. Implement, in R, a function for computing the mean of an aggregated dataset and a function for computing the variance of an aggregated dataset.

1.6   Extra In this last assignment we will take another look at the simulated dataset we considered in Problem 1.1. We used the following R code to produce this dataset:

```
> # Function for creating the data in demographics-synthetic.csv:
> create_data <- function(n, seed=10) {
+     set.seed(seed)
+
+     # Create:
+     gender <- rbinom(n, 1, .48)
+     height <- round(170 + gender*10 + rnorm(n, 0, 15), 1)
+     weight <- height / 2 + runif(n, -10, 10)
+     voting <- rbinom(n, 5, c(.3, .3, .2, .195, .005))
+     age <- round(23 + sqrt(rnorm(n, 0, 5)^2), 1)
+
+     # Recode:
+     voting <- ifelse(voting==4, 99, voting)
+     gender <- ifelse(gender==1, "Male", "Female")
+
+     # Return data frame:
+     data <- data.frame(
+       "Gender" = gender,
+       "Age" = age,
+       "Weight" = weight,
+       "Height" = height,
+       "Voting" = voting)
+
+     return(data)
+ }
>
> # Create the data and store it:
> n <- 500
> data <- create_data(n)
> write.csv(data, file="demographics-synthetic.csv")
```

Obviously, you should feel free to play around with this function. However, make sure you do the following:

1. Investigate the `rbinom()` function. What does it do?
2. Investigate the `rnorm()` function. What does it do?

3. What happens if you change the value of `seed`?
4. Explain what the `ifelse()` function does.

If you have time, you can always teach yourself more GGplot. For example, try to follow the tutorial at http://tutorials.iq.harvard.edu/R/Rgraphics/Rgraphics.html. It will certainly pay off in the rest of your data science career if you are able to quickly make informative (and cool-looking) plots!

Additional Material I: Installing and Running RStudio

RStudio is an integrated development environment (IDE) for R. It includes a console, a syntax-highlighting text editor that supports direct code execution, as well as tools for plotting, viewing your history, debugging and managing your workspace.

The Steps Involved When Installing RStudio

In order to run R and RStudio on your system, you need to follow the following three steps in the same order:

1. Install R
2. Install RStudio
3. (optionally) Install additional R-Packages

Installing R

Installing R is different for users of different operating systems:

- Windows users can download the latest version of R at https://cran.cnr.berkeley.edu/bin/windows/ and subsequently open the .exe file to install R.
- Mac users can get their version of R at https://cran.cnr.berkeley.edu/bin/macosx/ and open the downloaded .pkg file to install R.
- Linux users can follow the instructions on https://cran.cnr.berkeley.edu/bin/linux/. Users of Ubuntu with Apt-get installed can execute `sudo apt-get install r-base` in their terminal.

Install RStudio

After installing R, you will need to install RStudio. The different versions of RStudio can be found at https://www.rstudio.com/products/rstudio/download/#download. After installation you can open up RStudio; it should look like Fig. 1.19.

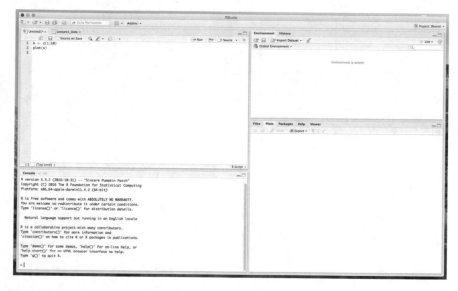

Fig. 1.19   Demonstrating `ggplot`. Shown is a scatterplot with a *trend line* for each gender

Installing Packages (Optional)

To install packages using RStudio click on the Packages tab in the bottom-right section and then click on install. A dialog box will appear. In the Install Packages dialog, write the package name you want to install under the Packages field and then click install. This will install the package you searched for or give you a list of matching package based on your package text.

You should now be good to go!

References

T. Baguley, M. Andrews, Handling missing data. *Modern Statistical Methods for HCI* (Springer, Berlin, 2016), pp. 57–82

M.J. Crawley, *The R Book* (Wiley, Hoboken, 2012)

L. Ippel, Getting started with [r]; a brief introduction. *Modern Statistical Methods for HCI* (Springer, Berlin, 2016), pp. 19–35

M.C. Kaptein, R. Van Emden, D. Iannuzzi, Tracking the decoy: maximizing the decoy effect through sequential experimentation. Palgrave Commun. **2**(1), 1–9 (2016)

G. Norman, Likert scales, levels of measurement and the "laws" of statistics. Adv. Health Sci. Educ. **15**(5), 625–632 (2010)

T. Rahlf, *Data Visualisation with R: 100 Examples* (Springer, Berlin, 2017)

D. Vidotto, J.K. Vermunt, M.C. Kaptein, Multiple imputation of missing categorical data using latent class models: state of the art. Psychol. Test Assess. Model. **57**(4), 542 (2015)

J. Young, J. Wessnitzer, Descriptive statistics, graphs, and visualisation. *Modern Statistical Methods for HCI* (Springer, Berlin, 2016), pp. 37–56

Chapter 2
Sampling Plans and Estimates

2.1 Introduction

In the previous chapter we computed descriptive statistics for the dataset on faces. The results showed that the average rating was 58.37 and that men rated the faces higher than women on average. If we are only interested in the participants in the study and we are willing to believe that the results are fully deterministic,[1] we could claim that the group of men rates higher than the group of women on average. However, if we believe that the ratings are not constant for one person for the same set of faces[2] or if we would like to know whether our statements would also hold for a larger group of people (who did not participate in our experiment), we must understand what other results could have been observed in our study if we had conducted the experiment at another time with the same group of participants or with another group of participants.

To be able to extend your conclusions beyond the observed data, which is called more technically *statistical inference*, you should wonder where the dataset came from, how participants were collected, and how the results were obtained. For example, if the women who participated in the study of rating faces all came from one small village in the Netherlands, while the men came from many different villages and cities in the Netherlands, you would probably agree that the comparison between the average ratings from men and women becomes less meaningful. In this situation the dataset is considered *selective* towards women in the small village. Selective means here that not all women from the villages and cities included in the study are represented by the women in the study, but only a specific subgroup of women have been included. To overcome these types of issues, we need to know about the concepts of *population*, *sample*, *sampling procedures*, and *estimation* of population

---

[1] Deterministic means here that the scores of the participants on the tested faces will always be the same: they are without any uncertainty. In reality though, scoring or rating is subjective and variable even for just one person.

[2] Not constant means that a rating of 60 for one face from one of the participants could also have been 50 or 65 if we had asked for the rating at another moment.

© Springer Nature Switzerland AG 2022
M. Kaptein and E. van den Heuvel, *Statistics for Data Scientists*, Undergraduate Topics in Computer Science, https://doi.org/10.1007/978-3-030-10531-0_2

characteristics, and also how these concepts are related to each other to be able to do proper statistical inference.

Figure 2.1 visualizes the relation between these concepts. On the left side we have a population of units (e.g., all men and women from the Netherlands) and on the right side we have a subset of units (the sample). Sampling procedures are formal *probabilistic* approaches to help collect units from the population for the sample. For the sample we like to use $x_1, x_2, ..., x_n$ for the observations of a certain variable (e.g., ratings on faces from pictures). The calculations on the sample data, which we have learned in Chap. 1, are ways of describing the sample. For the population the same notation $x_1, x_2, ..., x_N$ for all N units is used. Here we have used the same indices for the sample and the population, but this does not mean that the sample x_1, $x_2, ..., x_n$ is just the first n units from the population $x_1, x_2, ..., x_N$. Mathematically, we should have written $x_{i_1}, x_{i_2}, ..., x_{i_n}$ for the sample data, with $i_h \in \{1, 2, ..., N\}$ and $i_h \neq i_l$ when $h \neq l$, since any set of units $i_1, i_2, .., i_n$ from the population could have ended up in the sample. The values in the sample are referred to as a *realization* from the population.

If we know which sampling procedure was applied to collect the units for the sample, we would also know how close the calculations or statistics described in Chap. 1 would be to the theoretical value in the whole population.[3] Thus the sampling procedure and the choice of calculation on the sample data (an average, median, first quartile, standard deviation etc.) would make statistical inference mathematically precise and it would therefore help us when making statements beyond the sample data. In terms of statistical inference, the calculations on the sample data are referred to as *estimates* for the theoretical value in the whole population. Note that we often introduce a Greek symbol for a theoretical value in the population, like $\mu = \sum_{i=1}^{n} x_i/N$ and $\sigma^2 = \sum_{i=1}^{n} (x_i - \mu)^2/N$ for the mean and variance of the population, respectively (see Fig. 2.1).

By studying this chapter you will learn the following:

- The formal definitions of *population*, *units*, and *sample*.
- Different probability sampling approaches that one can use to select units from a population to create a *representative sample*.
- The concepts of *bias*, *standard error*, and *mean squared error* to evaluate the closeness of sample calculations to the population value and how these depend on the specific sampling approach.
- Methods of determining *bias*, *standard error*, and *mean square error* of a sampling procedure using R.

---

[3] The theoretical value is a calculation procedure applied to all units in the population. It is considered theoretical, since the values on each unit in the population may not exist but we believe that they could exist. The faces dataset makes this clear, since we only know the ratings of faces for participants in the study, and values beyond the study only exist if faces were to be rated by other units as well.

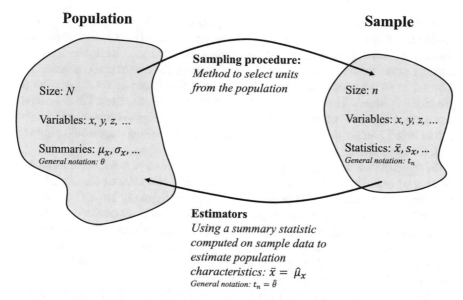

Fig. 2.1 Graphical representation of the material in this chapter: we are concerned with the relation(s) between population and sample, and the quality of sampling plans and estimators

2.2 Definitions and Standard Terminology

In this section we briefly introduce some definitions and standard terminology. Frequently, we wish to say something about a group of units other than just the ones we have measured. A *unit* is usually a concrete or physical thing for which we would like to measure its characteristics. In medical research and the social sciences units are mostly human beings, while in industry units are often products, but units can essentially be anything: text messages, financial transitions, sales, etc. The complete set of units that we would like to say something about is called the (target) *population*. The set of units for which we have obtained data is referred to as the *sample*. The sample is typically a subset of the population, although in theory the sample can form the whole population or the sample can contain units that are not from the target population. If we are interested in individuals in the age range of 25 years to 65 years, it could happen that a person with an age outside this range is accidently included in the sample.

Statistics is concerned with how we can say things, and what we can say, about a population given that we have only observed our sample data. As we mentioned before, we call this *statistical inference*: "Statistical inference is the process of deducing properties of an underlying population by analysis of the sample data. Statistical inference includes testing hypotheses for the population and deriving population estimates.", see e.g., Casella and Berger (2002).

In many situations it is unnecessary to specify the unit explicitly since it will be clear from the context, but it is not always easy to determine the unit. For instance, a circuit board contains many different components. Testing the quality of a circuit board after it has been produced requires the testing of all or a subset of the components. In this case it is not immediately clear whether the circuit board itself or whether the components are the units. In this setting the circuit board is sometimes referred to as the *sample unit*, since it is the unit that is physically taken from the production process. The components on the circuit board are referred to as *observation units*, since it is the unit that is measured. If the components were to be tested before being placed on the circuit board, however, the component would represent both the sample and observation unit.[4]

In principle we would expect that a population is always *finite*, since an infinite number of units does not exist in real life. However, populations are often treated as infinite. One reason is that populations can be really really large. Examples are *all* humans, *all* animals, *all* telephones, *all* emails, *all* internet connections, and *all* molecules in the world. It is mathematically often more convenient (as we will see later) to assume that such a population is infinite. However, in this chapter we are concerned solely with finite populations; we will discuss infinite populations in later chapters and discuss their relation with this chapter.

Properly defining or describing a population can be difficult. For instance, let's assume we would like to know whether students at Dutch universities are satisfied with their choice of study. At first the population could simply be all students at all Dutch Universities, but this is not necessarily well defined. Do we include all students and also students that have been registered abroad but are studying at a Dutch University for a short period of time? Do we include students of the University of Groningen who study at their campus in China? What about students who have chosen two programs (double bachelor or master) or are registered at multiple institutes? What about students that have recently changed their program? This simple example shows that we probably need to specify many more details before our population of interest is truly defined.[5] These details often require specification of a time period to eliminate changes in the population over time. A student at a Dutch University 10 years ago is probably different from a student today, but also the choices of program are different over these time periods. For example, Data Science did not exist as a bachelor program in the Netherlands before the academic year 2016–2017.

Furthermore, even if the population is established, measuring all units is often impossible or too elaborate. This means that information about the population can only be obtained by considering a subset of the population.[6] In the current era of

---

[4] Note that other terminology for sample and observation unit is used in different fields of science or even within statistics. For instance, units are sometimes called elements and an observation unit may be referred to as an elementary unit or element. The sample unit is sometimes referred to as enumeration or listing unit (see Levy and Lemeshow 2013).

[5] Unfortunately, many studies on human beings hardly ever make their target population explicit, which makes research claims somewhat fuzzy.

[6] Any subset of units from a population is called a *sample*, irrespective of how the sample has been obtained. We use the term *sampling* to refer to the activity of obtaining the units in a sample. This includes taking the measurements on the units and not just the physical collection of units.

big data[7] where we measure and store all kinds of information, we may think that sampling is redundant, since we can measure almost all units from the population. This is, however, not true for a number of reasons:

1. In many applications we really can't measure the complete population. For instance, one of the tests applied to aircraft engines is the "frozen bird test". A frozen dead bird is shot into one operating aircraft engine to determine if the engine can consume the bird (it goes through the engine) and still keeps running properly after the bird has gone through. Clearly, shooting a large solid projectile through an engine will cause damage to the engine, so this is not a test you would like to apply to each produced aircraft engine. In industry, many destructive tests are applied to determine product quality and this will only be done on a small subset of products.

2. Time, space, or budget restrictions often do not allow us to measure all units from a population. For instance, there is a legal obligation to verify the financial bookkeeping of all businesses that are registered at the stock exchange. It is impossible to verify each and every item in the bookkeeping and therefore sampling is necessary.

3. Big data itself may be an argument for sampling. If we have a very large sample or we have been able to measure all units from the population, the resulting dataset can be so large that it becomes impossible to analyze the full data at one computer. Sampling techniques can then help reduce the size of the data and still obtain the required information.

To be able to say something about a population by using a sample, the sample must be representative. A *representative sample* can be intuitively defined as a sample of units that has approximately the same distribution of characteristics as the population from which it was drawn. Although this is not an explicit definition, since it does not clarify how much "approximately" truly means, it at least intuitively presents a notion of the goal of representative sampling. Note that there have been alternative views on what a representative sample should be (see, e.g., Kruskal and Mosteller 1980). For instance, a representative sample is a miniature of the population or it should contain full coverage of the population, like Noah's Ark. However, it was Jerzy Neyman with his seminal paper in 1934 who made the definition of representative sampling precise (Neyman 1992).

There are many ways in which we can select units from a population. We often distinguish two groups of methods: non-representative sampling methods, and representative sampling methods. A non-representative sample implies that we do not know the exact process by which units in the population became part of the sample. The sample itself may describe the population appropriately or not, but we can never demonstrate or show this. For representative samples this is not the case: here we know how units have been included in the sample. More precisely, we would know the probability with which a unit of the population is being collected. Representative

---

[7] We have to admit that *Big Data* is a bit of a buzz word that is not very well defined. For now we will just go with the fuzzy notion of a very large dataset; one consisting of many rows and columns.

sampling is therefore also referred to as *random* or *probability sampling*. Thus the definition of a representative sample is related to how the data has been collected from the population and it requires random elements.

2.3 Non-representative Sampling

The most common non-representative sampling methods may be divided into three categories: *convenience sampling*, *haphazardly sampling*, and *purposive sampling*. Although these sampling methods are frequently in use, it is strongly recommended not to apply these methods, unless knowledge is available on how to adjust or correct the sample for inferential purposes.

2.3.1 Convenience Sampling

Convenience sampling collects only units from the population that can be easily obtained, such as the top layer of a pallet of boxes or trays with vials or the first cavity in a multi-cavity molding process. This may provide a *biased* sample, as it represents only one small part or time window of the whole processing window for a batch of products. The term *bias* indicates that we obtain the value of interest with a systematic mistake: we study bias in more detail later in this chapter. Convenience sampling is often justified by using the argument of population *homogeneity*.[8] This insinuates that either the population units are not truly different or the process produces the population of units in random order. Under these assumptions it is indeed irrelevant which set of units is collected, but these assumptions seem to contradict the need for sampling in the first place and are hardly ever justified.

2.3.2 Haphazard Sampling

Haphazard sampling is often believed to be an excellent way of collecting samples, because it gives a feeling or the impression that each unit was collected completely at *random*.[9] This way of sampling is best described by an example. If one stands in a library in front of a bookshelf and one is asked to collect an arbitrary book, then "just picking one" would be a haphazard sample. However, in practice it turns out that this procedure typically collects books in the center of the bookshelf and typically books

---

[8] *Homogeneity* means "being all the same or all of the same kind".

[9] *Random* is an often used word, but it is hard to properly define. In statistics we often use the term *uniformly random* for selection processes that are "governed by or involving equal chances for each unit".

that are larger or thicker. This is usually not what people feel or believe when they try to take an arbitrary book. Hence despite the feeling of randomness when performing haphazard sampling, often the resulting sample is not truly random. Another example is that human beings have the tendency to choose smaller digits when they are asked to choose digits from 1 to 6 (Towse et al. 2014).

2.3.3 Purposive Sampling

Purposive sampling or judgmental sampling tries to sample units for a specific purpose. This means that the collection of units is focused on one or more particular characteristics and hence it implies that only units that are more alike are sampled. In epidemiological research[10] purposive sampling can be very practical, since it may be used to exclude subjects with high risks for unrelated diseases. In clinical trials[11] inclusion (e.g., participants older than 65 years) and exclusion (e.g., no pregnant women) criteria are explicitly applied to make sure a sample has specific characteristics. This way of sampling is strongly related to the definition of the population, since deliberately excluding units from the sample is analogous to limiting the population of interest. Thus purposive sampling may be useful, but it is limited since it does not allow us in general to make statements about the whole population, and at best only about a limited part of the population (although we may not be sure either). In other words, it does most likely produce a biased sample with respect to the complete population.

2.4 Representative Sampling

All the sampling methods discussed above have the risk that some units are much more likely to be included in the sample than others, which can make statistics computed on the sample data bad estimates for the population parameters of interest. Even worse: with non-representative sampling some units are not only more likely to be included in the sample, we also do not actually know how likely units were included. Hence, even if we wanted to, we could not control for these systematic differences between units. When performing representative sampling we sample units in such a way that we do know how likely units are to be included in the sample (even if they will be different from unit to unit).

---

[10] The research field that is concerned with understanding health and disease.

[11] Experimental studies on human beings to determine if a new treatment is beneficial with respect to placebo or a currently available treatment.

To collect representative samples, statisticians have come up with the concept of *probability sampling* or *random sampling*. Random sampling is a sampling method that uses a random mechanism. This means that the probability of each unit in the population of becoming part of the sample is both positive and known. Note that we have not yet formally discussed the definition of probability: we will do this in the next chapter. However, for now the intuitive definition of probability that we all know from throwing a die suffices: if a die has six sides—which most of them have—then the probability of throwing a six is equal to $1/6$. We can easily generalize this idea to dice that have K sides, and thus the probability for each side is simply $1/K$. Hence, if we want to select a unit with probability $1/m$, we could create an m-sided die, roll it, and include the unit in the sample that belongs to the number on top of the die. Obviously this is not how samples are selected in practice, as we collect more than one unit and we use computers to generate (pseudo) random numbers. For the interested reader we have described how to generate random numbers using algorithms in the additional material at the end of this chapter.

The most common random sampling methods are: *simple random sampling*, *systematic sampling*, *stratified sampling*, and *cluster sampling*, we will discuss these below. Throughout this chapter we will assume the existence of a (finite) list of all units in the population. Such a list, however, will often not exist in practice. For instance, there is no complete list of all internet connections in the Netherlands nor is there a list of all dolphins in the ocean. We will however *assume* that such a list could be created, and discuss sampling mechanisms accordingly.

2.4.1 Simple Random Sampling

Simple random sampling is a way of collecting samples such that each unit from the population has the exact same probability of becoming part of the sample. Simple random sampling is a conceptually easy method of forming random samples but it can prove hard in practice. Because of its importance in statistical theory we discuss it in more detail.[12]

To illustrate simple random sampling, suppose that the entire population consists of six units ($N = 6$) only, numbered from 1 to 6. It is decided to collect a sample of three units ($n = 3$) from this population. For a simple random sample each of 20 combinations of three units could possibly form the sample S, i.e., the possible samples are:

---

[12] Simple random sampling is frequently combined with other choices or settings (see stratified and cluster sampling).

$$S_1 = (1, 2, 3) \quad S_2 = (1, 2, 4) \quad S_3 = (1, 2, 5) \quad S_4 = (1, 2, 6)$$
$$S_5 = (1, 3, 4) \quad S_6 = (1, 3, 5) \quad S_7 = (1, 3, 6)$$
$$S_8 = (1, 4, 5) \quad S_9 = (1, 4, 6)$$
$$S_{10} = (1, 5, 6)$$
$$S_{11} = (2, 3, 4) \quad S_{12} = (2, 3, 5) \quad S_{13} = (2, 3, 6)$$
$$S_{14} = (2, 4, 5) \quad S_{15} = (2, 4, 6)$$
$$S_{16} = (2, 5, 6)$$
$$S_{17} = (3, 4, 5) \quad S_{18} = (3, 4, 6)$$
$$S_{19} = (3, 5, 6)$$
$$S_{20} = (4, 5, 6)$$

The simple random sample can now be collected by generating one number k between 1 and 20 (using our $K = 20$-sided die) and then selecting S_k when k appears on top of the die. Note that each sample has the same probability (1/20) of being collected and that each unit has the same probability (1/2) of being collected.[13] This is a general property of simple random sampling: each unique sample has the same probability of being selected, and, as a result, each unit has the same probability of being selected (the numbers depend on the population and sample size). Hence, simple random sampling guarantees that each unit has the same probability of becoming part of the sample.

In R, this can be conducted by applying the function `sample`.[14] The function `sample` has (at least) three arguments: the data on the variable of interest (here we choose x), the number of samples drawn from the data (here we choose 1), and the indicator that tells us whether sampling is done *with replacement* (here we choose FALSE)[15]:

```
> x <- c(1:20)
> set.seed(575757)
> sample(x, 1, FALSE)
[1] 15
```

Thus, S_{15}, containing the units 2, 4, and 6, should be collected from the population.[16]

In practice, it might be cumbersome to create a list of all possible samples S_1, S_2, ..., S_K and subsequently select one of the samples because the number of samples K of size n that can be created with population size N rapidly increases with N and n. The number K of unique samples of size n is given by

$$K = N!/[n!(N - n)!],$$

---

[13] Each unit in the population appears in 10 out of 20 possible samples S_k. For instance, unit 3 appears in S_1, S_5, S_6, S_7, S_{11}, S_{12}, S_{13}, S_{17}, S_{18}, and S_{19}.

[14] Notice the argument FALSE that is supplied to the `sample` function is to indicate that there is no replacement. See also next page on sampling without replacement.

[15] Sampling with replacement implies that the units are put back into the population after being sampled and can be sampled again.

[16] Note that we have provided a seed number (initial value) to make the procedure reproducible. Every time the procedure is run, you will obtain sample 15.

with $x! = x \times (x-1) \times (x-2) \times \cdots \times 2 \times 1$. To see this, note that a set of numbers from $\{1, 2, 3, \ldots, N\}$ can be ordered in $N!$ ways.[17] These are all the possible permutations of the numbers of 1 to N. At the first position there are N units to choose from, at the second position there are $N - 1$ units left to choose from, then at the third position there are $N - 2$ units left to choose from, etc., until the last unit is fixed by the collection of all previous collected $N - 1$ units. This makes $N!$ possible options or permutations. Now assume that we take the first n positions as the sample. All permutations from $N!$ that would just permute the first n positions, which are $n!$ permutations, and/or permute the last $N - n$ positions, which are $(N - n)!$ permutations, would not lead to another set of units from the population. Thus given one permutation from the $N!$ permutations, there are $n! \times (N - n)!$ other permutations from the $N!$ permutations that result in the same sample. Thus the unique number of samples of size n from a population of size N is now $K = N!/[n!\,(N - n)!]$.[18]

A practically more feasible alternative approach of collecting a simple random sample is sampling the units sequentially from the population *without replacement*.[19] This leads to the *same* unique sample S defined by the permutations but without the need to list all the possible samples. To illustrate this consider again the example of $N = 6$ and $n = 3$. The probability that we would collect sample $S_{15} = (2, 4, 6)$, for instance, is calculated as follows. First we need to draw unit 2, which has a probability of $1/6$. Each unit in the population has the same probability of being collected. After unit 2 has been collected, the probability of collecting unit 4 is equal to $1/5$, since each of the remaining five units in the population has the same probability of being collected. Then finally, to collect unit 6, after units 2 and 4 have been taken out of the population, the probability is equal to $1/4$. To obtain the probability that $(2, 4, 6)$ occurs is just the product of probabilities (as we will learn in Chap. 3). Thus the probability of obtaining $(2, 4, 6)$ is equal to $1/120 = 1/6 \times 1/5 \times 1/4$. This is clearly not equal to $1/20$, which we indicated earlier as being the probability for $(2, 4, 6)$. This is because we have assumed in the calculation of the probability of $1/120$ that the order is strictly first unit 2, then unit 4, and finally unit 6. However, the order in which we collect units 2, 4, and 6 is irrelevant, and we ignored this order when computing our previous result of $1/20$. To resolve the difference, notice that the number of ways in which we can permute the three units is $3! = 3 \times 2 \times 1 = 6$ (namely, $(2, 4, 6)$; $(2, 6, 4)$; $(4, 2, 6)$; $(4, 6, 2)$; $(6, 2, 4)$; $(6, 4, 2)$). Each of these permutations has the same probability of being collected as permutation $(2, 4, 6)$. Thus the probability of collecting sample $(2, 4, 6)$, irrespective of the order in which

---

[17] The $x!$ notation is called "x-factorial".

[18] Note that it does not matter which n positions of a permutation of $\{1, 2, 3, \ldots, N\}$ you would choose. For our example there would be $6! = 720$ permutations of $\{1, 2, 3, \ldots, 6\}$. If we consider the permutations for which the first three elements remain $\{1, 2, 3\}$, there are 36 permutations ($3! = 6$ permutations of $\{1, 2, 3\}$ and $3! = 6$ permutations of $\{4, 5, 6\}$) leading to the same sample $\{1, 2, 3\}$. Thus there will be $20 = 720/36$ different samples (given by $S_1, S_2, \ldots, S_{20}$).

[19] Sampling without replacement means that the unit that is collected for the sample is not placed back in the population. This is common in medical science, marketing, psychology, etc. Sampling *with replacement* puts the unit back every time it is collected. For research on animals in the wild, units are of course being placed back.

they are collected, is now $6 \times 1/120 = 1/20$. Thus the approach of sequentially collecting units (without replacement) would give each triplet S_k above the same probability of $1/20$, as we found before!

In general, the probability of collecting sample S_k of size n from a population of N units, using sequential sampling, is equal to $n!\,(N-n)!/N!$ which is equal to one over the number of unique samples. The probability that a specific unit is part of the sample is given by n/N. To demonstrate this let's focus on the first unit ($i = 1$) of the population. In total there are $N!/[n!\,(N-n)!]$ possible samples of size n from a population of N units. The number of samples that does not contain unit 1 is equal to $(N-1)!/[n!\,(N-1-n)!]$, since they are all the samples from population $\{2, 3, \ldots, N\}$ without unit 1. Thus the probability that unit 1 is not contained in the sample is equal to $(N-1)!/[n!\,(N-1-n)!]\,/\,(N!/[n!\,(N-n)!]) = (N-n)/N$. The probability that unit 1 is contained in the sample is then $1 - (N-n)/N = n/N$. Since we can repeat this argument for each unit i in the population, the probability n/N holds true for every unit in the population.

2.4.2 Systematic Sampling

To obtain a sample of size n using systematic sampling, a few steps are required. First the population should be divided into n groups and the order of the units (if some order exists) should be maintained (or otherwise fix the order). Now suppose that each group consists of m units (thus the population size is $N = nm$) ordered from 1 to m in each group. From the first group one unit is randomly collected with probability $1/m$. Say the pth unit was the result. Then from each of the n groups the pth unit is collected too, forming the sample of n units. Note that systematic sampling provides only m possible sets of samples, i.e., $S_1, S_2, \ldots, S_m$.

Consider the population of six units again where we wish to collect a sample of three units. Splitting the population in to three groups for example provides the subgroups $(1, 2)$; $(3, 4)$; and $(5, 6)$. From the first group, which consists of only two units, one unit should be randomly collected with probability $1/m = 0.5$. Thus the sample can now only consist of $S_6 = (1, 3, 5)$ or $S_{15} = (2, 4, 6)$. Note that we have used the notation or index of the set of possible samples from simple random sampling. The possible samples from systematic sampling are quite different from the set of samples that can be obtained with simple random sampling. However, similar to simple random sampling, each unit in the population still has the same probability of being collected. The probability that a unit enters the sample is $p = 1/m$, which is the same as the probability of selecting one of the m possible sample sets.

The most important advantage of systematic sampling over simple random sampling is the ease with which the sample may be collected. Systematic sampling is often used in manufacturing in relation to a time period, for instance taking a unit every half hour. With a constant production speed a systematic sample is created if the first time point within the first half hour is taken randomly. This is clearly much easier than collecting a simple random sample at the end of production. It probably leads to fewer mistakes or to improper "short cuts" in sampling that would lead to a haphazard or convenience sample. Systematic sampling can also lead to more precise descriptive statistics than simple random sampling (see Cochran 2007). A clear disadvantage of systematic sampling is that the "period" for systematic sampling may coincide with particular patterns in the process or population.

2.4.3 Stratified Sampling

Simple random and systematic sampling methods implicitly assume that there is no particular group structure present in the population. At best they assume that the order of participants is aligned with an ordering in time. Structures in a population may be caused by particular characteristics. For instance, products may be manufactured on several different production lines, which form particular subpopulations within the population of products. Or, age, gender, and geographic area may form typical subgroups of people with different disease prevalence or incidence.[20]

When the numbers of units across these subpopulations are (substantially) different, simple random and systematic sampling may not collect units from each subgroup. Indeed, suppose that two production lines, say A and B, are used for the manufacturing of products and production line A produced 1,000 products, while production line B only produced 10 products. Then a simple random sample of 100 products may not necessarily contain any units from production line B.[21] Thus production line B would probably be under-represented. Stratified sampling is used to accommodate this issue by setting the sample size for each subpopulation (often called *strata*) to a fixed percentage of the number of units of the subpopulation. For instance, a 10% sample from the population of products from the two production lines A and B results in a simple random sample of size 100 units from line A and a simple random sample of size 1 unit from line B. By selecting 10% of the units from each stratum we are certain that each stratum is included in the sample.

Stratified sampling can also be applied to time periods, similar to systematic sampling. The population is then divided into n groups such that the order in units is maintained. From each group one unit is randomly collected with probability $1/m$, when each group contains m units. Note that this form of stratified sampling is not identical to systematic sampling (although this method of stratified sampling is sometimes referred to as systematic sampling). In the case of the ordered population of six units in the example for systematic sampling, the population is again split up into three periods, i.e., $(1, 2)$; $(3, 4)$; and $(5, 6)$. With stratified sampling the sample can now exist of the following possibilities: $S_6 = (1, 3, 5)$, $S_7 = (1, 3, 6)$, $S_8 = (1, 4, 5)$, $S_9 = (1, 4, 6)$, $S_{12} = (2, 3, 5)$, $S_{13} = (2, 3, 6)$, $S_{14} = (2, 4, 5)$, $S_{15} = (2, 4, 6)$, using

---

[20] Prevalence is the proportion of the population that is affected by the disease and incidence is the proportion or probability of occurrence within a certain time period.

[21] You will learn in the following chapter that this probability is equal to $\frac{1000}{1010} \frac{999}{1009} \frac{998}{1008} \cdots \frac{901}{911} = \frac{910}{1010} \cdots \frac{901}{1001} = 0.3508$.

the notation of the sampling sets for simple random sampling. Thus stratified sampling may lead to samples that are not possible with systematic sampling, but it does not produce all possible samples from simple random sampling.

The number of possible samples that can be collected with stratified random sampling is given by

$$K_S = \frac{N_1!}{n_1!(N_1 - n_1)!} \times \frac{N_2!}{n_2!(N_2 - n_2)!} \times \cdots \times \frac{N_M!}{n_M!(N_M - n_M)!}, \qquad (2.1)$$

with N_h the population size in stratum h, n_h the sample size taken from stratum h, and M the number of strata. Note that $N_1 + N_2 + \cdots + N_M = N$ and $n_1 + n_2 + \cdots + n_M = n$ represent the total population and sample size, respectively. In the example above we had $M = 3$ strata, a population size of $N_h = 2$ for each stratum, and a sample size $n_h = 1$ for each stratum. The number of possible samples is thus indeed $8 = 2 \times 2 \times 2$.

The probability of collecting any of the K_S samples is again $1/K_S$, with K_S given in Eq. (2.1), the same calculation principle as in simple and systematic sampling. The probability of collecting a unit now depends on the stratum the unit is part of. Using the probability for simple random sampling, the probability for any unit in stratum h is equal to n_h/N_h. If the ratio of the sample size n_h and stratum size N_h is the same across all strata, each unit in the population has the same probability of being collected in the sample. This type of stratified sampling is called *proportional stratified sampling*.

2.4.4 Cluster Sampling

Directly sampling units from populations is not always feasible. For instance, in several countries there are no complete or up-to-date lists of all houses in a certain geographic area. However, using maps of the region, groups or clusters of houses can be identified and these clusters can then be sampled. In other settings, economic considerations are used to form clusters of units that are being sampled. To determine how many hours per day children in the Netherlands play video games, it is logistically easier and financially cheaper to sample schools from the Netherlands and then contact (a random sample of) the children at these schools. Thus *cluster sampling* involves random sampling of groups or clusters of units in the population.

Cluster sampling can be less representative than sampling units directly. For instance, a random sample of 20,000 children from the Netherlands may cover the Netherlands more evenly than a random sample of 20 schools with on average 1,000 students. Additionally, cluster sampling introduces a specific structure in the sample which should also be addressed when the data is being analyzed. The cluster structure introduces two sources of variation in the data being collected. In the example of the number of hours per day that children play video games, children within one school may be more alike in their video game behavior than children from other schools.

These sources of variation need to be quantified to make proper statements on the population of interest.

Cluster sampling can be performed as *single-stage* or in *multiple stages*. A single-stage cluster sample uses a random sample of the clusters and then all units from these clusters are selected. In a *two-stage cluster* sample, the units from the sampled clusters are also randomly sampled instead of taking all units from the cluster. The number of stages can increase in general to any level, depending on the application. For instance, sampling children from the Netherlands can be done by sampling first a set of counties, then a set of cities within counties, then a set of schools within cities, and then finally a set of classes within schools (with or without sampling children from these classes). The sampling units for the first stage (e.g. counties) are referred to as *primary cluster units*. Sampling these different levels of clusters can be performed using simple random sampling, systematic sampling, or even stratified sampling, if certain cluster are put together on certain criteria.

Cluster sampling is in a way related to stratified sampling. For instance, in a two-stage cluster sample, the clusters may be viewed as strata, but instead of collecting units from each stratum, the strata themselves are first being randomly sampled. Since we deal with multiple levels of hierarchical clusters, the calculation of the probability of collecting one unit from the population and the probability of collecting one of the many sample sets is more cumbersome for cluster sampling. Therefore, we do not provide general formulae.

To illustrate the complexity of cluster random sampling, consider a population of children at six schools in a city in the Netherlands, with $N_1 = 500$, $N_2 = 300$, $N_3 = 700$, $N_4 = 1,500$, $N_5 = 1,100$, and $N_6 = 400$ children. Here we have six ($M = 6$) clusters in total. The municipality decided to use a two-stage cluster sampling approach, where first three ($m = 3$) schools are randomly collected and then one hundred ($n = 100$) children are randomly collected within the schools. To list all possible samples will be computationally impossible, since the number K of sample sets will be ridiculously large. For instance, the number of samples just from school 1 alone will already be equal to 2.0417×10^{107}. However, we may still calculate the probability for each unit to enter the sample.

The probability for a unit i in school h is determined by $p_h = n_h/N_h$, as we discussed for simple random sampling. It does not depend on the unit i, but it does depend on the school h. Thus the probabilities for collecting units in the six schools are $p_1 = 1/5$, $p_2 = 1/3$, $p_3 = 1/7$, $p_4 = 1/15$, $p_5 = 1/11$, and $p_6 = 1/4$, respectively. Since each school has the same probability ($1/6$) of being collected and we collect three schools, the probability for a unit being collected in the sample is now equal to $p_1 = 1/10$, $p_2 = 1/6$, $p_3 = 1/14$, $p_4 = 1/30$, $p_5 = 1/22$, and $p_6 = 1/8$, for units in the schools 1, 2, 3, 4, 5, and 6, respectively. Note that the units from the smallest school have the highest probability of being collected ($1/6$), while the probability for each unit in the population would be equal to $3 \times 100/[500 + 300 + 700 + 1500 + 1100 + 400] = 1/15$ for simple random sampling. Furthermore, the probability that school h is collected with simple random sampling is $p_h = N_h/[N_1 + N_2 + \cdots + N_6]$. Thus with simple random sampling the largest school has the highest probability of being represented in the sample, while in cluster random sampling each school has the same probability (1/6).

2.5 Evaluating Estimators Given Different Sampling Plans

In Chap. 1, several descriptive statistics were discussed, like the average, standard deviation, median, quartiles, quantiles, etc. They were introduced to summarize the collected data, but in the context of sampling they can be viewed as so-called *estimators*: quantities that we compute using the data in our sample to say something about the population. For example, can we determine how well the sample average $\bar{x}$, as defined in the previous chapter, estimates the population mean, μ? This means that we would like to determine the "closeness" of the sample average to the population mean. Whatever measure we would like to use for closeness, the sampling approach will influence the performance of the estimator. For instance, the sample average may generally be closer to the population mean under simple random sampling than under cluster random sampling. We will use *bias, mean square error* (MSE), and *standard error* (SE) as measures of closeness and we will illustrate these measures in this section for any type of statistic that we wish to calculate. To do this we will first provide a general framework for random sampling (Cochran 2007), for which simple random sampling, systematic sampling, stratified sampling, and cluster sampling are all a special case. In the third subsection we will illustrate the bias, mean squared error and standard error and in the fourth subsection we show how R can be used for evaluations.

2.5.1 Generic Formulation of Sampling Plans

A formal or mathematical definition for collecting a random sample of size n from a population of units indicated by $\Omega = \{1, 2, 3, \ldots, N\}$, $N \geq n$, can be described as follows: Let $S_1, S_2, \ldots, S_K$ be subsets of the population Ω, $S_k \subset \Omega, k = 1, 2, \ldots, K$, such that each subset S_k has n unique units from Ω and the union of all units from $S_1, S_2, \ldots, S_K$ forms the whole population Ω, i.e., $\Omega = \cup_{k=1}^{K} S_k$. Then each subset S_k is attached a probability π_k such that $\pi_k > 0$, for all $k = 1, 2, \ldots, K$, and $\pi_1 + \pi_2 + \cdots + \pi_K = 1$. A random sample of size n is obtained by drawing just one number from $1, 2, 3, \ldots, K$ using the probabilities $\pi_1, \pi_2, \pi_3, \ldots, \pi_K$. Subsets S_1, $S_2, \ldots, S_K$ can be assumed to be unique, $S_k \neq S_l$ when $k \neq l$, since otherwise we can create a unique set by adding the probabilities for the subsets that are equal. This does not mean that there is no overlap in units from different subsets, i.e., we do not require $S_k \cap S_l = \emptyset$. Note that simple random sampling, systematic sampling, stratified sampling, and cluster random sampling all satisfy this definition.

The set of samples $S_1, S_2, \ldots, S_K$ with their probabilities $\pi_1, \pi_2, \pi_3, \ldots, \pi_K$ is referred to as a *sampling plan*. Note that K can be very large and quite different for different sampling plans. It is also good to realize that the sets $S_1, S_2, \ldots, S_K$ and the probabilities $\pi_1, \pi_2, \pi_3, \ldots, \pi_K$ result in a set of probabilities $p_1, p_2, p_3, \ldots, p_N$ for units $1, 2, 3, \ldots, N$ in the population Ω, with $p_i > 0$.[22]

The sampling plan contains all the information necessary to analyze the quality of a sampling procedure. As long as we know $S_1, S_2, \ldots, S_K$ with their respective probabilities $\pi_1, \pi_2, \pi_3, \ldots, \pi_K$ we can use these in any further analysis. Hence, despite the differences between simple random sampling, systematic sampling, stratified sampling, and cluster sampling, our subsequent theory for judging the quality of a sampling plan can be solely stated in terms of the S_k's and π_k's.

2.5.2   Bias, Standard Error, and Mean Squared Error

Consider a population of N units and assume that we are interested in one characteristic or variable of the unit. For instance, the variable could represent height, weight, gender, hours of television watching per week, tensile strength, bacterial contamination, a face rating, etc. Each unit i in the population has a theoretical value x_i that may become available in the sample. Note that we consider numerical values only. The population parameter of interest can be defined by $\theta \equiv \theta(\mathbf{x})$, with $\mathbf{x} = (x_1, x_2, \ldots, x_N)$, as some kind of calculation on *all* theoretical values: for instance, the population mean $\mu = \sum_{i=1}^{N} x_i / N$ or the population variance $\sigma^2 = \sum_{i=1}^{N} (x_i - \mu)^2 / N$.

A sample S_k of size n can now be seen as the set of units, i.e., $S_k = \{i_1, i_2, \ldots, i_n\}$ with $i_h \in \{1, 2, 3, \ldots, N\}$ and all indices unique ($i_h \neq i_l$ when $h \neq l$). With every sample S_k we have observed a vector of observations $\mathbf{x}_k' = (x_{i_1}, x_{i_2}, \ldots, x_{i_n})$, with $'$ indicating the transpose.[23] Based on the observed data we compute the *descriptive statistic* $\hat{\theta}_k = T(\mathbf{x}_k)$ and use it as an *estimate* for the population parameter θ, with T a function applied to the observed data (i.e., some calculation procedure). In many cases the function T is identical to the calculation θ at the population level, but alternative functions may be used depending on the sampling plan. For instance, for estimation of the population mean, we may use average $\bar{x}_k = \sum_{h=1}^{n} x_{i_h} / n = \sum_{i \in S_k} x_i / n$, but we may also use a weighted average $\sum_{i \in S_k} w_i x_i / n$, with the weights adding up to one ($\sum_{i \in S_k} w_i = 1$); see Sect. 2.6. The function T is referred to as the *estimator*.

In general, the value $\hat{\theta}_k$ can be considered an estimate of the population parameter θ when sample S_k would be collected. The estimate $\hat{\theta}_k$ will most likely be different from the population parameter θ, because the sample is just a subset of the population. When the sample is representative the sample result should be "quite close" to the

---

[22] Note that these probabilities do not necessarily add up to one, i.e., $p_1 + p_2 + \cdots + p_N \neq 1$, since we allow $S_k \cap S_l \neq \emptyset$. Furthermore, the probabilities are not always the same for each unit.

[23] Vectors are mathematical entities. Here they are used to group the values $x_1, x_2, \ldots, x_n$ on variable x from sample S_k. They are typically denoted as bold face $\mathbf{x}_k$ to distinguish vectors from the single observation x_k. By definition, vectors are presented as columns and to be able to present them as rows we indicate this by $\mathbf{x}_k'$ and call this the *transposed* vector.

population parameter and then the sample result may be considered a good estimate of the population parameter.

2.5.2.1 Bias

The bias of an estimator T, given the sampling plan $S_1, S_2, S_3, \ldots, S_K$ with probabilities $\pi_1, \pi_2, \pi_3, \ldots, \pi_K$ for parameter θ is defined as

$$\text{bias} = \sum_{k=1}^{K} \hat{\theta}_k \pi_k - \theta.$$

In words: the bias is the difference between the weighted average—over all possible K samples—of the sample estimate $\hat{\theta}_k$'s and the true population parameter θ. The weights in this weighted average are provided by the probabilities π_k. Thus, if the bias of an estimator is zero, this means that, if we repeatedly take samples using our sampling plan and repeatedly compute our statistic of interest, the average over all of those statistics is equal to the true population parameter. If the bias is zero, the estimator, under the sampling plan that is being evaluated, is said to be *unbiased*. In statistics, "bias" is an objective statement about an estimator (in combination with a sampling plan), and while unbiased estimates are often preferred, it is not pejorative, unlike the ordinary English use of the term "bias" (hence, it is a bit of a bad term since bias does not necessarily mean "bad").

Note that we often define

$$\mathbb{E}(T) \equiv \sum_{k=1}^{K} \hat{\theta}_k \pi_k,$$

with T the estimator, and call this the *expected population parameter* for the estimator under the sampling plan.[24] The bias of an estimator is thus the difference between this estimator's expected value and the true population value.

2.5.2.2 Mean Squared Error

A small bias of an estimator under a sampling plan does not guarantee that individual sample results $\hat{\theta}_k$ are actually close to the population parameter θ; it just states that they are close on average, if we were to sample over and over again. However, we often only collect one sample, and thus the performance of an estimator on average is not our only concern. We are also concerned with the variability of the estimator

---

[24] The expectation operator $\mathbb{E}()$ is introduced here merely as a shorthand notation, but when we discuss probability and random variables more theoretically in Chap. 4 we will see that the expectation has a more formal interpretation.

across different samples. To capture the variability in the sample results $\hat{\theta}_1, \hat{\theta}_2, \ldots, \hat{\theta}_K$ with respect to the true value θ, we use the so-called mean squared error (MSE). This is defined as

$$\text{MSE} = \sum_{k=1}^{K} \left(\hat{\theta}_k - \theta \right)^2 \pi_k.$$

The MSE measures the weighted average squared distance of the sample results $\hat{\theta}_1, \hat{\theta}_2, \ldots, \hat{\theta}_K$ from the population parameter θ. The weights are again determined by the sampling probabilities. Often, the smaller the MSE the better the sampling plan. Sometimes the root mean square error (RMSE) is reported, which is simply $\text{RMSE} = \sqrt{\text{MSE}}$.

2.5.2.3 Standard Error

Another measure that is relevant to sampling plans, and closely related to the bias and MSE, is the standard error (SE). The standard error is defined as

$$\text{SE} = \sqrt{\sum_{k=1}^{K} \left(\hat{\theta}_k - \mathbb{E}(T) \right)^2 \pi_k}.$$

It represents the variability of the sampling plan with respect to the expected population parameter $\mathbb{E}(T)$ instead of using the true population parameter θ. Note that the standard error of an estimator is used as a measure to represent our uncertainty regarding an estimate. In many examples the standard error contains population parameters (see Sect. 2.6) that we do not know. To use standard errors in practice we have to estimate the standard error as well, and this is what researchers and professionals typically do. We will explore this in more detail below when we derive analytical expressions for the bias and standard error of the sample mean, sample variance, and sample proportion for simple random sampling, systematic sampling, stratified sampling, and one-stage and two-stage cluster random sampling, respectively.

Figure 2.2 shows how bias, MSE, and SE relate: if the bias is small, $\mathbb{E}(T)$ is close to the parameter value θ. On the other hand, if the bias is large, $\mathbb{E}(T)$ is not close to θ. If the MSE is small, the variability of the $\hat{\theta}_k$'s around θ is small, while if the MSE is large, the variability around θ is large. If the SE is small, the variability of the $\hat{\theta}_k$'s around $\mathbb{E}(T)$ is small. Finally, note that if the sampling plan is unbiased and thus $\mathbb{E}(T) = \theta$, the RMSE and the SE are identical. More generally, it can be demonstrated that

$$\text{RMSE} = \sqrt{\text{SE}^2 + (\mathbb{E}(T) - \theta)^2}.$$

Thus the RMSE is never smaller than the SE.

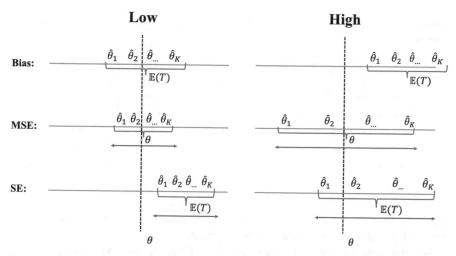

Fig. 2.2 A graphical illustration of the differences (and similarities) between bias, MSE, and SE. The top row of the plot shows that the bias of an estimator $\hat{\theta}$ is high if its expected value over repeated sampling $\mathbb{E}(T)$ is far away from the true value of θ. The middle row shows that the MSE quantifies the dispersion (or spread) of the estimator $\hat{\theta}$ over repeated sampling (i.e., $\hat{\theta}_k$ for each sample k) around the true value of θ. Finally, the standard error quantifies the dispersion of the estimator over repeated sampling around its expected value $\mathbb{E}(T)$

2.5.3 Illustration of a Comparison of Sampling Plans

Here we will assume that we have full knowledge about the population, so that we can evaluate the bias, standard error, and mean squared error for different sampling plans. Clearly, in practice we never have this information, otherwise the sampling becomes obsolete. However, we often do have some knowledge of the population in practice, using information from registries or historical data, and this information can be used to evaluate different strategies, often in combination with simulations (see Sect. 2.5.4).

Our population of interest is provided in Table 2.1, which is taken from Table 2.3 of Levy and Lemeshow (2013). The population consists of six schools in a community with in each school the total number of students and the number of students that were not immunized for measles. In total there are 314 students, of which 30 students are not immunized for measles. The population parameter of interest is $\theta = 30/314 = 0.09554$, the proportion of students not being immunized for measles. We assume that schools 1, 3, and 4 are located in the north of the community and the schools 2, 5, and 6 are located in the south. Two sampling approaches are being considered: a single-stage cluster sample with a simple random sample of two clusters (schools)

Table 2.1 Schools with total number of students and students not being immunized for measles

School	Number of students		Proportion
	Total number	Not immunized	
1	59	4	0.068
2	28	5	0.179
3	90	3	0.033
4	44	3	0.068
5	36	7	0.194
6	57	8	0.140

and a single-stage cluster sample with stratified sampling of two clusters. For the stratified sampling, the strata are north and south.

For the single-stage cluster sampling with simple random sampling, there are $K = 6!/[2! \times 4!] = 15$ possible samples of two schools: $S_1 = (1, 2)$; $S_2 = (1, 3)$; ...; $S_6 = (1, 6)$; $S_7 = (2, 3)$; ...; $S_{10} = (2, 6)$; ....; $S_{15} = (5, 6)$. Each pair of schools has the same probability of being collected, i.e., $\pi_k = 1/K = 1/15$ for $k = 1, 2, \ldots, 15$. The expected population parameter for this sampling approach is given by

$$
\begin{aligned}
\mathbb{E}(T) = \sum_{k=1}^{K} \hat{\theta}_k \pi_k \\
= \left[\frac{9}{87} + \frac{7}{149} + \frac{7}{103} + \frac{11}{95} + \frac{12}{116} + \frac{8}{118} + \frac{8}{72} + \frac{12}{64} \right. \\
\left. + \frac{13}{85} + \frac{6}{134} + \frac{10}{126} + \frac{11}{147} + \frac{10}{80} + \frac{11}{101} + \frac{15}{93} \right] \times \frac{1}{15} \\
= 0.10341.
\end{aligned}
$$

The bias is therefore determined by bias $= 0.10341 - 0.09554 = 0.00787$. Thus the simple random sample of two schools (single stage cluster sample) is not fully unbiased, but the bias is relatively small.

The MSE is

$$
\text{MSE} = \left[\left(\frac{9}{87} - \frac{30}{314} \right)^2 + \left(\frac{7}{149} - \frac{30}{314} \right)^2 + \cdots + \left(\frac{15}{93} - \frac{30}{314} \right)^2 \right] \times \frac{1}{15} = 0.00167.
$$

This implies a root mean square error of RMSE $= \sqrt{0.00167} = 0.04087$. From the bias and the MSE we can determine the SE as well: SE $= \sqrt{\text{MSE} - \text{bias}^2} = 0.04010$.

For the single-stage cluster sample with stratified sampling, there are $K = 9$ possible samples of two schools: $S_1 = (1, 2)$; $S_2 = (1, 5)$; $S_3 = (1, 6)$; $S_4 = (2, 3)$; $S_5 = (3, 5)$; $S_6 = (3, 6)$; $S_7 = (2, 4)$; $S_8 = (4, 5)$; $S_9 = (4, 6)$. Each pair has the same probability of being collected, i.e., $\pi_k = 1/K = 1/9$ for $k = 1, 2, \ldots, 9$. The expected population parameter for this sampling approach is given by

$$\mathbb{E}(T) = \sum_{k=1}^{K} \hat{\theta}_k \pi_k$$
$$= \left[\frac{9}{87} + \frac{11}{95} + \frac{12}{116} + \frac{8}{118} + \frac{10}{126} + \frac{11}{147} + \frac{8}{72} + \frac{10}{80} + \frac{11}{101} \right] \times \frac{1}{9}$$
$$= 0.09886.$$

The bias is therefore determined at bias $= 0.09886 - 0.09554 = 0.00331$. Again the sampling approach is not fully unbiased, but it is closer to zero than the simple random sample of two schools.

The MSE is determined at

$$\text{MSE} = \left[\left(\frac{9}{87} - \frac{30}{314} \right)^2 + \left(\frac{11}{95} - \frac{30}{314} \right)^2 + \cdots + \left(\frac{11}{101} - \frac{30}{314} \right)^2 \right] \times \frac{1}{9} = 0.00036.$$

This implies a root mean square error of RMSE $= \sqrt{0.00036} = 0.01910$, which is smaller than the RMSE for the simple random sample of two schools. The SE of this stratified sampling plan is given by SE $= 0.01881$.

Clearly, the single-stage cluster sample with stratification is better than the simple random sample of schools, since the bias is closer to zero and the MSE is lower. The reason is that the proportion of immunization is quite different between the strata. Schools 1, 3, and 4 have a high proportion of immunization, while schools 2, 5, and 6 have a lower proportion of immunization. The stratified sampling approach makes sure that both proportions are being represented in the sample.

We could have considered alternative sampling plans, like simple random sampling and stratified sampling from the complete population of school data. In that case we would have needed the information on student level instead of the aggregated data at the school level. Thus we need a list of all students telling us whether the student would have been immunized for measles or not and then calculate the criteria for all possible sample sets as we just illustrated. However, for these sampling plans there are many sample sets (i.e., the S_k's) and it would be tedious to do so. On the other hand, for these sampling plans there exist generic and explicit mathematical formulas for the bias, MSE, and SE if we wish to estimate a population mean, proportion, or variance. Thus the calculations in these cases are relatively easy: we present these theoretical results in Sects. 2.6, 2.7 and 2.8.

2.5.4 Comparing Sampling Plans Using R

In practice we may use the computer to help us evaluate, in particular when we want to use more complex sampling plans, to study other population parameters, or when we want to investigate other estimators. For instance, we may want to select a sample of high-school children again, similar to the data we have in our dataset high-school.csv, to investigate if television watching behavior has changed since the year 2000. We could take the same sampling approach as before, but we could also study alternative sampling procedures that may be more complex and could not be studied previously, since a historical dataset was not available then. Improving the sampling plan, i.e., reducing MSE, can help us lower the number of samples and costs. Based on the historical data we may investigate if stratification on grades helps us reduce the standard error, or maybe we can reduce the number of schools with an increase in the number of students within schools. These types of evaluations may be mathematically difficult, in particular if several other stratifications and cluster samples are already involved, but it may not be so difficult to study these plans with the computer (now that some data are available). In other cases we may even generate our own data or extent the historical data to help evaluate sampling plans and estimators.

Here we will provide a generic approach with R to be able to investigate sampling plans and estimators. The R codes are relatively straightforward, but they can be extended and made more complex to address specific situations. The goal is to understand the structure, and not the possible sophistication of sampling and programming. Recall that this type of computer approach requires historical data or some knowledge of the population to be able to mimic data from the population. The general structure is the following:

1. generate population data,
2. execute a sampling plan and generate sample data,
3. compute a statistic using the sample data, and
4. execute these steps a large number of times and compute bias, MSE and SE using the results.

Under the assumption that we have appropriately created the population data, this procedure will give us (approximately) the values for bias, MSE, and SE when we repeat the procedure many times. To illustrate this it may be easiest to think about a simple random sample. Each time we draw a sample from the population with the computer we draw in principle from the set $S_1, S_2, \ldots, S_K$ with their probabilities $\pi_1, \pi_2, \pi_3, \ldots, \pi_K$. Thus if we repeat this procedure many times, we will see all sample sets $S_1, S_2, \ldots, S_K$ appear in the proper proportions $\pi_1, \pi_2, \pi_3, \ldots, \pi_K$. With the computer we can draw samples repeatedly, which we cannot do in practice, but the results depend on the quality of the simulated population data. If we are less sure about the appropriateness of the population data, it may be better to introduce variations in the population data to determine if the results of the sampling plan are robust against other possible realistic populations.

We will now provide the R code for the steps listed above. Each time we will create our own functions. If you want to explore different sampling plans, or different estimators, you only need to change the code within that specific function and the program can be executed again. Note that—merely as an example—we will focus on computing the bias, MSE, and SE of the sample mean under simple random sampling.

2.5.4.1   Generating Population Data

We start our R script by writing a function to generate our population data, although in some cases this step can be omitted if we have historical data available. We want to generate a data.frame object that contains the indices of the units i, and the value of the variable x. Furthermore, we will add the option to also include a group membership, which might be useful if, in the future, we want to evaluate cluster sampling. In this case the grouping is independent of the units, but this can be changed if certain structures need to be incorporated. The code looks as follows:

```
> ### Function to generate population data of size N
> ### with variable of interest X and a possible grouping
> ### The function returns a data frame with N rows and
> ### columns i (units), x (variable), and, possibly,
> ### groups (the grouping)
> generate_population_data <- function(N, groups=1) {
+
+    # Generate index:
+    i <- c(1:N)
+
+    # Generate N random numbers uniformly between 0 and 10:
+    x <- runif(N, min=0, max=10)
+
+    # If groups, generate group membership with equal
+    # probability:
+    group <- sample(1:groups, size=N, replace=TRUE)
+
+    return(data.frame(i=i, x=x, group=group))
+
+ }
```

The comments make clear what each line of code does. Calling this function using generate_population_data(100) returns a data.frame consisting of 100 rows. This number can be increased if a larger population is required.

2.5.4.2   Implementing a Sampling Plan

Now that we have a way of creating a population of a certain size, we need to implement our sampling method. Simple random sampling, where each unit has the same probability of being selected, can be implemented as follows:

```
> ### Select a number of units from a population
> ### The function returns a vector of selected units
> sample_from_population <- function(P, n) {
+
+    # Create a vector of selected units
+    # currently for simple random sample
+    selected <- sample(P$i, size=n)
+
+    return(selected)
```

```
+
+ }
```

Again, the comments make clear what each line of code does. Obviously it is easy to change this function to implement different sampling methods but this is outside the scope of the book.

2.5.4.3   Computing an Estimator

After selecting the units in the sample, we need to compute an estimator. Obviously we could use the R functions mean, mode, etc. that we have seen in the previous chapter. However, to make our R script flexible we create a separate function to compute the estimator; we can then just change the content of this function if we want to consider different estimators. The function takes a vector of sample data $x_1, \ldots, x_n$ and returns the value of the estimator:

```
> ### Compute the value of an estimator given
> ### a vector of sample data x
> ### The function returns a scalar
> compute_estimator <- function(x) {
+
+    # Compute the estimator and return
+    est <- sum(x)/length(x)
+    return(est)
+
+ }
```

2.5.4.4   Computing Bias, MSE, and SE

That almost wraps up all the functions we need: we can now create a population, sample from it, and compute estimators. However, we also need a function which, given a whole vector of estimators $\hat{\theta}_1, \ldots, \hat{\theta}_K$ and the true population value θ computes the bias, MSE, and SE. This can be implemented as follows:

```
> print_summary <- function(theta, estimators) {
+
+    # Compute Bias
+    bias <- mean(estimators) - theta
+
+    # Compute MSE
+    mse <- sum( (estimators - theta)^2 ) / length(estimators)
+
+    # Compute SE
+    se <- sqrt(sum( (estimators - mean(estimators))^2 ) /
+           length(estimators))
+
+    # Print nicely by putting into data frame
+    result <- data.frame(bias=bias, mse=mse, se=se)
```

```
+    print(result)
+
+ }
```

Note that this function prints the bias, MSE, and SE to the screen of the user.

2.5.4.5 Putting It All Together

The functions above give us all the ingredients we need to start simulating the results of repeatedly sampling from a population using a specific sampling plan. By executing the sampling plan over and over again, we can find its performance on average. The following code implements our simulation:

```
> ### Putting it all together and running the simulation:
>
> # Set seed, N, n, and the number of simulations
> set.seed(12345)
> N <- 1000
> n <- 10
> sims <- 100000
>
> # Generate the population data:
> P <- generate_population_data(N=N)
>
> # Generate an empty vector to store results
> estimators <- rep(NA, times=sims)
>
> # For each simulation....
> for (j in 1:sims) {
+
+    # Select sample from P...
+    select <- sample_from_population(P=P, n=n)
+    sample_data <- P[select, ]
+
+    # Compute estimator and store result.
+    estimators[j] <- compute_estimator(sample_data$x)
+
+ }
>
> # Compute the population parameter:
> theta <- mean(P$x)
>
> # Compute bias, mse, se and print to screen:
> print_summary(theta=theta, estimators=estimators)
            bias        mse         se
1 -0.0001635409 0.7963648 0.8923927
```

Again, the comments make clear what each line of code does. For this run of the code—and this will differ each time you run the code depending on the random numbers that the computer generates—we find that the bias is −0.0001635409; this is quite close to the value of zero that we know is the analytical solution to our question (see Sect. 2.6). If you try a larger sample you will see that the MSE and SE will diminish in size (see Sect. 2.6 again). Note that we ran this code after setting the so-called seed—basically the starting point of the computers random number generator—using `set.seed(12345)`. If you do the same, you should get the exact same result (for more details on random number generation see the additional material at the end of this chapter).

2.6   Estimation of the Population Mean

For estimators of the population mean $\mu = \sum_{i=1}^{N} x_i / N$ in the form of weighted averages, the bias, MSE, and SE can be formulated mathematically when the sampling plan is simple random sampling, systematic sampling, stratified sampling, and cluster sampling. When we obtain the values $x_1, x_2, \ldots, x_n$ from sample S_k, we may "average" them in different ways. We may feel that some observations are more important or reliable than other observations and we may want to use this in averaging. This can be done using a *weighted average*, where the weights would help quantify how much more one observation is valued over other observations. A weighted average for the data observed from sample S_k is now defined as $\bar{x}_{w,k} = \sum_{i \in S_k} w_{ik} x_i$, with $\sum_{i \in S_k} w_{ik} = 1$. Note that the weights need to add up to one and that the weights may in principle depend on the sample set S_k, although we will restrict ourselves to weights that are independent of S_k, i.e., $w_{ik} = w_i$. If we choose weight $w_1 = 2/n$ and weight $w_2 = 1/(2n)$, the first observation is four times as important as the second observation. If every observation has the same weight, we obtain the arithmetic average $\bar{x}_k = \sum_{i \in S_k} x_i / n$. In practice, weights can depend on other variables like sex and age, in particular in stratified sampling. In this section we will describe the bias, MSE, and SE for weighted averages under the four sampling plans. The SE and MSE depend on population variances, which we will define in the following subsections. Since we have the calculation rule MSE = bias^2 + SE^2 we will mainly focus on bias and MSE. A summary of the theory is provided in Table 2.2 (Cochran 2007).

2.6.1   Simple Random Sampling

Recall that for simple random sampling we are drawing a sample of size n from a population of size N where the number of possible samples K is given by $N! / [n! (N - n)!]$ and each sample S_k is selected with probability $1/K$ (see Sect. 2.4.1). In this sampling plan, it can be demonstrated that the arithmetic average is the only unbiased estimator for the population mean in the class of weighted averages (see Cochran 2007). Thus we will focus on the arithmetic average $\bar{x}_k$ for simple random sampling.

Table 2.2 Bias and MSE for estimators of the population mean

Sampling plan	Estimator	Bias	MSE
Simple random sampling	$\frac{1}{n}\sum_{i=1}^n x_i$	0	$\sigma^2\left(\frac{N}{N-1}\right)\left(1-\frac{n}{N}\right)\frac{1}{n}$
Systematic sampling	$\frac{1}{n}\sum_{i=1}^n x_i$	0	$\sigma^2 - \frac{1}{N}\sum_{h=1}^n\sum_{i=1}^m (x_{hi}-\bar{x}_h)^2$
Stratified sampling	$\sum_{h=1}^M \left(\frac{N_h}{N}\right)\left(\frac{1}{n_h}\sum_{i=1}^{n_h} x_{hi}\right)$	0	$\sum_{h=1}^M\left[\left(\frac{N_h^2(N_h-n_h)}{n_h(N_h-1)N^2}\right)\sigma_h^2\right]$
Single-stage cluster sampling	$\sum_{h=1}^m \left(N_h/\sum_{h=1}^m N_h\right)\bar{x}_h$	≥ 0 [a]	$\sim \frac{M^2}{m(M-1)N^2}\left(1-\frac{m}{M}\right)\sum_{h=1}^M N_h^2(\mu_h-\mu)^2$
	$\frac{M}{mN}\sum_{h=1}^m N_h\bar{x}_h$	0	$\frac{M^2}{m(M-1)N^2}\left(1-\frac{m}{M}\right)\sum_{h=1}^M (N_h\mu_h - N\mu/M)^2$
Two-stage cluster sampling	$\sum_{h=1}^m \left(N_h/\sum_{h=1}^m N_h\right)\bar{x}_h$	≥ 0 [a]	$\sim \frac{M}{mN^2}\left[\left(1-\frac{m}{M}\right)\frac{1}{M-1}\sum_{h=1}^M N_h^2(\mu_h-\mu)^2 + \sum_{h=1}^M\left(\frac{N_h^2(N_h-n_h)}{n_h(N_h-1)}\sigma_h^2\right)\right]$
	$\frac{M}{mN}\sum_{h=1}^m N_h\bar{x}_h$	0	$\frac{M}{mN^2}\left[\left(1-\frac{m}{M}\right)\frac{1}{M-1}\sum_{h=1}^M (N_h\mu_h - N\mu/M)^2 + \sum_{h=1}^M\left(\frac{N_h^2(N_h-n_h)}{n_h(N_h-1)}\sigma_h^2\right)\right]$

[a] Unbiasedness is obtained only when the cluster sizes N_h are all equal. If the cluster sizes are not all equal, the bias is close to zero but positive

To demonstrate that $\bar{x}_k$ is unbiased, recall that each sample S_k has the same probability $\pi_k = 1/K$ of being collected and that $\mathbb{E}(T) = \sum_{k=1}^{K} \bar{x}_k/K = \sum_{k=1}^{K} \sum_{i \in S_k} x_i/(nK)$. Since each unit i has a probability of n/N to occur in the sample, the unit i will be present in nK/N of the samples $S_1, S_2, \ldots, S_K$. Thus $\sum_{k=1}^{K} \sum_{i \in S_k} x_i$ becomes equal to $(nK/N) \sum_{i=1}^{N} x_i$, which implies that $\mathbb{E}(T) = \sum_{i=1}^{N} x_i/N = \mu$ which is the definition of the population mean. This makes the bias equal to bias $= \mathbb{E}(T) - \mu = 0$.

Now that we know that the arithmetic average is an unbiased estimator for the population mean, we also know that the standard error (SE) of the arithmetic average is equal to the root mean squared error (RMSE). The MSE of the arithmetic average is given by (see also Table 2.2)

$$\text{MSE}(\bar{x}_k) = \sigma^2 \left(\frac{N}{N-1} \right) \left(1 - \frac{n}{N} \right) \frac{1}{n}, \tag{2.2}$$

with $\sigma^2 = \sum_{i=1}^{N} (x_i - \mu)^2/N$ the population variance.

The expression of the MSE clearly shows that it becomes equal to zero when the sample size n becomes equal to the population size N. Thus the arithmetic average becomes equal to the population mean when the sample size is equal to the population size. This may seem intuitively obvious, but it is not. The MSE will not become zero when the estimator is biased, even if the sample size is equal to the population size. This may become clear if we use a constant times the arithmetic average as an estimator for estimation of the population mean, i.e., $c\bar{x}_k$, with c some constant. This estimator has a bias $(c-1)\mu$ and a squared standard error of $\text{SE}^2 = c^2\sigma^2(N/(N-1))(1-n/N)/n$.[25] The MSE is now equal to MSE $= [(c-1)\mu]^2 + c^2\sigma^2(N/(N-1))(1-n/N)/n$ and this becomes equal to the squared bias $[(c-1)\mu]^2$ when the sample size is equal to the population size. Although we would favor the arithmetic average $\bar{x}_k$ over $c\bar{x}_k$, the MSE for the estimator $c\bar{x}_k$ can actually be lower than the MSE of the arithmetic average $\bar{x}_k$ for sample sizes smaller than the population, but this depends heavily on μ and c.

2.6.1.1 Estimation of the MSE

Reporting the MSE or RMSE is important in practice, since it helps us evaluate the closeness of the estimator to the population parameter: the smaller the MSE, the closer the estimate would be to the population parameter. However, the expression for the MSE in Eq. (2.2) shows that we cannot use this MSE in practice, as we do not know the population variance σ^2. Thus, the MSE must be estimated too to be able to use it next to the estimator for μ. Section 2.8 shows that an unbiased estimator for σ^2 is given by $[(N-1)/N]s_k^2$, with s_k^2 the sample variance $s_k^2 = \sum_{i \in S_k} (x_i - \bar{x}_k)^2/(n-1)$ that we have seen in Chap. 1. Using this sample variance estimator, the MSE we may

---

[25] For any estimator T we have the following convenient calculation rules: $\mathbb{E}(cT) = c\mathbb{E}(T)$ and $\text{SE}(cT) = c\text{SE}(T)$, where c is a constant.

use in practice becomes

$$\hat{MSE}(\bar{x}_k) = (N - n)s_k^2/(Nn).$$ (2.3)

This is an unbiased estimator of the MSE of the arithmetic average in Eq. (2.2) and it may also be used when the observations are binary. Note that the standard error of $\bar{x}_k$ is just the square root of the $\hat{MSE}$ in Eq. (2.3).

2.6.2 Systematic Sampling

Recall that for systematic sampling we split our population into n groups consisting of m units, where from each group we select the kth unit ($k \in \{1, 2, \ldots, m\}$). Here we will assume that the ratio of the population size N and sample size n is an integer, i.e., $N/n = m \in \{1, 2, 3, \ldots, N\}$ to keep calculations mathematically more simple. The population is now perfectly split into n groups of size m. The sampling plan is S_1, $S_2, \ldots, S_m$, with $S_k = \{k, k + m, k + 2m, \ldots, k + (n - 1)m\}$, and each sample S_k having probability $1/m$ of being collected. The sample average can now be written as $\bar{x}_k = \sum_{i=1}^{n} x_{k+m(i-1)}/n$.

If the population can be perfectly split up into n groups of m units, the sample average $\bar{x}_k$ is an unbiased estimator of the population mean $\mu = \sum_{h=1}^{m} \sum_{i=1}^{n} x_{h+m(i-1)}/N$, with $N = mn$. The mean square error is given by (see Table 2.2)

$$\sigma^2 - \frac{1}{N} \sum_{h=1}^{n} \sum_{i-1}^{m} \left(x_{h+m(i-1)} - \bar{x}_h\right)^2,$$ (2.4)

with σ^2 the population variance given by $\sigma^2 = \sum_{k=1}^{m} \sum_{i=1}^{n} (x_{k+m(i-1)} - \mu)^2/N$ and with $\bar{x}_h$ the sample average for sample set S_h. It is obvious that the MSE of the sample average under systematic sampling is different from the MSE of the sample average under simple random sampling (just compare Eq. (2.4) with Eq. (2.2)). Systematic sampling can be more efficient than simple random sampling, in particular when the variance in the systematic samples is larger than the population variance (which is impossible to verify in practice).

2.6.2.1 Estimation of the MSE

In the general setting for systematic sampling, an unbiased estimation of the MSE in Eq. (2.4) is not possible. The literature has discussed solutions (see, e.g., Were et al. 2015), but this will be outside the scope of this book. However, given that there is no systematic difference between units based on their position in the groups (and the ratio of sample and population size is an integer), the MSE of the sample average under systematic sampling becomes equal to the MSE of the sample average under simple random sampling. In this case, the sample variance multiplied by $(N - 1)/N$, i.e., $[(N - 1)/N]s_k^2$, with $s_k^2 = \sum_{i=1}^{n}(x_{k+m(i-1)} - \bar{x}_k)^2/(n - 1)$, is an unbiased estimator of the MSE in Eq. (2.2).

2.6.3 Stratified Sampling

To discuss the properties of the weighted sample average under stratified sampling we will change the notation for the index of units. Instead of using index i for a unit in the population, we will use the indices (h, i) to indicate the unit $i \in \{1, 2, \ldots, N_h\}$ for units in stratum $h \in \{1, 2, \ldots, M\}$, and with $N_1 + N_2 + \cdots + N_M = N$ the total number of population units. Thus the variable x_{hi} represents the value of unit i in stratum h. The population mean can then be rewritten into

$$\mu = \frac{1}{N} \sum_{h=1}^{M} \sum_{i=1}^{n} x_{hi} = \sum_{h=1}^{m} w_h \mu_h \tag{2.5}$$

with $w_h = N_h/N$ and $\mu_h = \sum_{i=1}^{N_h} x_{hi}/N_h$ the population average in stratum h or the strata mean. Note that the weights add up to one, i.e., $w_1 + w_2 + \cdots + w_M = 1$. Thus the population mean is a weighted mean of the strata means μ_h.

The variance σ_h^2 in stratum h is defined as $\sigma_h^2 = \sum_{i=1}^{N_h}(x_{hi} - \mu_h)^2/N_h$. The relationship between the population variance and the strata variances is given by

$$\sigma^2 \equiv \frac{1}{N} \sum_{h=1}^{M} \sum_{i=1}^{N_h}(x_{hi} - \mu)^2 = \sum_{h=1}^{M} w_h \sigma_h^2 + \sum_{h=1}^{M} w_h(\mu_h - \mu)^2. \tag{2.6}$$

Thus the population variance is the sum of two parts. The first part represents a weighted mean of the within strata variances and the second part represents a weighted mean of the squared distances of the strata means to the population mean. They may be referred to as *within* and *between (strata) variances*.

Now let's assume that we have determined in some way the sample size n_h in stratum h, such that the sum is equal to the total sample size $n = n_1 + n_2 + \cdots + n_m$. In case we draw a simple random sample in each stratum, the possible samples for stratum h are now denoted by $S_{h,1}, S_{h,2}, \ldots, S_{h,K_h}$, with $K_h = N_h!/[n_h!(N_h - n_h)!]$. $S_{h,k}$ is the collected sample in stratum h, the sample mean is given by $\bar{x}_{h,k} = \sum_{i \in S_{h,k}} x_{hi}/n_h$. The sample variance in stratum h is then denoted by $s_{h,k}^2 = \sum_{i \in S_{h,k}}(x_{hi} - \bar{x}_{h,k})^2/(n_h - 1)$.

The bias and standard error within each stratum h now follow the theory of simple random sampling in Sect. 2.6.1. This implies that the bias of the sample average $\bar{x}_{h,k}$ in stratum h is zero for the stratum mean μ_h. Thus the standard error of the sample average in stratum h can now be estimated with $\hat{SE}(\bar{x}_{h,k}) = \sqrt{(1 - f_h)}s_{h,k}/\sqrt{n_h}$, with $f_h = n_h/N_h$ the sample fraction in stratum h.

To obtain an estimate of the overall population mean we need to combine the sample statistics from the different strata. Since the population mean is equal to the weighted strata means with weights $w_h = N_h/N$, the most logical way is to use the weighted strata sample means with the exact same weights. Thus the population mean can be estimated by $\bar{x}_k = \sum_{h=1}^{M} w_h \bar{x}_{h,k}$. This estimate is unbiased. This can be seen easily, since $\bar{x}_{h,k}$ is unbiased for the stratum mean μ_h and the weights w_h are the same weights that are needed to combine the strata means into the population mean in Eq. (2.5).

The MSE of the weighted sample average $\bar{x}_k = \sum_{h=1}^{M} w_h \bar{x}_{h,k}$ can now be determined using simple algebraic calculations. First recall that the SE of an estimator that is multiplied by a constant is the SE of the estimator multiplied by the constant (i.e., $\text{SE}(cT) = c\text{SE}(T)$). Secondly, the simple random samples from the different strata are completely unrelated, which implies that the sum of the squared standard errors of $w_h \bar{x}_{h,k}$ form the squared standard error of the sample average $\bar{x}_k$. Finally, the squared standard error of $\bar{x}_{h,k}$ is equal to its MSE since $\bar{x}_{h,k}$ is unbiased. Thus the MSE of $\bar{x}_k$ is now equal to (see Table 2.2)

$$\text{MSE}(\bar{x}_k) = \sum_{h=1}^{M} w_h^2 \text{MSE}(\bar{x}_{h,k}) = \sum_{h=1}^{M} N_h(1 - f_h)w_h^2 \sigma_h^2 / [(N_h - 1)n_h].$$

2.6.3.1 Estimation of the MSE

Since the strata variance $s_{h,k}^2$ can be used to estimate $\text{MSE}(\bar{x}_{h,k})$, see Sect. 2.6.1, an estimator of the MSE of $\bar{x}_k$ is now given by

$$\hat{\text{MSE}}(\bar{x}_k) = \sum_{h=1}^{M} (1 - f_h)w_h^2 s_{h,k}^2 / n_h. \tag{2.7}$$

An estimate of the standard error of $\bar{x}_k$ is now obtained by taking the square root of the estimated MSE in Eq. (2.7).

2.6.4 Cluster Sampling

For single-stage cluster sampling, we just perform a simple random sample of m clusters from a total of M available clusters and then evaluate all units in the collected clusters. If cluster h is being collected in our single-stage cluster sample, we would collect or observe the mean $\bar{x}_h$ on all N_h units in this cluster. Thus here we sample the cluster mean $\bar{x}_h = \mu_h$. To estimate the population mean $\mu = \sum_{h=1}^{M} \sum_{i=1}^{N_h} x_{hi}/N$, with x_{hi} the value on variable x for unit i in cluster h and $N = \sum_{h=1}^{M} N_h$, there are several possibilities. Initially, we may be inclined to use the theory from simple random sampling, since we draw a simple random sample of clusters. However, this may not be the best choice. Indeed, for simple random sampling we would calculate the arithmetic average of the cluster means that we have collected, i.e., $\bar{x} = \sum_{h=1}^{m} \bar{x}_h/m$. This estimator is an unbiased estimator of $\sum_{h=1}^{M} \mu_h/M = \sum_{h=1}^{M} \sum_{i=1}^{N_h} x_{hi}/(MN_h)$, but this mean is not equal to the population mean. This is obvious, since the arithmetic average of cluster means does not weigh the size of the clusters in the calculation. A large cluster will have the same weight in the arithmetic average $\bar{x}$ as a very small cluster. Thus only when the cluster sizes are all equal ($N_1 = N_2 = \cdots = N_M$) will the average $\bar{x}$ be an unbiased estimator of the population mean μ.

Two commonly used approaches in practice for estimation of the population mean under single-stage cluster sampling are listed in Table 2.2. Both approaches calculate the total sum of observations $\sum_{h=1}^{m} N_h\bar{x}_h$ from the units in the collected clusters, but one approach considers the cluster sizes N_h as being fixed numbers, while the other approach treats the cluster sizes as being random numbers. If the cluster sizes are considered fixed, the total sum of observations is divided by the total number of units $\sum_{h=1}^{m} N_h$ that were collected in the single-stage cluster sample, giving the estimator $\bar{x}_F = \sum_{h=1}^{m} N_h\bar{x}_h/\sum_{h=1}^{m} N_h$. If the cluster sizes are considered random, the total sum of observations is divided by the expected number of units $(m/M) \cdot N$ in the single-stage cluster sample, with $N = \sum_{h=1}^{M} N_h$ being the population size. This gives the estimator $\bar{x}_R = [M/(mN)] \sum_{h=1}^{m} N_h\bar{x}_h$ and it would require knowledge of the number of clusters and the total population size. Note that $\bar{x}$, $\bar{x}_F$, and $\bar{x}_R$ are all equal when the cluster sizes are all equal.

In Sect. 2.5.3 we illustrated the bias and MSE for a single-stage cluster sample on a binary variable for a population of children who were clustered by schools (see the data in Table 2.1). We applied the first or fixed cluster size approach ($\bar{x}_F$) for estimating the population proportion and demonstrated that it was somewhat biased. It is unbiased only when the cluster sizes are all equal, but the bias is often very small and can be ignored in practice. On the other hand, the second approach ($\bar{x}_R$), where we would assume the population sizes are random, is unbiased in all cases. To illustrate this, consider the data on non-immunized children in Table 2.1. We draw randomly one pair of schools for our single-stage cluster and there are in total 15 possible pairs of schools that could be collected. Each pair has a probability of $1/15$ to be collected. The total population size is $N = 314$ children with in total 30 events, which gives a population mean of $15/257 = 0.09554$. The expected value of the statistic $\bar{x}_R$, using our generic theory in Sect. 2.5.2, is now given by

$$\mathbb{E}(\tfrac{M}{mN} \textstyle\sum_{h=1}^{m} N_h\bar{x}_h) = \frac{6}{628}[9 + 7 + 7 + 11 + \cdots + 11 + 15]\frac{1}{15}$$

$$= \frac{15}{257} = 0.09554,$$

which implies that there is no bias, since the population proportion is identical to our expected value of the estimator $\bar{x}_R$. According to the formula for MSE in Table 2.2, the MSE for this estimator can be calculated as

$$\frac{6^2}{2(6-1)314^2}\left(1-\frac{2}{6}\right)\left[(4-5)^2+(5-5)^2+\cdots+(8-5)^2\right]=0.0005355,$$

since the mean number of events for one school $N\mu/M$ is equal to 5. Note that this MSE is approximately one third of the MSE that we calculated in Sect. 2.5.3 for estimator $\bar{x}_F$, but this does not mean that we should always use $\bar{x}_R$. If the cluster means do not vary a lot, the estimator $\bar{x}_F$ would be preferred over $\bar{x}_R$, but when there is relatively large variation across cluster means, $\bar{x}_R$ would be preferred over $\bar{x}_F$. In other words, when the cluster means are truly *heterogeneous*, $\bar{x}_R$ provides the smallest standard error. In the example with the schools, the cluster proportions of non-immunized children vary substantially or are heterogeneous.

2.6.4.1 Estimation of the MSE

In practice we cannot know the MSE, since it involves all cluster means and the population mean, but we can estimate it from a single-stage cluster sample when we draw more than one cluster. The estimator for the MSE of $\bar{x}_R$ is equal to

$$\hat{\text{MSE}}(\bar{x}_R)=\frac{M-m}{N^2Mm(m-1)}\sum_{h=1}^{m}(MN_h\bar{x}_h-N\bar{x}_R)^2. \tag{2.8}$$

Knowing that the estimator $\bar{x}_R$ is unbiased for the population mean μ, we can estimate the standard error of $\bar{x}_R$ by $\sqrt{(\hat{\text{MSE}}(\bar{x}_R))}$.

Estimators $\bar{x}_F$ and $\bar{x}_R$ for the population mean in a two-stage cluster sample are the same as for the single-stage cluster sampling. The only difference is that the observed average per cluster $\bar{x}_h$ is now based on a simple random sample from cluster h and it is (most likely) not equal to the cluster mean μ_h any more. Based on the theory of simple random sampling we know that the cluster sample average $\bar{x}_h$ is an unbiased estimator of the cluster mean μ_h. This implies that the bias of $\bar{x}_F$ and $\bar{x}_R$ for two-stage cluster sampling would be equal to the bias of $\bar{x}_F$ and $\bar{x}_R$ for single-stage cluster sampling. However, the MSE of $\bar{x}_F$ and $\bar{x}_R$ for two-stage cluster sampling is different from the MSE of $\bar{x}_F$ and $\bar{x}_R$ under single-stage cluster sampling.

In single-stage cluster sampling the MSE of $\bar{x}_F$ and $\bar{x}_R$ is only determined by differences in the cluster means and the population mean. This term describes the variability across cluster means. This can be seen best in the MSE of $\bar{x}_F$, where the distance of μ_h to μ is squared. In two-stage cluster sampling the MSE of $\bar{x}_F$ and $\bar{x}_R$ would also include squared distances of the unit values to their cluster means, or in other words it includes within-cluster variability. That is why there is a term with the cluster variance σ_h^2 in the MSE additional to the term that quantifies the variability among the cluster means. The MSE of $\bar{x}_F$ and $\bar{x}_R$ in two-stage cluster sampling can also be estimated from one two-stage cluster sample when more than one cluster is collected, but this is outside the scope of our book. More information can be found in the literature (see Cochran 2007; Levy and Lemeshow 2013).

2.7   Estimation of the Population Proportion

Population proportions are very much similar to population means. The population values $z_1, z_2, \ldots, z_N$ are now represented by binary values, i.e., $z_i \in \{0, 1\}$, and the population proportion η is obtained by the fraction of units with the value 1, i.e., the population mean $\eta = \sum_{i=1}^{N} z_i / N$. Binary variables occur if the unit has only two possible outcomes, like having a disease or not, but a binary variable is often created from other variables. For instance, having hypertension or high blood pressure is often defined by a systolic blood pressure of 140 mm Hg or higher. Thus the binary variable z_i in this case is created from a continuous variable x_i using $z_i = 1_{[C,\infty)}(x_i)$, with $1_A(x)$ an indicator variable being equal to 1 if $x \in A$ and 0 otherwise. In the example of hypertension, the population proportion represents the fraction or proportion of people with a systolic blood pressure equal to or higher than 140 mm Hg, i.e., $\eta = \sum_{i=1}^{N} 1_{[C,\infty)}(x_i) / N$.

When the characteristic of the unit is binary, the variability in the values z_1, $z_2, \ldots, z_N$ is determined by the proportion η. The population variance σ^2 for the binary values is $\sum_{i=1}^{N}(z_i - \eta)^2 / N$, but this can be rewritten into

$$\sigma^2 = \sum_{i=1}^{N}(z_i - \eta)^2 / N = \sum_{i=1}^{N} z_i^2 / N - \eta^2 = \sum_{i=1}^{N} z_i / N - \eta^2 = \eta(1 - \eta).$$

Thus the population mean and population variance for binary variables are perfectly related. Knowing the population mean implies that you can calculate the population variance, but also the other way around. Knowing the population variance implies that you can calculate the population mean.

This relationship between the population mean and population variance for binary variables is slightly different for sampled data. If $z_1, z_2, \ldots, z_n$ is a sample of binary data from sample S_k (recall the notation in Sect. 2.5.1), the sample average is defined as $\hat{\eta}_k = \sum_{i \in S_k} z_i / n$, but the sample variance is $s_k^2 = \sum_{i \in S_k}(z_i - \hat{\eta}_k)^2 / (n - 1) = n \hat{\eta}_k (1 - \hat{\eta}_k) / (n - 1)$. The difference in relations for samples and populations is caused by the use of $n - 1$ in the sample variance, but N in the population variance. We could decide to use n for the sample variance instead of $n - 1$, but then the sample variance $s_k^2 = \sum_{i \in S_k}(x_i - \bar{x}_k)^2 / n$ for the population variance σ^2 for any variable x becomes biased (see Sect. 2.8). So, we wish to use $n - 1$ in the sample variance. Alternatively, we could use $N - 1$ in the population variance instead of N, and this is often done in literature to simplify the estimate of the MSE for the sample average in Eq. (2.3), but then the population variance for binary variables would contain the population size. So, here there is a trade-off and we choose to define the population variance by $\sigma^2 = \sum_{i=1}^{N}(x_i - \mu)^2 / N$, with μ the population mean and N the total population size. This is applied for any variable x, including binary variables.

Since population proportions can be seen as population means, the theory from Sect. 2.6 can be applied on the binary samples $z_1, z_2, \ldots, z_n$. Practically, this means that we just replace the variable x by the binary variable z in each formula for sample statistics used in Sect. 2.6. This applies to all the discussed sampling plans. Some formulas can be rewritten making use of the relation between (population and sample) means and variances, but this is left to the reader, as in practice there is no need to rewrite the formulas. We just use the same statistics, whether the variable is binary or not.[26]

2.8 Estimation of the Population Variance

Reporting an estimate for the population mean is much more common in society than reporting an estimate for the population variance or standard deviation, because governmental institutes, marketing bureaus, and other agencies are typically more interested in means or totals, like unemployment rates, economic growth, number of sales, number of views per website per day, etc. However, the variability among units is also relevant. A statement by the government that consumer spending is (most likely) increasing by 5% next year is a statement about the population mean, but it does not provide the full picture. If on average the consumer spending is increasing, this does not mean that all groups increase in consumer spending. For instance, elderly people or students may not increase in their spending even though the mean increases. Thus differences between units or groups of units is also important. These differences are often quantified by the variance or standard deviation. In this section we will only focus on the population variance for *simple random sampling*, since variances for other sampling plans are more complicated, in particular the derivation of the MSE and SE.

In Sect. 2.6 we have defined the population variance by $\sigma^2 = \sum_{i=1}^{N}(x_i - \mu)^2/N$, with μ the population mean and N the total population size. If we draw a simple random sample S_k (recall the notation in Sect. 2.5.1) of n units, the sample variance was defined in Sect. 2.4.1 as $s_k^2 = \sum_{i \in S_k}(x_i - \bar{x}_k)^2/(n-1)$, with $\bar{x}_k = \sum_{i \in S_k} x_i/n$ the sample average. This sample variance is not an unbiased estimator of the population variance, as it estimates $N\sigma^2/(N-1)$ (see also Sect. 2.7). Again, when the population size is large this ratio is very close to one and s_k^2 would certainly be a proper estimate (which is frequently used in practice), but to obtain an unbiased estimate of σ^2 in all cases it is better to use $(N-1)s_k^2/N$.

In Sect. 2.7 we also mentioned that we should use $n-1$ in the calculation of the sample variance instead of n to obtain an unbiased estimate of the population variance. Indeed, if we were to use $(N-1)\sum_{i \in S_k}(x_i - \bar{x}_k)^2/(Nn)$ as the sample estimator, the expected value becomes $(n-1)\sigma^2/n$ and the bias becomes $-\sigma^2/n$. This bias cannot be neglected if the ratio of population variance and sample size is relatively large. Investigating the variability of systolic blood pressure in a sample of only 15 people with a population standard deviation of 12 mm Hg gives a bias of almost -10. Clearly, when sample sizes increase the bias rapidly decreases and

---

[26] If the variable x is ordinal or nominal the theory in Sect. 2.6 cannot just be used, since ordinal and nominal values cannot be interpreted numerically (see Chap. 1). The binary variable (being either ordinal or nominal) is the exception as long as we code the two possible outcomes as 0 and 1.

there would no longer be a strong argument anymore to use $n - 1$. An intuitive reason for the use of $n - 1$ in the sample variance is that we first have to estimate the population mean μ with the sample average $\bar{x}_k$ and therefore we loose "one degree of freedom" from all n observations or available degrees of freedom. The use of the sample average makes one observation redundant. The divisor $n - 1$ is referred to as the number of *degrees of freedom* used to calculate the sample variance.

The MSE of the estimator $(N - 1)s_k^2/N$, with $s_k^2 = \sum_{i \in S_k}(x_i - \bar{x}_k)^2/(n - 1)$, is complicated to determine and it contains the population kurtosis. It is equal to

$$\text{MSE}\left(\frac{Ns_k^2}{N-1}\right) = \sigma^4\left[\frac{(N-n)(N-1)(Nn-N-n-1)}{N(N-2)(N-3)n(n-1)}\gamma_2 + \frac{2(N-n)}{(N-2)(n-1)}\right],$$
(2.9)

with γ_2 the population excess kurtosis defined by $\gamma_2 = \sum_{i=1}^{N}(x_i - \mu)^4/(N\sigma^4) - 3$. When the excess kurtosis is close to zero and the population is large with respect to the sample size, the MSE of the estimator $(N - 1)s_k^2/N$ is close to $2\sigma^4/(n - 1)$. More generally, when the population size is large, the MSE is approximately $\sigma^4(\gamma_2/n + 2/(n - 1))$, which is often used as an approximation to describe the MSE in Eq. (2.9).

2.8.1   Estimation of the MSE

The MSE in Eq. (2.9) can be estimated by substituting estimates for the population parameters σ^4 and γ_2. An obvious and common estimator of the squared variance σ^4 is of course $(N - 1)^2s_k^4/N^2$, even though this estimator is biased. An estimator for the population kurtosis γ_2 was provided in Chap. 1 by g_2 in Eq. (1.5). This estimator is unfortunately biased for estimation of the population kurtosis, but it is considered an appropriate estimator for the kurtosis. Using these two estimators into Eq. (2.9) and taking the square root, an estimator of the precision of the variance estimator $(N - 1)s_k^2/N$ is obtained. Thus an estimator of the MSE in Eq. (2.9) is now

$$\frac{(N-1)^2s_k^4}{N^2}\left[\frac{(N-n)(N-1)(Nn-N-n-1)}{N(N-2)(N-3)n(n-1)}g_2 + \frac{2(N-n)}{(N-2)(n-1)}\right]$$

and in case we deal with large populations we may use the simplified version $s_k^4[g_2/n + 2/(n - 1)]$.

2.9 Conclusions

In this chapter we discussed how we may collect and use the sample data such that it may say something about the population parameters of interest. This process is called statistical inference. We have shown that the quality of inference depends on the estimators used, as well as the way in which units are sampled from the population. The nice and surprising part of probability sampling is that the collected or sampled values can be used to estimate the population parameters and their precision.

The results presented in this chapter provide a good feel for what is called sampling variability: by taking a sample from a population we always introduce uncertainty in our resulting inferences. However, to take these ideas further, we need to extend our theoretical knowledge: we need more understanding of probability theory. In the next two chapters we will develop this theory, and subsequently we will return to the topic of estimation.

Problems

2.1 Random sampling of $n = 3$ units from a population $\{1, 2, 3, 4, 5, 6, 7, 8, 9\}$ with $N = 9$ units. The units represent three periods. Period one contains units $\{1, 2, 3\}$, period two contains units $\{4, 5, 6\}$, and period 3 contains units $\{7, 8, 9\}$.

1. How many subsets of three units can be formed ignoring the time structure?
2. Considering the time structure, how many systematic subsets of size 3 can be formed?
3. Considering the time structure, how many stratified subsets of size 3 can be formed?
4. For assignments (1), (2), and (3) separately, use R to select one random subset from all possible subsets that you have formed.
5. Use R to draw directly from the population, without creating any subsets, for simple random sampling, systematic sampling, and stratified sampling.

2.2 To determine the proportion of children who are not immunized with measles, a single stage cluster sampling of $n = 3$ schools from the population of $N = 6$ schools described in Table 2.1 is considered. Recall that the population parameter is $\theta = 30/314 = 0.09554$.

1. How many possible simple random samples of three schools are there? Make a collection of all possible simple random samples of three schools and calculate the bias, standard error, and mean square error of this sampling approach.
2. Now stratify the schools in three strata: stratum 1 is $\{1, 3\}$, stratum 2 is $\{4, 6\}$, and stratum 3 is $\{2, 5\}$. How many possible stratified samples of 3 schools are there given these strata? Make a collection of all possible stratified samples of three schools and calculate the bias, standard error, and mean square error of this stratified sampling approach.

3. Next, stratify the schools as follows: stratum 1 is {1, 2}, stratum 2 is {3, 5}, and stratum 3 is {4, 6}. Make a collection of all possible stratified samples of three schools and calculate the bias, standard error, and mean square error of this alternative stratified sampling approach.
4. Which sampling procedure would you prefer if you have to choose from the three options in (1), (2), or (3)?
5. Set your seed to 19,670,912 and subsequently draw one simple random sample of three schools. Given only the data in this sample, what are the estimates of the bias, SE, and MSE?

2.3 Consider the dataset `high-school.csv` with 50,069 high-school children from the Netherlands. The data contains 13 different variables as we have discussed in the beginning of this book. We will use this dataset to answer the following questions. Use seed number 19,670,912.

1. For the numerical variables, what are the population means and population variances?
2. What are the population proportions of children that do not spend time on sports, watch television, or, computer (determine this for the three variables separately)?
3. For the categorical variables, what are the population proportions?
4. Use R to draw a simple random sample of $n = 1,200$ children.

   a. Estimate the means of the numerical variables and provide an estimate of the corresponding standard errors.
   b. Estimate the proportions of children that do not spend time on sports, watch television or computer and provide the standard errors.
   c. Estimate the proportion of female and male children that prefer mathematics and calculate their standard errors.

5. Use R to draw a stratified sample from the 12 provinces.

   a. Explain the choice of sample sizes within each of the provinces.
   b. Estimate the means of the numerical variables and provide an estimate of the corresponding standard errors.
   c. Estimate the proportion of female and male children that prefer mathematics and calculate their standard errors.

6. Which sampling approach in (4) and (5) for the separate variables would you prefer? Why?
7. What would be your recommendation for the sampling approach if you wish to estimate the proportion of children that prefer mathematics?

2.4 In this assignment we will explore the R code presented in Sect. 2.5.4. Make the following changes to the R code and interpret the results (make sure to execute `set.seed(12345)` before every simulation run):

1. Plot the distribution of the estimators of the population mean and add a vertical line at the population mean to the plot.

2. Increase the sample size n to 100, rerun the simulation, and plot the distribution of the estimators again. Do the results from the `print_summary()` function change? Does the distribution of the estimators look different?

3. Use the sample median (instead of the sample mean) as the estimator for the population mean and rerun the simulation. Compare the results from the `print_summary()` function for the two different estimators (use a sample size of $n = 10$ in both simulations). Which estimator performs better in the simulations?

4. Next, our interest focuses on the population variance instead of the population mean. Change the R code such that it uses the sample variance as an estimator for the population variance. Make sure that you use the correct denominator when calculating the sample and population variances. Run the simulation again with a sample size of $n = 10$ and plot the distribution of the estimators. Is the distribution symmetric or is it skewed? Can you think of a reason why the distribution is (not) symmetric?

5. Now we are interested in estimating the population mean using single-stage cluster sampling. Change the R code such that there are 50 clusters (or groups) in our population. In the simulation, draw a simple random sample of five clusters out of these 50 clusters and use all units in the sampled clusters to estimate the population mean.

6. The R code in Sect. 2.5.4 does not use the probability of obtaining a sample π_k explicitly. Why not?

Additional Material I: Generating Random Numbers

Suppose we want to randomly select one out of N numbers with equal probability. In the old days, before the use of computers, we would generate random numbers by (e.g.,) printing out cards numbered 1 to N and subsequently shuffling these cards, with N the number of units in our population. If the shuffling was done well, this would ensure that each number had the same probability of being selected. Nowadays, however, we use computers to generate random numbers.[27]

Most computer *algorithms*[28] that produce random numbers, produce numbers that are determined by the previous number the algorithm generated. By definition such a series of numbers cannot be truly random, since if the algorithm is known the sequence of numbers is known exactly. However, they do *appear* to be random and therefore they are called *pseudorandom*. An effective (pseudo-, but we will drop that) random number generator is characterized by the following properties (see Sheskin 2003):

---

[27] Note that it is not at all easy to produce "real" random numbers; this is true both when we use computers or when we shuffle cards. There is actually a scientific literature on how to shuffle cards such that they are really randomly ordered.

[28] An algorithm is a set of instructions designed to perform a specific task.

1. Each digit or number $0, 1, 2, \ldots, 9$ has the same probability $(1/10)$ of occurring in the sequences of numbers.
2. The random number generator should have a long period before it starts repeating the same sequence.
3. The sequence of numbers should never degenerate, which means that the algorithm should not produce the same number over and over again after a certain period.
4. The algorithm should be able to reproduce the same set of numbers, which is often guaranteed by providing a starting or initial value (the so called *seed*, which we already encountered in the simulation and the assignments of this chapter).
5. The algorithm should be efficient and not utilize lots of computer memory.

Many such algorithms have been developed. One of the earliest algorithms was the *midsquare procedure*. This procedure starts with an initial value $m_0 > 0$, chosen by the user. The value is squared and then the middle r digits are used as the first random number. This process is then repeated to produce the set of numbers. To illustrate this we use the example from Sheskin (2003). The initial value is $m_0 = 4{,}931$ and r is taken equal to 4. Squaring m_0 results into $m_0^2 = 24{,}314{,}761$. The middle four digits are $m_1 = 3{,}147$, which is the first random number in the sequence. To continue this process we obtain $m_2 = 360$ as the next number, since $(3147)^2 = 9{,}903{,}609$. Note that we could have taken 9,036, since there is no exact middle number of length 4. The third random number would be $m_3 = 2{,}960$ since 129,600 is the square of 360. The midsquare method is not used frequently, since the procedure seems to have a short period or may degenerate; hence, it is not a really good pseudo-random number generator.[29]

An improvement over the midsquare procedure is the *midproduct procedure*. The procedure starts with two initial numbers m_0 and m_1 of length r. The numbers are multiplied and the middle r digits are taken for number m_2. Then the numbers m_1 and m_2 are multiplied to generate m_3. This continues to generate the set of pseudo random numbers $m_2, m_3, m_4, \ldots$. In some cases a constant multiplier is used. This means that the numbers m_k, $k = 2, 3, 4, \ldots$, are determined from the middle four digits of the product $m_{k-2} \times m_{k-1}$. Although the midproduct procedure is better than the midsquare procedure it still has a short period or it may degenerate.

A much better approach is the so-called *linear congruential method*. This method produces numbers in the range 0 to $n - 1$ and it uses a recursive relationship. The relationship is given by $m_k = (a \times m_{k-1} + c) \bmod n$, for $k = 1, 2, \ldots$. The parameter a is a constant multiplier, the parameter c is an additive constant or increment, and n is the modulus.[30] The value m_0 is the initial value or seed value. The following example is again from Sheskin (2003). Consider $m_0 = 47$, $a = 17$, $c = 79$, and $n = 100$. This results into $m_1 = (17 \times 47 + 79) \bmod 100 = (878) \bmod 100 = 78$. Continuing the procedure, the sequence of pseudo random numbers would become

---

[29] So, why did we discuss it? Well, because it provides an easy start!.

[30] The operation $x \bmod n$ provide the rest value when x is divided by n. For instance, 26 mod 7 is equal to 5.

$m_1 = 78$, $m_2 = 5$, $m_3 = 64$, $m_4 = 67$, $m_5 = 18$, $m_6 = 85$, etc. To generate a high-quality sequence of random numbers with the linear congruential method, the value n must be large. It has been suggested to take $n = 2^{31} - 1 = 2{,}147{,}483{,}647$.

Instead of producing a set of integer values, the random numbers can be represented as a uniform set of values in between $[0, 1)$, also denoted as RND. This is done by dividing the random numbers by 10^r, with r being the maximum number of digits of the random numbers. For the example of the midsquare procedure r was 4 and the range for the random numbers was 0 to 9999. Thus the value 0 would be theoretically possible, but the value 1 could not be attained, since $9999/10^4 < 1$. For the linear congruential method, we could divide the random numbers by the modulus number n, thus by $n = 100$ in the specific example. Again, the value 1 is not possible, but the value 0 would be attainable. To generate integers from a list of say 1 to N, both boundaries being included, we could just take the integer part of the value $1 + N \times$ RND. For instance, if we want to draw one number from 1 to 6 (to simulate the throwing of a die) and we use the example of the linear congruential method, the value RND would become equal to $RND = m_1/n = 78/100 = 0.78$. The value $1 + N \times RND = 1 + 6 \times 0.78 = 5.68$. The integer value is then 5.

Modern statistical software packages apply substantially more sophisticated recursive algorithms. The Mersenne-Twister algorithm is currently applied in most statistical software packages, like R, Python, SPSS, SAS, etc. The algorithm is outside the scope of this book, but it can be found in Matsumoto and Nishimura (1998).

References

G. Casella, R.L. Berger, *Statistical Inference*, vol. 2 (Duxbury Pacific Grove, CA, 2002)

W.G. Cochran, *Sampling Techniques* (Wiley, Hoboken, 2007)

W. Kruskal, F. Mosteller, Representative sampling, iv: the history of the concept in statistics, 1895–1939, in *International Statistical Review/Revue Internationale de Statistique* (1980), pp. 169–195

P.S. Levy, S. Lemeshow, *Sampling of Populations: Methods and Applications* (Wiley, Hoboken, 2013)

M. Matsumoto, T. Nishimura, Mersenne twister: a 623-dimensionally equidistributed uniform pseudo-random number generator. ACM Trans. Model. Comput. Simul. (TOMACS) 8(1), 3–30 (1998)

J. Neyman, On the two different aspects of the representative method: the method of stratified sampling and the method of purposive selection. *Breakthroughs in Statistics* (Springer, Berlin,1992), pp. 123–150

D.J. Sheskin, *Handbook of Parametric and Nonparametric Statistical Procedures* (CRC Press, Boca Raton, 2003)

J.N. Towse, T. Loetscher, P. Brugger, Not all numbers are equal: preferences and biases among children and adults when generating random sequences. Front. Psychol. 5, 19 (2014)

F. Were, G. Orwa, R. Odhiambo, A design unbiased variance estimator of the systematic sample means. Am. J. Theor. Appl. Stat. 4(3), 201–210 (2015)

Chapter 3
Probability Theory

3.1 Introduction

Statistics is a science that is concerned with principles, methods, and techniques for collecting, processing, analyzing, presenting, and interpreting (numerical) data. Statistics can be divided roughly into descriptive statistics (Chap. 1) and inferential statistics (Chap. 2), as we have already suggested. Descriptive statistics summarizes and visualizes the observed data. It is usually not very difficult, but it forms an essential part of reporting (scientific) results. Inferential statistics tries to draw conclusions from the data that would hold true for part or the whole of the population from which the data is collected. The theory of probability, which is the topic of the next two theoretical chapters, makes it possible to connect the two disciplines of descriptive and inferential statistics.

We have already encountered some ideas from probability theory in the previous chapter. To start with, we discussed the probability of selecting a specific sample π_k and we briefly defined the notion of probability based on the throwing of a dice. In this chapter we work out these ideas more formally and discuss the probabilities of events; we define probabilities and discuss how to calculate with probabilities. In the previous chapter, when discussing bias, we have also encountered the expected population parameter $\mathbb{E}(T)$, but we have not yet detailed what expectations are exactly; this is something we cover in Chap. 4.

To summarize, descriptive statistics only get us so far. If we want to do more interesting things we need to have a formal, theoretical, understanding of probability. This is exactly what we cover in the next two chapters. However, despite being primarily theoretical, we introduce practical examples of each of the concepts we introduce throughout the chapters.

In this chapter we will study:

- Basic principles and terminology of probability theory
- Calculation rules for probability; the probability axioms
- Conditional probability
- Measures of risk, and their association with study designs
- Simpson's Paradox

© Springer Nature Switzerland AG 2022
M. Kaptein and E. van den Heuvel, *Statistics for Data Scientists*, Undergraduate Topics
in Computer Science, https://doi.org/10.1007/978-3-030-10531-0_3

3.2 Definitions of Probability

In daily life, probability has a subjective interpretation, because everyone may have his or her own intuition or ideas about the likelihood of particular events occuring. An *event* is defined as something that happens or it is seen as a result of something. For instance, airplane crashes around the world or congenital anomalies in newborn babies can be considered events. These two types of events are considered rare because their frequency of occurrence is considered small with respect to the possible number of opportunities, but nobody knows exactly what the probability of such an event is. Using information or empirical data, the probability of an event can be made more quantitative. One could assign the ratio of the frequency of an event and the frequency of opportunities for the event to occur as the probability of this event: for instance, the yearly number of newborn babies with a congenital anomaly as a ratio of all yearly newborns or the number of airplane crashes in the last decade as a ratio of the number of flights in the same period.

The possible opportunities for an event to occur can also be viewed as a population of units (e.g., newborns or flights in a particular period of time) and the events can be seen as the population units with a specific characteristic (e.g., congenital anomalies or airplane crashes). In this context the definition of the probability of an event A for a finite population can be given by

$$\Pr(A) = \frac{N_A}{N},\tag{3.1}$$

with N_A the number of units with characteristic A and N the size of the population.

The definition in Eq. (3.1) is only correct if each opportunity for the event to occur is as likely to produce the event as any other opportunity. Indeed, if for instance, congenital anomalies may occur more frequently for older women than for younger women or airplane crashes might occur more frequently for intercontinental flights than for continental flights, the definition in Eq. (3.1) is inappropriate. Of course, in such cases we may reduce the population into a smaller set of units or divide it into several subsets and then apply the definition in Eq. (3.1) to these subsets, but this can only be performed if the number of units in the subsets does not become too small. If we have to create very many subsets, it may happen that the probabilities for these subsets are only equal to 1 or 0, which would make the definition less useful. Another limitation of this definition is that it is defined for finite populations only. In the case of tossing a coin until a head appears, the sequence of tosses can in theory be infinitely long and the definition in Eq. (3.1) seems unsuitable.

This brings us to an alternative, and more theoretical, approach to the definition of probability of an event, which assigns to this probability the proportion of the occurrence of an event obtained from infinitely many repeated and identical trials or experiments under similar conditions. This definition has its origin in gambling; thus we will explain it by considering dice throwing once again: if a die is thrown n times and the event A is the single dot facing up, then the probability $\Pr(A)$ of the event A can be *approximated* by the ratio of the number of throws n_A with a single dot facing up and the total number of throws n, i.e.,

$$\Pr(A) \approx \frac{n_A}{n}.$$

When the number of repeated trials n is increased it is expected that the proportion n_A/n converges to some value p (which would be equal to $1/6$ if the die is *fair*). Thus an alternative definition of probability can now be given by

$$\Pr(A) = \lim_{n \to \infty} \frac{n_A}{n} = p. \tag{3.2}$$

Clearly, this definition is only appropriate when repeated and identical trials can be conducted under almost similar conditions (e.g., gaming and gambling). Thus definition Eq. (3.2) is more conceptual for real situations, since it is impossible to conduct infinitely many or even many of these trials in practice. Indeed, in the case of congenital anomalies, it would be extreme to have each mother deliver (infinitely) many babies just to be able to approximate or apply the definition of probability in Eq. (3.2).

This fact, however, does not imply that the definition in Eq. (3.2) is useless. On the contrary, it merely shows where particular assumptions about probabilities are introduced. For instance, for each pregnant woman we could assign an individual probability parameter of reproducing a newborn baby with a congenital anomaly. If we are willing to assume that the probabilities for these women are all equal, the probability in Eq. (3.1) is an approximation to the probability in Eq. (3.2), but alternatively we could also assume that equal probabilities only exist for pregnant women of certain age. In the Additional material at the end of this chapter we provide a brief overview of the history of probability, highlighting different ways in which people have thought about probabilities and probability theory over the years.

Definition of probability. A formal mathematical framework of probability can be constructed (see the Additional material at the end of this chapter). In this textbook and in context with the above-mentioned definitions Eq. (3.1) and Eq. (3.2), we simply define the probability $\Pr(A)$ of an event A as an (unknown) value between zero and one, $0 \le \Pr(A) \le 1$, where both boundaries are allowed, which could either be approximated by collecting appropriate and real data or by the limit of a proportion of repeated and identical trials.[1] To operationalize probability we also need some calculation rules; we discuss these in the next section.

---

[1] It should be noted here that a probability of zero does not necessarily mean that the event will never occur. This seems contradictory, but we will explain this later. On the other hand, if the event can never occur, the probability is zero.

3.3 Probability Axioms

There are several calculation rules for probabilities, but before we discuss some of them we need to introduce some standard notation on events.

- The *complement* of event A is denoted by A^c and it indicates that event A does not occur. Thus the complement of having a female baby is having a male baby (although there exists literature that suggests that gender or sex is much more fluent).
- The occurrence of two events A and B at the same time is denoted by $A \cap B$. This is often referred to as the *joint* or *mutual* event. If event A represents a congenital anomaly in a newborn baby and event B represents the gender male of the baby, then $A \cap B$ represents the event that the baby is both male and has a congenital anomaly.
- The event that either A or B (or both) occurs is denoted by $A \cup B$. Thus in the example of newborn babies, $A \cup B$ means that the baby is either male (with or without an anomaly) or is female with a congenital anomaly. This is the complement of the event of having a female baby without an anomaly (i.e., $(A \cup B)^c = A^c \cap B^c$).

We also have to provide some additional definitions relevant for probabilities

- The probability of no event must be zero. Not having events is indicated by the empty set $\emptyset$ and the probability is $\Pr(\emptyset) = 0$. For instance, if two events A and B can never occur together (mutually exclusive events), then it follows that $A \cap B = \emptyset$ and $\Pr(A \cap B) = \Pr(\emptyset) = 0$. The mutual event that a newborn baby has an anomaly in its uterus and is also a boy does not exist. This should have probability zero of occurring.
- The probability that event A occurs is one minus the probability that the event A does not occur; thus $\Pr(A) = 1 - \Pr(A^c)$. This rule is based on the assumption that either event A occurs or event A^c occurs. This means that $\Pr(A \cup A^c) = 1$, since we will see either A or A^c.
- We call two events A and B *independent* if and only if the probability of the mutual event is equal to the product of the probabilities of each event A and B separately. Thus the independence of events A and B (denoted by $A \perp B$) is equivalent with $\Pr(A \cap B) = \Pr(A) \cdot \Pr(B)$. Using products of probabilities when independence is given or assumed is applied frequently throughout the book. Note that any event A with the non-event $\emptyset$ is independent: $\Pr(A \cap \emptyset) = \Pr(\emptyset) = 0 = 0 \cdot \Pr(A) = \Pr(\emptyset)\Pr(A)$. Alternatively, if two events with a positive probability ($\Pr(A) > 0$ and $\Pr(B) > 0$) that are also mutually exclusive can never be independent: $0 = \Pr(\emptyset) = \Pr(A \cap B) < \Pr(A)\Pr(B)$. We will discuss dependencies in more detail in Chap. 6.

Using the above definitions we can define the following calculation rules:

1. If the events A and B are independent, then the events A and B^c, the events A^c and B, and the events A^c and B^c are also independent. Demonstrating this fact is left to the reader and part of the assignments.

2. The probability of the occurrence of either event A or B or both is equal to the sum of the probabilities of these events separately minus the probability that both events occur at the same time, i.e., $\Pr(A \cup B) = \Pr(A) + \Pr(B) - \Pr(A \cap B)$. Note that mutually exclusive events A and B imply that $\Pr(A \cup B) = \Pr(A) + \Pr(B)$.

3. The probability of an event A is the sum of the probability of both events A and B and the probability of both events A and B^c, thus $\Pr(A) = \Pr(A \cap B) + \Pr(A \cap B^c)$. This is sometimes referred to as the *law of total probability* and is frequently applied throughout the book. Note that this rule follows directly from the second rule.

Our definition of probability, combined with these calculation rules, jointly compose the core of our theoretical discussion of probabilities. In essence, all of the material in this chapter and the next can be derived from these simple rules. However, you will be surprised by the many interesting results we can find merely based on these simple rules!

3.3.1 Example: Using the Probability Axioms

To explain the calculation rules using a practical example we will make use of a deck of cards. A deck of cards contains 52 playing cards: 13 clubs, 13 diamonds, 13 hearts, and 13 spades. Diamonds and hearts have color red and clubs and spades are black. Each suit has the same 13 different values: 2, 3, 4, 5, 6, 7, 8, 9, 10, jack, queen, king, and ace. Now suppose one card is randomly collected from the deck, then we can answer the following questions:

- What is the probability that this card is a heart?
- What is the probability that this card is not a heart?
- What is the probability that it is a heart and a king?
- What is the probability that the card is a heart or a king?
- Are the events that the card is a heart and is a king independent?

Note that by virtue of our random selection of the card we are actually investigating the probabilities of a specific simple random sample containing a single unit (as discussed in Chap. 2).

The probability that the card is a heart is equal to 1/4. This can be deduced in at least two ways. First of all, there are 52 cards in total (the population of cards) and each card is as likely to be drawn as any other card. The 13 hearts are all favorable for the event or outcome of drawing a heart. Thus using definition Eq. (3.1), the probability is given by $13/52 = 1/4$. An alternative approach is to define the population by the four suits, and only one suit would provide the appropriate event of drawing a card of hearts, leading to 1/4 directly.

The probability that the card is not heart is now 3/4, using the definition that the probability of the complementary event is one minus the probability of the event.

There is only one king of hearts, which leads to the probability of $1/52$ of the event that the card is both heart and a king, using definition Eq. (3.1).

The probability of drawing a king is $4/52 = 1/13$, since there are four kings among the 52 cards. The probability that the randomly collected card is hearts or a king is equal to $1/4 + 1/13 - 1/52 = 4/13$, using the second calculation rule.

Since the product of probabilities of hearts and a king is $(1/4) \cdot (1/13) = 1/52$, which is equal to the probability of drawing the king of hearts, the definition of independence implies that the events are independent.

3.4   Conditional Probability

In some situations probability statements are of interest for a particular subset of outcomes. For instance, what is the probability that the newborn baby has a congenital anomaly *given* that the baby is a boy. This question is of interest because it could be possible that congenital anomalies are more frequent for boys than for girls. This probability is generally not the same as the probability that both events congenital anomaly and gender male occur, because we have excluded events of the type female. If event A represents a congenital anomaly and B represents the event of a male newborn baby, then the probability of interest is the so-called *conditional probability* denoted by $\Pr(A|B)$. We refer to this conditional probability as the probability of event A given event B. It is defined by

$$\Pr(A|B) = \frac{\Pr(A \cap B)}{\Pr(B)}, \text{ when } \Pr(B) > 0. \tag{3.3}$$

Clearly, if event B could never occur (i.e., $\Pr(B) = 0$), there is no reason to define the conditional probability in Eq. (3.3). However, for calculation purposes discussed hereafter we will need to define the conditional probability $\Pr(A|B) \equiv 0$ when $\Pr(B) = 0$.

In the case of newborn babies, one may think that the conditional probability is equal to the probability of a congenital anomaly, but this is only true when the two events congenital anomaly and gender male are independent. Indeed, if the events A and B are independent (and $\Pr(B) > 0$), then $\Pr(A \cap B) = \Pr(A)\Pr(B)$. Using the alternative formulation $\Pr(A \cap B) = \Pr(A|B)\Pr(B)$ of formula Eq. (3.3), the independence of the events A and B results in $\Pr(A|B) = \Pr(A)$. Thus the conditional probability of a congenital anomaly given the child is a boy is then equal to the probability of having a congenital anomaly irrespective of gender.

If we deal with two events A and B, the relevant probabilities can be summarized in the a 2×2 *contingency table* given in Table 3.1. In column A and row B, the probability of the occurrence of both events A and B at the same time is given

Table 3.1 Conditional probabilities in a 2 × 2 contingency table

	A	A^c	
B	$\Pr(A \cap B) = \Pr(A\|B)\Pr(B)$	$\Pr(A^c \cap B) = \Pr(A^c\|B)\Pr(B)$	$\Pr(B)$
B^c	$\Pr(A \cap B^c) = \Pr(A\|B^c)\Pr(B^c)$	$\Pr(A^c \cap B^c) = \Pr(A^c\|B^c)\Pr(B^c)$	$\Pr(B^c)$
	$\Pr(A)$	$\Pr(A^c)$	1

by $\Pr(A \cap B)$. Using the conditional relation Eq. (3.3) it can also be expressed by $\Pr(A|B)\Pr(B)$.[2] For the other three cells the same type of probabilities are presented.

For row B in Table 3.1 the two probabilities in column A and A^c add up to the probability $\Pr(B)$ of event B, since $\Pr(A|B) + \Pr(A^c|B) = 1$. This last relation is due to the fact that if event B has already occurred, the probability for the occurrence of event A or event A^c are given by $\Pr(A|B)$ and $\Pr(A^c|B)$ (see also the second probability rule). Indeed, if we know that the new-born baby is a boy (event B), only the conditional probabilities of congenital anomalies $\Pr(A|B)$ and $\Pr(A^c|B)$ in row B can be observed. If on the other hand we know that the newborn baby is a girl, only the conditional probabilities $\Pr(A|B^c)$ and $\Pr(A^c|B^c)$ in row B^c can be observed. Also note that $\Pr(A|B)\Pr(B)$ and $\Pr(A|B^c)\Pr(B^c)$ add up to $\Pr(A)$, which also follows from the second probability rule after applying definition Eq. (3.3).

3.4.1 Example: Using Conditional Probabilities

Conditional probabilities play an important role in medical testing where the medical test can produce false negative and false positive test results. In this context, the *sensitivity* of a medical test is the probability of a positive test results (disease indicated) when the patient truly has the disease and the *specificity* of the test is the probability of a negative test result (disease not indicated) for patients without this particular disease.

If event A represents a positive test result (and thus A^c represents an event with a negative test result) in a patient and event B represents the presence of the disease (and thus B^c represents the event with no disease), then the sensitivity and specificity of the medical test are given by the conditional probabilities $\Pr(A|B)$ and $\Pr(A^c|B^c)$, respectively (see Table 3.1).

Suppose that a diagnostic test for the detection of a particular disease has a sensitivity of 0.95 and a specificity of 0.9. Assume further that the proportion of patients with the disease (a priori probability) in the target population is equal to 0.7. Thus in this

---

[2] Using definition Eq. (3.3) we can write $\Pr(A \cap B)$ as $\Pr(A|B)\Pr(B)$, as we did in Table 3.1, but also as $\Pr(B|A)\Pr(A)$. Which one to use mostly depends on the practical situation. In Table 3.1 we could have used $\Pr(B|A)\Pr(A)$ as well.

example we have received the following information in Table 3.1: $\Pr(A|B) = 0.95$, $\Pr(A^c|B^c) = 0.9$, and $\Pr(B) = 0.7$. This particular example is for instance discussed by Veening et al. (2009, Chap. 13) for the detection of a hernia. Possible questions of interest are the size of the diagnostic probabilities:

- What is the probability of having the disease when the medical test has provided a positive test result: $\Pr(B|A)$?
- What is the probability of not having the disease when the medical test has provided a negative test result: $\Pr(B^c|A^c)$?

The calculation of these diagnostic probabilities can be obtained by the theorem of Bayes, or just *Bayes Theorem*, which is given by

$$\Pr(B|A) = \frac{\Pr(A \cap B)}{\Pr(A)} = \frac{\Pr(A|B)\Pr(B)}{\Pr(A)},$$

and is easily derived using relation Eq. (3.3).

Using relations $\Pr(A) = \Pr(A|B)\Pr(B) + \Pr(A|B^c)\Pr(B^c)$, see Table 3.1, $\Pr(B^c) = 1 - \Pr(B)$, and $\Pr(A|B^c) = 1 - \Pr(A^c|B^c)$, the conditional probability $\Pr(B|A)$ can be rewritten as

$$\Pr(B|A) = \frac{\Pr(A|B)\Pr(B)}{\Pr(A|B)\Pr(B) + [1 - \Pr(A^c|B^c)][1 - \Pr(B)]}. \tag{3.4}$$

The last expression in Eq. (3.4) contains only the terms $\Pr(A|B)$, $\Pr(A^c|B^c)$, and $\Pr(B)$ which were all provided in the example. Thus it becomes possible to calculate the conditional probability $\Pr(B|A)$. The answer is $0.9568 \approx (0.95 \cdot 0.7) / (0.95 \cdot 0.7 + 0.1 \cdot 0.3)$.

For the probability $\Pr(B^c|A^c)$ a similar formula can be established. It is the same formula Eq. (3.4), but with A replaced by A^c, B replaced by B^c, A^c replaced by A, and B^c replaced by B. If we again apply $\Pr(B^c) = 1 - \Pr(B)$, then we find for the conditional probability $\Pr(B^c|A^c)$ the following formula

$$\Pr\left(B^c|A^c\right) = \frac{\Pr(A^c|B^c)[1 - \Pr(B)]}{\Pr(A^c|B^c)[1 - \Pr(B)] + [1 - \Pr(A|B)]\Pr(B)}. \tag{3.5}$$

Thus the answer to the second question above is $0.8852 \approx (0.9 \cdot 0.3) / (0.9 \cdot 0.3 + 0.05 \cdot 0.7)$.

In the general public, there is little intuition about these conditional probabilities $\Pr(B|A)$ and $\Pr(B^c|A^c)$, when $\Pr(A|B)$, $\Pr(A^c|B^c)$, and $\Pr(B)$ are given, due to the role of $\Pr(B)$. If in our example above the probability of the disease is rare, say $\Pr(B) = 0.005$, the probability of having the disease given a positive test $\Pr(B|A)$ would become approximately equal to 0.046, which is still very low. Without the positive test, the probability for an arbitrary person to have the disease would be equal to $\Pr(B) = 0.005$, but if this person receives a positive test, this probability increases to 0.046. Clearly, there is a huge increase in probability (more than 9 times higher), but since the a priori probability of the disease was very small, the probability on the disease after a positive test remains low.

3.4.2 Computing Probabilities Using R

In Chaps. 1 and 2 we have seen how R can be used to perform descriptive analyses of data and simulate different sampling procedures. R is also a handy calculator for performing simple calculations such as those presented in the examples in Sects. 3.3.1, 3.4.1, and 3.5.4. For instance, consider the example of a medical test from Sect. 3.4.1, where the event A (A^c) represents a positive (negative) test result and the event B (B^c) represents the presence (absence) of a particular disease in a patient. The following probabilities are given: $\Pr(A|B) = 0.95$ (sensitivity of the test), $\Pr(A^c|B^c) = 0.9$ (specificity of the test), and $\Pr(B) = 0.7$ (proportion of patients in the target population having the disease). What we would like to know is the probability of having the disease given that the test is positive, $\Pr(B|A)$, and the probability of not having the disease given that the test is negative, $\Pr(B^c|A^c)$. These probabilities can be calculated using the formulas in Eqs. (3.4) and (3.5), respectively. We can use R to carry out these calculations as follows:

```
> # Specify known probabilities:
> P_A_given_B <- 0.95
> P_notA_given_notB <- 0.9
> P_B <- 0.7
>
> # Calculate Pr(B/A):
> P_A_given_B*P_B/(P_A_given_B*P_B+(1-P_notA_given_notB)*(1-P_B))
[1] 0.9568345
>
> # Calculate Pr(B^c/A^c):
> P_notA_given_notB*(1-P_B)/(P_notA_given_notB*(1-P_B)+
+    (1-P_A_given_B)*P_B)
[1] 0.8852459
```

Note that in the R code above P_A_given_B is the sensitivity $\Pr(A|B)$ and P_notA_given_notB is the specificity $\Pr(A^c|B^c)$.

3.5 Measures of Risk

The example of conditional probabilities for medical tests is a relevant example of probability theory, but it is of course not the only application of probability theory. Conditional probabilities also play a dominant role in the investigation of associations between events (see, for example Jewell 2003; Rothman et al. 2008; White et al. 2008). Two events or variables are considered *associated* when they are not independent. For instance, the probability of a congenital anomaly in newborn babies may be different for boys and girls. If independence is violated, a comparison of the two conditional probabilities of congenital anomalies given boys and girls can be used to quantify the strength of the association. Different *association measures* or *measures of risk* exist for this purpose, such as the *risk difference*, *relative risk*, and *odds ratio*. We discuss each of these in turn.

To discuss these measures, it is convenient to change our notation slightly so that it is easier to see what is an *outcome* and what is an *explanatory variable*. Hence, instead of using A and B for events we will use D to denote the event of interest (outcome, result, *disease*) and E to denote the event that may affect the outcome (risk factor, explanatory event, *exposure*). In our example D is the event of the congenital anomaly and E is the event of being male, but one may consider many other examples:

- Lung cancer (D) with smoking (E) or non-smoking (E^c).
- Product failure (D) with automation (E) or manual processing (E^c).
- Passing the data science exam (D) with the use of our book (E) or the use of other books or no books at all (E^c).

3.5.1 Risk Difference

The *risk difference* or *excess risk* is an absolute measure of risk, since it is nothing more than the difference in the conditional probabilities, i.e.

$$ER = \Pr(D|E) - \Pr\left(D|E^c\right).$$

The risk difference is based on an additive model, i.e., $\Pr(D|E) = ER + \Pr(D|E^c)$. It always lies between -1 and 1.

If the risk difference is positive ($ER > 0$), there is a greater risk of the outcome when exposed (E) than when unexposed (E^c). A negative risk difference ($ER < 0$) implies that the exposure (E) is protective for the outcome. If the risk difference is equal to zero ($ER = 0$), the outcome (D) is independent of the exposure (E).

To see this last statement we will assume that $0 < \Pr(E) < 1$ and then use the definition of conditional probability and the calculation rules from Sect. 3.3.

$$
\begin{aligned}
ER = 0 &\Longleftrightarrow \Pr(D|E) = \Pr(D|E^c) \\
&\Longleftrightarrow \Pr(E^c)\Pr(D \cap E) = \Pr(E)\Pr(D \cap E^c) \\
&\Longleftrightarrow [1 - \Pr(E)]\Pr(D \cap E) = \Pr(E)[\Pr(D) - \Pr(D \cap E)] \\
&\Longleftrightarrow \Pr(D \cap E) = \Pr(D)\Pr(E).
\end{aligned}
$$

Many researchers feel that the risk difference is the most important measure, since it can be viewed as the excess number of cases (D) as a fraction of the population size. If the complete population (of size N) were to be exposed, the number of cases would be equal to $N \cdot \Pr(D) = N \cdot \Pr(D|E)$. If the complete population were unexposed the number of cases would be equal to $N \cdot \Pr(D) = N \cdot \Pr(D|E^c)$. Thus the difference in these numbers of cases indicates how the number of cases were to change if a completely exposed population would change to a completely unexposed population.

3.5.2 Relative Risk

The *relative risk* would compare the two conditional probabilities $\Pr(D|E)$ and $\Pr(D|E^c)$ by taking the ratio, i.e.

$$RR = \frac{\Pr(D|E)}{\Pr(D|E^c)}.$$

It is common to take as denominator the risk of the outcome D for the unexposed group. Thus a relative risk larger than 1 ($RR > 1$) indicates that the exposed group has a higher probability of the outcome (D) than the unexposed one. A relative risk equal to one ($RR = 1$) implies that the outcome and exposure are independent. A relative risk less than one ($RR < 1$) indicates that the unexposed group has a higher probability of the outcome. The relative risk is based on a multiplicative model, i.e., $\Pr(D|E) = RR \cdot \Pr(D|E^c)$.

3.5.3 Odds Ratio

The third measure of risk is also a relative measure. The *odds ratio* compares the odds for the exposed group with the odds for the unexposed group. The *odds* is a measure of how likely the outcome occurs with respect to not observing this outcome. The odds comes from gambling, where profits of bets are expressed as 1 to x. For instance, the odds of 1 to 6 means that it is six times more likely to loose than to win. The odds can be defined mathematically by $O = p/(1 - p)$, with p the probability of winning. The odds of the exposed group is $O_E = \Pr(D|E)/[1 - \Pr(D|E)]$ and the odds for the unexposed group is $O_{E^c} = \Pr(D|E^c)/[1 - \Pr(D|E^c)]$. The odds ratio is now given by

$$OR = \frac{O_E}{O_{E^c}} = \frac{\Pr(D|E)[1 - \Pr(D|E^c)]}{\Pr(D|E^c)[1 - \Pr(D|E)]} = \frac{\Pr(D^c|E^c)}{\Pr(D^c|E)} \times RR. \tag{3.6}$$

Similar to the relative risk, it is common to use the unexposed group as reference group, which implies that the odds of the unexposed group O_{E^c} is used in the denominator.

An odds ratio larger than one ($OR > 1$) indicates that the exposed group has a higher odds than the unexposed group, which implies that the exposed group has a higher probability of outcome D. An odds ratio of one ($OR = 1$) indicates that the outcome is independent of the exposure, and an odds ratio smaller than one ($OR < 1$) indicates that the unexposed group has a higher probability of the outcome.

Note that the odds ratio and relative risk are always ordered. To be more precise, the odds ratio is always further away from 1 than the relative risk, i.e., $1 < RR < OR$ or $OR < RR < 1$. To see this, we will only demonstrate $1 < RR < OR$, since the other ordering $OR < RR < 1$ can be demonstrated in a similar way. If $RR > 1$, we have that $\Pr(D|E) > \Pr(D|E^c)$, using its definition. Since $\Pr(D^c|E) = 1 - \Pr(D|E)$ and $\Pr(D^c|E^c) = 1 - \Pr(D|E^c)$, we obtain that $\Pr(D^c|E) < \Pr(D^c|E^c)$. Combining this inequality with the relation in Eq. (3.6), we see that $OR > RR$. Note that the odds ratio and relative risk are equal to each other when $RR = 1$ (or $OR = 1$).

3.5.4   Example: Using Risk Measures

To illustrate the calculations of the different risk measures we will consider the example of Dupuytren disease (outcome D) and discuss whether gender has an influence on this disease (E is male). Dupuytren disease causes the formation of nodules and strains in the palm of the hand that may lead to flexion contracture of the fingers. Based on a random sample of size $n = 763$ from the population of Groningen, a contingency table with Dupuytren disease and gender was obtained (Table 3.2). More can be found in Lanting et al. (2013).

The probabilities in the middle four cells in Table 3.2 (thus not in the bottom row nor in the last column) represent the probabilities that the two events occur together. Thus the probabilities $\Pr(D|E)$ and $\Pr(D|E^c)$ can be obtained by $\Pr(D|E) = \Pr(D \cap E) / \Pr(E) = 92/348 = 0.2644$ and $\Pr(D|E^c) = \Pr(D \cap E^c) / \Pr(E^c) = 77/415 = 0.1855$, respectively.

Given this information we can compute the risk difference, the relative risk, and the odds ratio:

- The risk difference is now $ER = 0.0788$, which implies that males have 7.88% absolute higher risk of Dupuytren disease than females.
- The relative risk is $RR = (92/348) / (77/415) = 1.4248$. This implies that males have a risk of Dupuytren disease that is almost 1.5 times larger than the risk for females.

Table 3.2 2×2 contingency table for Dupuytren disease and gender

Exposure	Disease outcome		Total
	Dupuytren (D)	No Dupuytren (D^c)	
Male (E)	$92/763 = 0.1206$	$256/763 = 0.3355$	$348/763 = 0.4561$
Female (E^c)	$77/763 = 0.1009$	$338/763 = 0.4430$	$415/763 = 0.5439$
Total	$169/763 = 0.2215$	$594/763 = 0.7785$	1

- The odds ratio of Dupuytren disease for males is $OR = (92 \times 338)/(77 \times 256) = 1.5775$, indicating that the odds for Dupuytren disease in males is more than 1.5 time larger than the odds for females. Thus males have a higher risk for Dupuytren disease.

In the next section we discuss what measures of association or risk we can calculate if we sample from the population in three different ways. Each of the sampling approaches will provide a 2×2 contingency table, just like the one in Table 3.2, but they may not provide estimates of the population proportions.

3.6 Sampling from Populations: Different Study Designs

The odds ratio is often considered more complex than the relative risk, in particular because of the simplicity of interpretation of the relative risk. The odds ratio is, however, more frequently used in practice than the relative risk. An important reason for this is that the odds ratio is symmetric in exposure E and outcome D. If the roles of the exposure and outcome are interchanged the odds ratio does not change, but the relative risk does.

To see this, we will again use the data presented in Table 3.2. Interchanging the roles of E and D results in a relative risk of $\Pr(E|D) / \Pr(E|D^c)$. This relative risk is equal to $(92/169)/(256/594) = 1.2631$ and it is quite different from the relative risk $\Pr(D|E) / \Pr(D|E^c) = 1.4248$. When the roles of E and D are interchanged, the odds ratio becomes $[\Pr(E|D)/(1 - \Pr(E|D))]/[\Pr(E|D^c)/(1 - \Pr(E|D^c))]$. Calculating this odds ratio results in the odds ratio of 1.5775 (as we already calculated in Sect. 3.5.4), since $[(92/169)/(77/169)]/[(256/594)/(338/594)] = (92 \times 338)/(77 \times 256)$. These results can be proven mathematically, we ask you to do so in the assignments.

As we will see hereafter, the symmetry of the odds ratio makes it possible to investigate the association between D and E irrespective of the way that the sample from the population was collected. This would not be the case for the risk difference and the relative risk. There are many ways in which we can select a sample from a population, but three of them are particularly common in medical research: *population-based* (cross-sectional), *exposure-based* (cohort study), and *disease-based* (case-control study). We discuss these three in turn, and also discuss their limitations with respect to calculating the different risk or association measures if there are any.

3.6.1 Cross-Sectional Study

In a cross-sectional study a simple random sample of size n is taken from the population (see Chap. 2). For each unit in the sample both the exposure and outcome are being observed and the units are then summarized into the four cells (E, D), (E, D^c),

(E^c, D), and (E^c, D^c). The 2×2 contingency table would then contain the number of units in each cell, just like we saw in Table 3.2 for Dupuytren disease.

This way of sampling implies that the proportions in the last row ($\Pr(D)$ and $\Pr(D^c)$) and the proportions in the last column ($\Pr(E)$ and $\Pr(E^c)$) of Table 3.1 would be unknown before sampling and they are being determined by the probability of outcome and exposure in the population. The example of Dupuytren disease in Table 3.2 was actually obtained with a population-based sample. Thus the observed probabilities in Table 3.1 obtained from the sample represent unbiased estimates of the population probabilities .

Here we have applied the theory of simple random sampling of Sect. 2.7 for estimation of a population proportion. For instance, if we define the binary variable x_i by 1 if unit i has both events E and D (thus $E \cap D$) and it is zero otherwise, the estimate of the population proportion $\Pr(E \cap D)$ would be the sample average of this binary variable. This sample average is equal to the number of units in cell (E, D) divided by the total sample size n; see also Table 3.2. This would also hold for any of the other cells, including the cells in the row and column totals ($\Pr(D)$, $\Pr(D^c)$, $\Pr(E)$, and $\Pr(E^c)$). Thus if we also want to express the mean squared error (MSE) for estimating any of the probabilities in Table 3.1, we could apply the MSE from Table 2.2 (see Sect. 2.2). Since the 2×2 contingency table with sample data provides proper estimates of the population proportions, the measures of risk that would use these estimates from the sampled contingency table are estimates of the population measures of risk. Thus calculation of the risk difference, the relative risk, and the odds ratio are all appropriate for cross-sectional studies.

3.6.2 Cohort Study

In a cohort study, a simple random sample is taken from the population of units who are exposed and another simple random sample is taken from the population of units who are unexposed. Thus this way of sampling relates directly to stratified sampling discussed in Chap. 2 with the strata being the group of exposed (E) and the group of unexposed (E^c). In each sample or stratum the outcome D is noted and the contingency table in Table 3.1 is filled. In this setting, the probabilities $\Pr(E)$ and $\Pr(E^c)$ are preselected before sampling and are fixed in the sample, whatever they are in the population. Thus the sample and the population may have very different probabilities.

To illustrate this we consider the example of newborn babies again. The outcome will be the occurrence of congenital anomalies (D) and the exposure would represent the gender of the child, with male being the event (E). If we select a random sample of 500 male newborn babies and 1,000 female newborn babies, the observed contingency table would have twice as many girls as boys, but in practice this ratio is approximately one. Thus a consequence is that the probabilities in the cells of the contingency tables are no longer appropriate estimates for the population probabilities, since we have destroyed the ratio in probabilities for E and E^c. This also implies that the probability $\Pr(D \cap E)$ in Table 3.1 does not reflect the true probability in the population either. This would become even more obvious if we assume that the exposure E is really rare in the population ($\Pr(E) \approx 0$) and all units with the exposure also have the outcome D. In the sample we would observe that $\Pr(D \cap E)$ is equal to one, while in the population this probability is close to zero since the exposure hardly occurs.[3]

Despite the fact that we cannot use the joint probabilities in the contingency table as estimates for the population probabilities, the risk difference, the relative risk, and the odds ratio in the sample are all appropriate estimates for the population when a cohort study is used. The reason is that these measures use the conditional probabilities only, where conditioning is done on the exposure. The $\Pr(D|E)$ and $\Pr(D|E^c)$ in the sample do represent the conditional population probabilities.

3.6.3 Case-Control Study

In a case-control study a simple random sample is taken from the population of units having the outcome and from the population of units without the outcome. Thus this way of sampling relates also directly to stratified sampling discussed in Chap. 2 with the strata being the group with outcome (D) and the group without outcome (D^c). In each sample or stratum the exposure of each unit is noted. Thus for disease-based sampling the probabilities $\Pr(D)$ and $\Pr(D^c)$ are known before sampling and are fixed in the sample. This means that the observed probabilities in the sample are inappropriate as estimates for the same probabilities in the population. Thus we cannot estimate how many units in the population have the outcome. Similar to the cohort study, we cannot estimate the joint probabilities $\Pr(D \cap E)$, $\Pr(D \cap E^c)$, $\Pr(D \cap E)$, and $\Pr(D \cap E)$ in the population from the sample. This is similar to the discussion in cohort studies.

The problem with case-control studies is that the conditional probabilities $\Pr(D|E)$ and $\Pr(D|E^c)$ cannot be determined either. To illustrate this, assume the following probabilities in the population $\Pr(D \cap E) = 0.08$, $\Pr(D \cap E^c) = 0.02$, $\Pr(D^c \cap E) = 0.12$, and $\Pr(D^c \cap E^c) = 0.78$. Thus in the population we have $\Pr(D|E) = 0.4$ and $\Pr(D|E^c) = 0.025$, which gives a relative risk of $RR = 16$. Now let us assume that the sample size in the outcome D group is equal to 900 and it is the same as in the D^c group. We would expect the following numbers in the 2×2 contingency Table 3.3, because $\Pr(E|D) = 0.8$ and $\Pr(E|D^c) = 0.133$.

The conditional probabilities $\Pr(D|E)$ and $\Pr(D|E^c)$ are now equal to $\Pr(D|E) = 720/840 = 0.8571$ and $\Pr(D|E^c) = 180/960 = 0.1875$. The relative risk for Table 3.3 is now given by $RR = 4.5714$, which is substantially lower than the relative risk in the population. The odds ratio, though, can still be properly estimated, due to the symmetry of the odds ratio, which does not change if the roles of D and E interchange.

---

[3] If, in this case, the population size(s) were known, we could calculate weighted averages to estimate the population parameters as we did in Chap. 2.

Table 3.3 2×2 contingency table for an artificial case-control study

Exposure	Disease outcome		Total
	D	D^c	
Male	720	120	840
Female	180	780	960
Total	900	900	1800

3.7 Simpson's Paradox

In Sect. 3.5 we discussed three different measures of risk for two types of events (outcome D and exposure E). In Sect. 3.6 we discussed three different observational study designs and demonstrated that not all three measures can be determined in each of these observational studies. However, in practice it is even more complicated, since we should always be aware of a third event C that may change the conclusions if the event data of D and E are split for C and C^c. This issue is best explained through an example. In Table 3.4 we report the numbers of successful removal of kidney stones with either percutaneous nephrolithotomy or open surgery (see Charig et al. 1986 for more details).

The relative risk of removal of kidney stones for percutaneous nephrolithotomy with respect to open surgery (which can be calculated from this dataset, as it is a cohort study) is determined by $RR = (289/350)/(273/350) = 1.0586$. This means that percutaneous nephrolithotomy increases the "risk" of successful removal of the kidney stones with respect to open surgery. However, if the data are split by size of kidney stone, a 2×2 contingency table for stones smaller and larger than 2 cm in diameter can be created. Let's assume that we obtain the two 2×2 contingency tables in Table 3.5. Note that if we combine the two tables into one 2×2 contingency table we obtain Table 3.4. The relative risks for the two sizes of kidney stones separately are determined at $RR_{\leq 2} = 0.9309$ and $RR_{>2} = 0.9417$. Thus it seems that open surgery has a higher success of kidney stone removal than percutaneous nephrolithotomy for both small and large stones. This seems to contradict the results from Table 3.4 and this contradiction is called Simpson's paradox (Simpson 1951), named after Edward Hugh Simpson.

Table 3.4 2×2 contingency table for removal of kidney stones and two surgical treatments

Exposure	Kidney stones outcome		Total
	Removal (D)	No removal (D^c)	
Nephrolithotomy (E)	289	61	350
Open surgery (E^c)	273	77	350
Total	562	138	700

Table 3.5 2×2 contingency table for removal of kidney stones and two surgical treatments by size of kidney stones

Exposure	Kidney Stones Outcome $\leq$ 2cm (C)		Total
	Removal (D)	No Removal (D^c)	
Nephrolithotomy (E)	234	36	270
Open Surgery(E^c)	81	6	87
Total	315	42	357

Exposure	Kidney Stones Outcome > 2cm (C^c)		Total
	Removal (D)	No Removal (D^c)	
Nephrolithotomy (E)	55	25	80
Open Surgery(E^c)	192	71	263
Total	247	96	343

Simpson (1951) demonstrated that the association between D and E in the collapsed contingency table is preserved in the two separate contingency tables for C and C^c whenever one or both of the following restrictions hold true

$$\Pr(D \cap E \cap C) \Pr(D \cap E^c \cap C^c) = \Pr(D \cap E^c \cap C) \Pr(D \cap E \cap C^c)$$

$$\Pr(D \cap E \cap C) \Pr(D^c \cap E \cap C^c) = \Pr(D^c \cap E \cap C) \Pr(D \cap E \cap C^c)$$

The first equation implies that the odds ratio for having the exposure E for the presence or absence of C in the outcome group D is equal to one, i.e.

$$OR_{EC|D} = \frac{\Pr(E|C, D)[1 - \Pr(E|C^c, D)]}{\Pr(E|C^c, D)[1 - \Pr(E|C, D)]} = 1.$$

Thus E and C must be independent in the outcome group D, which means that $\Pr(E \cap C|D) = \Pr(E|D) \Pr(C|D)$. The second equation implies that the odds ratio for the outcome D in the presence or absence of C in the exposed group E is equal to one, i.e.

$$OR_{DC|E} = \frac{\Pr(D|C, E)[1 - \Pr(D|C^c, E)]}{\Pr(D|C^c, E)[1 - \Pr(D|C, E)]} = 1.$$

Thus this means that D and C are independent in the exposed group E, which means that $\Pr(D \cap C|E) = \Pr(D|E) \Pr(C|E)$.

In the example of kidney stones, we see that $\Pr(E \cap C|D) = 234/562$, $\Pr(E|D) = 289/562$, and $\Pr(C|D) = 315/562$. The product of probabilities $\Pr(E|D) \Pr(C|D) = 0.2882$, which is substantially lower than $\Pr(E \cap C|D) = 0.4164$. Additionally, we also obtain $\Pr(D \cap C|E) = 234/350$, $\Pr(D|E) = 289/350$, and $\Pr(C|E) = 270/350$. This shows that $\Pr(D|E) \Pr(C|E) = 0.6370$, which is lower than $\Pr(D \cap C|E) = 0.6686$.

If the two independence requirements are violated, the event C is called a *confounder*. In this case we should report the stratified analysis. Thus for the example of kidney stone removal, the analysis should be conducted on the data in Table 3.5. This means that open surgery has a slightly higher success rate than percutaneous nephrolithotomy.[4]

Simpson's paradox also shows that data analysis is far from trivial and care should be taken when making bold statements about associations of events in populations.

3.8 Conclusion

In this chapter we started our exploration of the theory of probability. To do so, we defined probabilities, and we gave the basic computation rules to work with probabilities. We discussed probabilities (in terms of events) and several derived quantities that are used in practice to summarize data, such as distinct risk measures. Also, we discussed how sampling (or study design) and appropriate risk measures are closely related.

The probability rules we discuss in this chapter provide the foundation for discussing more probability theory; namely the theory of random variables. We will do so in the next chapter. In the additional materials for this chapter you will find a short history of probability. It is interesting to see that the same rules have originated multiple times, using different definitions of probabilities. For now we will continue using our definitions presented here. However, in Chap. 8 we will get back to some of the fundamental discussions and discuss the important role that Eq. (3.4) has in thinking about probabilities.

Problems

3.1 Two fair dice are thrown one by one.

1. What is the probability that the first die shows an odd number of eyes facing up?
2. What is the probability that the sum of the eyes of the two dice is eleven?

3.2 A card is randomly drawn from an incomplete deck of cards from which the ace of diamonds is missing.

1. What is the probability that the card is "clubs"?
2. What is the probability that the card is a "queen"?
3. Are the events "clubs" and "queen" independent?

3.3 In a group of children from primary school there are 18 girls and 15 boys. Of the girls, 9 have had measles. Of the boys, 6 have had measles.

---

[4] Note that Simpson's Paradox, and its solutions, are still heavily debated (see, Armistead 2014 for examples).

1. What is the probability that a randomly chosen child from this group has had measles?
2. If we randomly choose one person from the group of 18 girls, what is the probability that this girl has had measles?
3. Are the events "boy" and "measles" in this example independent?

3.4 In a Japanese cohort study, 5,322 male non-smokers and 7,019 male smokers were followed for four years. Of these men, 16 non-smokers and 77 smokers developed lung cancer.

1. What is the probability that a randomly chosen non-smoker from this group developed lung cancer?
2. What is the probability that a randomly chosen smoker from this group developed lung cancer?
3. Are the events "smoking" and "lung cancer" in this example independent?
4. What is the conditional probability that the patient is a smoker if he has developed lung cancer?

3.5 Prove mathematically that $A \perp B^c$, $A^c \perp B$, and $A^c \perp B^c$ if $A \perp B$.

3.6 In a life table the following probabilities are provided. Females can expect to live to an age of 50 years with a probability of 0.898. The probability drops to 0.571 for females with a life expectancy of 70 years. Given that a woman is 50 years old, what is the probability that she lives to an age of 70 years?

3.7 Suppose a particular disease is prevalent in a population with 60%. The sensitivity and specificity of the medical test for this disease are both 0.9. A patient from this population is visiting the physician and is tested for the disease.

1. What is the probability that the patient has the disease when the patient is tested positively?
2. If the sensitivity is 0.9, what is the minimum required specificity of the medical test to know with at least 95% certainty that the patient has the disease when tested positively?
3. If the specificity is 0.9, what is the minimum required sensitivity of the medical test to know with at least 95% certainty that the patient does not have the disease when tested negatively?

3.8 Use R to carry out the calculations presented in Sect. 3.5.4. First, use the `matrix()` function to store the numbers presented in Table 3.2 in a 3×3 matrix. Use the `dimnames` argument of the `matrix()` function to give the matrix meaningful row and column names. If you do not know how to use the `dimnames` argument try running `?matrix()` or search the Internet. Use the numbers stored in the matrix to calculate the risk difference, the relative risk, and the odds ratio. Give an interpretation of the results.

3.9 Consider the following 2×2 contingency table for the removal of kidney stones using two different treatments:

Treatment	Removal of kidney stones		Total
	Successful	Not successful	
Open surgery	273	77	350
Small incision	289	61	350
Total	552	148	700

1. What do you think is the study design that the researchers of the removal of kidney stones have selected?
2. Calculate the risk difference, the relative risk, and the odds ratio for a successful removal of kidney stones for a small incision with respect to open surgery. Based on these results, formulate your conclusion.
3. Prove mathematically that the odds ratio for outcome D with and without exposure E is the same as the odds ratio for the exposure E with and without the outcome D.

Additional Material I: A Historical Background of Probability

The theory of probability was inspired by and has its origin in gaming and gambling. In the 16th century, the Italian mathematician Girolamo Gardano (1501–1576) is considered the first to have calculated probabilities by theoretical arguments and possibly started the development of modern probability (see David (1955)). Also Galileo-Galilei (1564–1642) discussed probabilities, in particular for throwing three dice, but he may have thought that the problem was of little interest. The two French mathematicians Blaise Pascal (1623–1662) and Pierre de Fermat (1601–1665) discussed more complex calculations of popular dice games in a set of letter correspondences. They are often credited with the development of the first fundamental principles for probability theory. Their correspondence was probably initiated by a seeming contradiction presented by the French nobleman Chevalier de Méré, who believed that scoring a six once in four throws of a die is equal to scoring a double six simultaneously in 24 throws of two dice (see Sheynin (1977)). As we now know, these probabilities are approximately 0.5177 and 0.4914, respectively.

The Dutch scientist Christiaan Huygens (1629–1695) is known to be the first to write a book solely on the subject of probability in 1657, in which a systematic manner of probability calculations was set out for gambling questions. Although he might not have met Pascal or Fermat, he was probably introduced to the theory of probability when he spent time in Paris in 1655. It is believed that the manuscript of Huygens initiated much more interest in the theory of probability. It profoundly influenced two other important contributors to this theory, namely the Swiss mathematician James Bernoulli (1654–1705) and the French mathematician Abraham de

Moivre (1667–1754). They both contributed to probability theory by introducing more complicated calculations in gambling questions, but Bernoulli also provided a philosophical foundation that would make probability suitable for broader applications.

The French mathematician and astronomer Pierre-Simon Laplace (1749–1827) took this development much further and applied probability theory to a host of new applications. It can be claimed that Laplace is responsible for the early development of mathematical statistics (see Stigler (1975)). In 1812 he published the first edition of a book on probability theory with a wide variety of analytical principles. For instance, he presents and applies the theorem of Bayes. This important calculation rule in probability, which plays a role in, for instance, diagnostic tests, as we discussed before, is credited to the English mathematician Thomas Bayes (1702–1761). It was published posthumously in 1764, but it did not receive much attention until Laplace published it in his book. It is unknown if Laplace was aware of the publication in 1764.

A more fundamental mathematical formulation of the definition of probability was developed by the Russian mathematician Andrey Nikolaevich Kolmogorov (1903–1987), who built upon theoretical results of other mathematical scientists. One could claim that the work of Kolmogorov ended the search for a precise mathematical definition of probability that is also comprehensive enough to be useful to describe a large set of practical phenomena.

Additional Material II: A Formal Definition of Probability

The first step is to introduce an outcome space Ω of elementary events (Grimmett et al. 2001). This is the set of outcomes that we may (theoretically) observe. For example, the outcome space Ω can be equal to $\Omega = \{1, 2, 3, 4, 5, 6\}$ if we through a die and each side of the die can finish on top. Then in the second step we need to define what is called a σ-field (or σ-algebra) $\mathscr{F}$. This is a set of subsets of the outcome space Ω. It needs to satisfy the following conditions:

1. The empty set $\emptyset$ must be an element of $\mathscr{F}$. Thus $\emptyset \in \mathscr{F}$.
2. The union of any number of subsets of $\mathscr{F}$ should be part of $\mathscr{F}$. If $A_1, A_2, \ldots \in \mathscr{F}$ then $\cup_{i=1}^{\infty} A_i \in \mathscr{F}$.
3. The complement of any subset in $\mathscr{F}$ is part of $\mathscr{F}$. If $A \in \mathscr{F}$ then $A^c \in \mathscr{F}$.

Note that we may define different σ-fields on the same outcome space. For instance, for the outcome space $\Omega = \{1, 2, 3, 4, 5, 6\}$ we could define $\mathscr{F} = \{\emptyset, \{6\}, \{1, 2, 3, 4, 5\}, \Omega\}$ as the σ-field. This σ-field shows that we are interested in the event of throwing a six. We may also be interested in the event of throwing an odd number, which would imply that the σ-field is equal to $\mathscr{F} = \{\emptyset, \{1, 3, 5\}, \{2, 4, 6\}, \Omega\}$. Alternatively, we may be interested in throwing any outcome, which means that the σ-field would contain any possible subset of Ω. Thus $\mathscr{F}$ would be equal to

$$\mathscr{F} = \{\emptyset, \{1\}, \{2\}, \ldots, \{6\}, \{1, 2\}, \{1, 3\}, \ldots, \{5, 6\}, \{1, 2, 3\}, \{1, 2, 4\}, \ldots, \{4, 5, 6\}, \ldots, \{1, 2, 3, 4, 5, 6\}\}.$$

Note that all three σ-fields satisfy the conditions listed above. Thus the σ-field determines what sort of events we are interested in. The σ-field relates to the question or to the probability of interest.

Now the final step is to define the probability measure Pr on $(\Omega, \mathscr{F})$ as a function from the σ-field $\mathscr{F}$ to the interval $[0, 1]$, i.e., Pr $: \mathscr{F} \to [0.1]$, that satisfies:

1. $\Pr(\emptyset) = 0$.
2. If $A_1, A_2, \ldots$ is a collection of disjoint members of $\mathscr{F}$, i.e., $A_i \in \mathscr{F}$ and $A_i \cap A_j = \emptyset$ for $i \neq j$, then $\Pr\left(\cup_{i=1}^{\infty} A_i\right) = \sum_{i=1}^{\infty} \Pr(A_i)$.

The triplet $(\Omega, \mathscr{F}, \Pr)$ is called a *probability space* and Pr is called the *probability measure*.

Thus if we believe that throwing a six with one die is equal to $1/6$, the probability space may be written as $(\{1, 2, 3, 4, 5, 6\}, \{\emptyset, \{6\}, \{1, 2, 3, 4, 5\}, \Omega\}, \Pr)$ with $\Pr(\{6\}) = 1/6$. Alternatively, if we do not know the probability of throwing a 6, we may introduce an unknown probability θ for the probability of throwing a 6, i.e., $\Pr(\{6\}) = \theta$. The probability space remains what it is, but Pr is changed now.

References

T.W. Armistead, Resurrecting the third variable: a critique of pearl's causal analysis of Simpson's paradox. Am. Stat. **68**(1), 1–7 (2014)

C.R. Charig, D.R. Webb, S.R. Payne, J.E. Wickham, Comparison of treatment of renal calculi by open surgery, percutaneous nephrolithotomy, and extracorporeal shockwave lithotripsy. Br. Med. J. (Clin. Res. Ed.) **292**(6524), 879–882 (1986)

G. Grimmett, D. Stirzaker et al., *Probability and Random Processes* (Oxford University Press, Oxford, 2001)

N.P. Jewell, *Statistics for Epidemiology* (Chapman and Hall/CRC, Boca Raton, 2003)

R. Lanting, E.R. Van Den Heuvel, B. Westerink, P.M. Werker, Prevalence of dupuytren disease in the Netherlands. Plast. Reconstr. Surg. **132**(2), 394–403 (2013)

K.J. Rothman, S. Greenland, T.L. Lash et al., *Modern Epidemiology*, vol. 3 (Wolters Kluwer Health/Lippincott Williams & Wilkins, Philadelphia, 2008)

E.H. Simpson, The interpretation of interaction in contingency tables. J. Roy. Stat. Soc.: Ser. B (Methodol.) **13**(2), 238–241 (1951)

E.P. Veening, R.O.B. Gans, J.B.M. Kuks, *Medische Consultvoering* (Bohn Stafleu van Loghum, Houten, 2009)

E. White, B.K. Armstrong, R. Saracci, *Principles of Exposure Measurement in Epidemiology: Collecting, Evaluating and Improving Measures of Disease Risk Factors* (OUP, Oxford, 2008)

F.N. David, Studies in the History of Probability and Statistics I. Dicing and Gaming (A Note on the History of Probability). *Biometrika*, **42**(1/2), 1–5 (1955)

O.B. Sheynin, Early history of the theory of probability. Archive for History of Exact Sciences, **17**(3), 201–259 (1977)

S.M. Stigler, Studies in the History of Probability and Statistics. XXXIV: Napoleonic statistics: The work of Laplace. *Biometrika*, **62**(2), 503–517 (1975)

Chapter 4
Random Variables and Distributions

4.1 Introduction

In the first chapter we discussed the calculation of some statistics that could be useful to summarize the observed data. In Chap. 2 we explained sampling approaches for the proper collection of data from populations. We demonstrated, using the appropriate statistics, how we may extend our conclusions beyond our sample to our population. Probability sampling required reasoning with probabilities, and we provided a more detailed description of this topic in Chap. 3. The topic of probability seems distant from the type of data that we looked at in the first chapter, but we did show how probability is related to measures of effect size for binary data. We will continue discussing real-world data in this chapter, but to do so we will need to make one more theoretical step. We will need to go from distinct events to dealing with more abstract *random variables*. This allows us to extend our theory on probability to other types of data without restricting it to specific events (i.e., binary data).

Thus, this chapter will introduce random variables so that we can talk about continuous and discrete data. Random variables are directly related to the data that we collect from the population; a relationship we explore in depth. Subsequently we will discuss the *distributions of random variables*. Distributions relate probabilities to outcomes of random variables. We will show that distributions may be considered "models" for describing variables from populations. We will discuss separately distributions for *discrete* random variables and for *continuous* random variables. In each case we will introduce several well-known distributions. In both cases we will also discuss properties of the random variables: we will explain their *expected value, variance*, and *moments*. These properties provide summaries of the population. They are closely related to the mean, variance, skewness, and kurtosis we discussed in Chaps. 1 and 2. However, we will only finish our circle—from data to theory to data—in the next chapter.

© Springer Nature Switzerland AG 2022
M. Kaptein and E. van den Heuvel, *Statistics for Data Scientists*, Undergraduate Topics in Computer Science, https://doi.org/10.1007/978-3-030-10531-0_4

In this chapter we will discuss:

- Populations and density functions
- The definition of random variables and probability distributions
- Probability distributions for continuous random variables
- Probability distributions for discrete random variables
- Formal definitions of means, variances, standard deviations, and other moments
- Examples of parametric probability distributions (Bernoulli, binomial, Poisson, normal, lognormal, uniform, and exponential)
- Using R to work with probability distributions.

4.2   Probability Density Functions

In Chap. 1 we introduced the histogram to visualize our data and we gave an example of a density plot, or in other words a *density function* (see Fig. 1.10). The density function may be viewed as a smooth version of the histogram if we standardize the frequencies on the vertical axis to proportions. It may be viewed as an approximation of the histogram on all units from the population if the population is very large (say million's and million's of units). The density function characterizes the occurrence of values for a specific variable (as depicted on the x-axis) on all units from the population. Since in practice all populations are finite, the density function is an abstract formulation of, or a "model" for, the "frequencies" of all population values. In statistics the density function is typically denoted by the small letter f and it is typically referred to as *probability density function* (PDF).

Since we have assumed that the PDF f is some kind of smooth approximation of the histogram, the PDF must satisfy two important conditions or properties. The first condition is that it cannot be negative, i.e., $f(x) \geq 0$ for every value x that is present in the population. Clearly, we cannot observe a negative frequency or proportion in histograms. We often extend the domain of this PDF to the whole real line $\mathbb{R}$, even though the values from the population may be restricted to a smaller domain. For values of x outside this domain, the PDF can then be defined equal to zero: $f(x) = 0$. For instance, measuring the amount of hours per week that school children watch television ranges theoretically from 0 to 168 hours. A PDF f would then be considered equal to zero for any negative value of x and for values larger than 168 hours (and possibly also for values inside this interval, but that depends on the behavior of all children in the population). The second condition for a PDF is that we assume that the "area under the curve" is equal to one, i.e.,

$$\int_{\mathbb{R}} f(x)dx = 1.$$

This essentially means that 100% of all unit values together form the population. If we were able to observe all values from the population we must have considered or obtained all units from the population. This property makes it possible to relate PDFs to proportions of units in the population (or, as we will see later, to probability statements), as we have already indicated in Chap. 1. For instance, the proportion of school children that watches television for less than or equal to 2 hours per week can now be written as

$$0 \leq \int_{-\infty}^{2} f(x)dx = \int_{0}^{2} f(x)dx \leq \int_{\mathbb{R}} f(x)dx = 1.$$

Indeed, if all children watch television for less than two hours per week, then the integral on the left side would be equal to 1, since watching for less than two hours per week still represents the whole population of school children, but if all school children watch television for more than two hours per week, the integral would be equal to 0, since no child will watch less then two hours per week. In practice the proportion will be somewhere in between 0 and 1, since there will be children who hardly watch any television and those who watch a lot. Thus the integral indicates what proportion of school-children watch television for less than or equal to two hours a week. By studying these proportions (or integrals) for any type of interval, say $[a, b] \subset \mathbb{R}$, we would know or be able to retrieve the shape of the PDF f, or in other words, we would know exactly what proportion of the population would have what set of values. We will discuss this later in more detail.

Many different PDFs exist and they have been proposed over a period of more than two centuries to be able to describe populations and data in practical settings. These functions are often parametric functions, i.e., the PDF is known up to a set of parameters. The PDF is then often denoted by f_{θ}, where θ represents the set or vector of m density parameters: $\theta = (\theta_1, \theta_2, \ldots, \theta_m)^T$.[1] Many books have been written on PDFs, so it would go too far to provide a long list here, but we do want to provide information about the *normal, log normal, uniform,* and *exponential* PDFs to give a flavor of the differences.

4.2.1 Normal Density Function

The normal PDF is very special within statistics, both for theoretical and for practical reasons. We will learn in Chaps. 4, 5, and 6 that it can be used to approximate other PDFs when the sample size or the size of the data is getting large. This has the advantage that important features of the normal density function can be transferred to other densities when the approximation is quite close. Beside these theoretical aspects, the normal PDF has been used often to analyze all kinds of datasets and it is

---

[1] Although the subscript notation that we introduce here is often used, in some contexts the notation $f(\cdot|\theta)$ is preferred to make explicit that the distribution function is *conditional* on the parameters (for example in Chap. 8 of this book). In the current chapter we will, however, use the subscript notation.

Fig. 4.1 Three normal density curves for different choices of the parameters

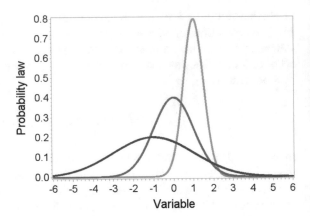

an underlying assumption of several of the modeling approaches that is outside the scope of this book.

The normal PDF was probably first introduced explicitly as a PDF by Carl Friedrich Gauss, and it is therefore often referred to as the *Gauss curve*. He used the normal PDF to describe random errors in measuring orbits (Sheynin 1979), in particular for the calculation of the orbit of the dwarf planet Ceres. In that period the topic was referred to as the "theory of errors". It was an important research area to determine how to deal with measurement errors in calculations. Today the normal PDF is still frequently used for describing data, since many types of measurements, like physical dimensions, are often properly described by the normal PDF.

The normal PDF has just two parameters: μ and σ.[2] The parameter μ indicates the mean value of the population of the variable of interest and the parameter σ indicates the standard deviation. These parameters represent the exact same two population parameters that we discussed in Chap. 2. The shape of the normal PDF is equal to the famous "bell-shape" curve that we have all seen somewhere before. Three different curves are represented in Fig. 4.1.

The normal PDF is mathematically formulated by

$$f_{\mu,\sigma}(x) = \frac{1}{\sigma\sqrt{2\pi}} \exp\left(-\frac{(x-\mu)^2}{2\sigma^2}\right), \tag{4.1}$$

with $\mu \in \mathbb{R}$ and $\sigma^2 > 0$. It is obvious that the normal PDF satisfies the first condition: $f_{\mu,\sigma}(x) > 0$ for all $x \in \mathbb{R}$, but it is not straightforward to show that the integral of this density is equal to one (but it really is).

When we choose $\mu = 0$ and $\sigma = 1$ in Eq. (4.1), we refer to this normal PDF as the *standard normal density function* or *standard normal PDF* and we rather prefer the

---

[2] Although in general we like to denote the parameters of a PDF with $\boldsymbol{\theta} = (\theta_1, \theta_2, \ldots, \theta_m)^T$, for many specific PDFs other notation is used. For the normal PDF we should have used $\boldsymbol{\theta} = (\theta_1, \theta_2)^T$, with $\theta_1 = \mu$ and $\theta_2 = \sigma$, but μ and σ are more common in the literature.

notation ϕ instead of $f_{0,1}$, i.e., $\phi(x) = \exp\{-x^2/2\}/\sqrt{2\pi}$. This means that the normal PDF $f_{\mu,\sigma}(x)$ in Eq. (4.1) can now also be written as $f_{\mu,\sigma}(x) = \phi((x - \mu)/\sigma)/\sigma$.

Some well-known characteristics of the normal PDF are the areas under the curve for specific intervals. For instance, 95.45% of all the population values fall within the interval $[\mu - 2\sigma, \mu + 2\sigma]$, or formulated in terms of the integral:

$$\int_{\mu-2\sigma}^{\mu+2\sigma} \phi((x - \mu)/\sigma)dx = \int_{-2}^{2} \phi(x)dx = 0.9545.$$

Alternatively, 95% of the values fall within $[\mu - 1.96\sigma, \mu + 1.96\sigma]$ and 99.73% of all the population values fall within the interval $[\mu - 3\sigma, \mu + 3\sigma]$.

4.2.1.1 Normally Distributed Measurement Errors

Putting these characteristics in practice we follow the ideas of Gauss on measurement errors. We will assume that the normal PDF can be used to describe random errors in measuring some kind of quantity η (e.g., blood pressure of a human being, the diameter of a planet, the tensile strength of one plastic tube, etc.). We would expect that the population of all possible measurement errors, that we may obtain when we measure the quantity,[3] are on average equal to zero ($\mu = 0$), since it would be as likely to measure higher as lower values than the true value η that we are trying to capture. Based on the shape of the standard normal PDF, it is much more likely to obtain random errors closer to zero than random errors that will be far away from zero. Moreover, approximately 95.45% of all the possible random errors that we may obtain will fall within plus or minus twice the standard deviation away from zero, i.e., $[-2\sigma, +2\sigma]$, and 99.73% will fall within $[-3\sigma, +3\sigma]$. The standard deviation σ is here a measure of the precision of the measurement system.

For instance, the standard deviation of measuring blood pressure with oscillometric devices in human beings is approximately equal to 4.4 and 3.4 mmHg for systolic and diastolic blood pressure, respectively (Liu et al. 2015). Thus 95.45% of the random systolic blood pressure errors fall within $[-8.8, 8.8]$ mmHg and 99.73% will fall within $[-13.2, 13.2]$ mmHg. Thus if we measure a person with a systolic blood pressure $\eta = 120$ mmHg, our blood pressure reading will fall within 111.2 mmHg and 128.8 mmHg with 95.45% certainty and within 106.8 mmHg and 133.2 mmHg with 99.73% certainty. Something similar can be determined for diastolic blood pressure.

---

[3] Here we assume the existence of an infinite population of measurement errors having the normal PDF from which one error e is randomly sampled when we conduct one measurement of the quantity. This error is then added to the true value η to obtain a measurement $x = \eta + e$ of the quantity or a reading of the unit.

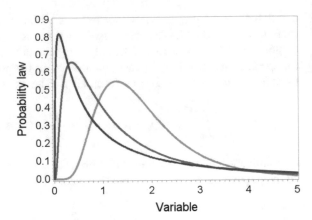

Fig. 4.2 Three lognormal population densities for different values of its parameters

4.2.2   *Lognormal Density Function*

Although the origin of the lognormal PDF comes from a more theoretical setting, the log normal PDF became very popular at the beginning of the 20th century, when the log normal PDF was being used for biological data (Kotz et al. 2004). It has been used in many different applications, ranging from agriculture to economics and from biology to the physical sciences. The lognormal PDF has some very nice properties, which makes it useful in many applications. As we will see, the lognormal PDF describes populations with positive values. This would make more sense than the normal PDF, which describes both positive and negative values, when quantities like particle size, economic growth, duration of games and activities, and measures of size are being studied. Furthermore, the relative standard deviation, which was formulated in Chap. 1, is constant for the lognormal PDF. This means that larger values demonstrate larger variability, but the ratio with variability is constant whether we observe smaller or larger values. Finally, the lognormal PDF is not symmetric like the normal PDF (see Fig. 4.2), which makes sense when values are limited from below, but not from above.

On the other hand, the lognormal PDF is closely related to the normal PDF. If the population values can be described by a lognormal PDF, the normal PDF would then describe the logarithmic transformation of the population values (using the natural logarithm). Thus, the relationship between the normal and lognormal PDFs is based on a log transformation. In practice, we often make use of this transformation, so that we can borrow the normal PDF characteristics in the log scale and then transform the results back to the original scale (using the inverse of the logarithm: $\exp\{x\}$).

The mathematical formulation of the lognormal density is given by

$$f_{\mu,\sigma}\left(x\right) = \frac{1}{x\sigma\sqrt{2\pi}}\exp\left\{-\frac{(\log(x)-\mu)^2}{2\sigma^2}\right\}, \tag{4.2}$$

with $x > 0$, $\log(x)$ the natural logarithm, $\mu \in \mathbb{R}$, and $\sigma^2 > 0$.[4] This PDF has been formulated only on positive values $x > 0$, while we have indicated that the domain of PDFs is typically formulated on the whole real line $\mathbb{R}$. For the part that is left out ($x \leq 0$), the density is then automatically assumed equal to zero. Thus the lognormal density is equal to zero ($f_{\mu,\sigma}(x) = 0$) for values of $x \leq 0$. In Fig. 4.2 a few examples of the log normal PDF are visualized.

Thus the lognormal PDF is also always non-negative for all values of $x \in \mathbb{R}$. Knowing that the integral of the normal PDF is equal to one, helps us to demonstrate that the integral of the lognormal PDF is also equal to one:

$$\int_0^\infty \frac{1}{x\sigma\sqrt{2\pi}} \exp\left\{-\frac{(\log(x) - \mu)^2}{2\sigma^2}\right\} dx = \int_{-\infty}^\infty \frac{1}{\sigma\sqrt{2\pi}} \exp\left\{-\frac{(x - \mu)^2}{2\sigma^2}\right\} dx = 1$$

Thus the lognormal density function also satisfies the two criteria for a PDF.

The parameters μ and σ have a different meaning in the lognormal PDF than in the normal PDF. They do represent the population mean and standard deviation, but only for the logarithmic transformed values of the population. Their meaning in relation to the mean and standard deviation of the population values in the original scale is now more complicated. The population mean and standard deviation in the original scale are now functions of both μ and σ. They are given by $\exp\{\mu + \sigma^2/2\}$ and $(\exp\{\sigma^2\} - 1)\exp\{2\mu + \sigma^2\}$, for the mean and standard deviation, respectively, see the additional material at the end of this chapter. The relative standard deviation (i.e., the standard deviation divided by the mean) is now a function of the parameter σ only: $\sqrt{\exp\{\sigma^2\} - 1}$. This property makes the lognormal density very useful for chemical measurements, where it is often assumed that the measurement error is a fixed percentage of the value that is being measured, i.e., they have a constant relative standard deviation. Indeed, in chemical analysis, the measurement error for higher concentrations is larger than for lower concentrations.

4.2.3 Uniform Density Function

We saw that a random measurement error that could be described by a normal PDF is more likely to be closer to zero than to be further away from zero (due to the bell shape of the density). For a uniform PDF this is different. If the random measurement error would be described by a uniform PDF, being close to or far away form zero would be equally likely. However, the uniform PDF has a finite domain, which means that the density is positive on an interval, say $[\theta_0, \theta_1]$, with $\theta_0 < \theta_1$ and $\theta_0, \theta_1 \in \mathbb{R}$, but zero everywhere else.

---

[4] Note that we use the same notation $f_{\mu,\sigma}$ for the normal PDF and lognormal PDF. This does not mean that the normal and lognormal PDFs are equal, but we did not want to use a different letter every time we introduce a new PDF. We believe that this does not lead to confusion, since we always mention which PDF we refer to.

The mathematical formulation of the uniform PDF is therefore given by

$$f_{\theta_0, \theta_1}(x) = \frac{1}{\theta_1 - \theta_0}, \quad x \in [\theta_0, \theta_1]. \tag{4.3}$$

It is obvious that the density is non-negative on the real line $\mathbb{R}$, since it is positive on $[\theta_0, \theta_1]$ and zero everywhere else. Furthermore, the area under the curve is equal to one, which can easily be determined using standard integral calculations:

$$\int_{\mathbb{R}} f_{\theta_0, \theta_1}(x)\, dx = \int_{\theta_0}^{\theta_1} \frac{1}{\theta_1 - \theta_0} dx = \frac{\theta_1 - \theta_0}{\theta_1 - \theta_0} = 1.$$

Thus the uniform PDF satisfies the two conditions for a PDF. The *standard uniform density* is given by the density in Eq. (4.3) with $\theta_0 = 0$ and $\theta_1 = 1$.

As can be seen from the density function in Eq. (4.3), the uniform PDF has two parameters (θ_0 and θ_1), similar to the normal and lognormal PDF, but the parameters θ_0 and θ_1 have a truly different interpretation. They indicate the lowest and highest values present in the population, or in other words, they represent the minimum and maximum values in the population. The mean and standard deviation for a population that is described by a uniform PDF are equal to $(\theta_0 + \theta_1)/2$ and $(\theta_1 - \theta_0)/\sqrt{12}$, respectively.

As an example, consider that the random measurement error for systolic blood pressure follows a uniform density symmetric around zero and has a population standard deviation of 4.4 mmHg. In this case, the parameters θ_0 and θ_1 would be equal to -7.62 and 7.62, respectively. A systolic blood pressure reading for a person with an actual systolic blood pressure of 120 mmHg would fall within the interval [112.38, 127.62] mmHg with 100% confidence and any value would be as likely as any other value in this interval. Although the uniform PDF is probably no longer used for measurement errors, it was suggested as possible PDFs for the theory of error before the normal density for measurement errors was introduced (Sheynin 1995).

The uniform PDF has also some nice theoretical properties that we will use later in this chapter. We would be able to simulate a population described by any density function through the use of the standard uniform density. If we draw a population using the standard uniform density, we would be able to make a proper transformation of these uniform values such that the transformed values would describe another density. Drawing values according to the uniform density with a computer using a pseudo-random number generator has been discussed in the additional material of Chap. 2.

4.2.4   Exponential Density Function

The exponential PDF has become a very popular density function in practice in the field of reliability, representing the failure times of complex equipment (Barlow

1984). For instance, it may describe the life time of a population of phones that were bought in 2019. Some phones may live for many years, while others may break-down or stop working within months. One important characteristic of the exponential PDF is its lack of memory of failure times. The occurrence of a failure of a machine (like a phone) in, say, week 52, assuming it survives week 51, is the same as the probability that this machine will fail this week (assuming it survived last week). Thus the exponential PDF describes populations with positive values, similar to the lognormal PDF.

The exponential PDF has only one parameter, which is different from the log-normal density, and it is mathematically described for positive x by the following function

$$f_\lambda(x) = \lambda \exp\{-x\lambda\}, \tag{4.4}$$

with $x > 0$ and $\lambda > 0$. For values of x less than or equal to zero, the PDF is equal to zero (as we already mentioned).

It is interesting to note that the exponential PDF is closely related to the double exponential PDF, which had been applied at least a century earlier in relation to the theory of errors (Hald 2008). The double exponential is symmetric, like the normal PDF, but the exponential PDF is skewed to the right like the lognormal PDF (see Fig. 4.3). The double exponential PDF can easily be determined using the exponential PDF. The double exponential PDF, for any value $x \in \mathbb{R}$, is defined by $0.5 f_\lambda(|x|)$, with $|\cdot|$ the absolute function.

Both the exponential and double exponential satisfy the conditions for a PDF. Clearly, they are non-negative on the real line $\mathbb{R}$ and it can be easily shown that the area under the PDF is equal to one. For the exponential PDF we have

$$\int_\mathbb{R} f_\lambda(x)dx = \int_0^\infty \lambda \exp\{-\lambda x\} = [-\exp\{-\lambda x\}]_0^\infty = 1.$$

Since the double exponential PDF is half the exponential PDF on positive values of x and half the exponential PDF on the negative values of x, the area under the double exponential PDF is also equal to one.

A few choices of the exponential PDF are visualized in Fig. 4.3. Visualizing the double exponential is just half the exponential PDF together with its mirror image on the left side of $x = 0$.

For the exponential PDF, the parameter λ is related to the population mean, since λ^{-1} represents the population mean. Thus the smaller the value of λ, the larger the population mean. If the life time of a phone (in years) is described by the exponential density with $\lambda = 0.25$, the mean life-time of a phone is then equal to 4 years. The population standard deviation is also equal to λ^{-1} and therefore depends on the mean of the population, but the relative standard deviation is then independent of the parameter and it is equal to 1. For the double exponential PDF, the mean is equal to zero, which makes it an attractive PDF for measurement errors. The standard deviation of the double exponential is equal to $\sqrt{2}/\lambda$. Thus, the larger the value λ, the closer the measurement errors will be.

Fig. 4.3 Three exponential
density curves

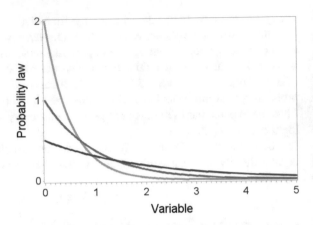

4.3   Distribution Functions and Continuous Random Variables

In the discussions on the theory of errors, PDFs were used to help describe the random measurement errors to be able to come to a proper calculation (often the arithmetic average or median) of the collected observations, like the calculation of the orbit of a celestial body using several measured positions. Whether a calculation of the observations would be better or more precise than just one of these observation was a topic of study.[5] The random errors were considered imperfections of the measurement process (often the human eye) that was trying to capture the true value of interest. These random errors were in that period not always seen as random variables (Sheynin 1995), while the concept of a random variable was already in use long before the theory of errors was discussed. Indeed, random variables were already used when the fundamentals of probability were developed many years earlier.

  An intuitive definition of a *random variable* or *random quantity* is a variable for which the value or outcome is unknown and for which the outcome is influenced by some form of random phenomenon. There exists a more formal or mathematical definition, but it is outside the scope of this book.[6] A random variable is typically denoted by a capital letter, say X, Y, or T, to indicate that we do not know the value yet. A *realization* or an outcome of the random variable is then indicated by the same, but lower case, letter x, y, or t. This is in line with our definition of realization in Chap. 2. Indeed, a random variable may be seen as a variable that can in principle

---

[5] Now we know, with all our knowledge on probability and statistics, that a calculation of the observations like the arithmetic average is in most cases better than just selecting one of them.

[6] There is actually, and perhaps surprisingly, quite an active debate surrounding the definition of a random variable. A definition that is more mathematical but might still be accessible is the following: "A random variable is a mapping (i.e., a function) from events to the real number line". This definition allows us to mathematically link the material in Chap. 3—where we discussed events—to the material presented in this chapter. However, this definition is sometimes perceived as confusing as it does not contain any reference to random processes or outcomes.

be equal to any of the values in the population, and after probability sampling the outcome(s) will become known. Before sampling the outcomes are unknown, so we use capital X, and after sampling the outcomes would become known, so we use x. The probability sampling approach makes the variable of interest a random variable, as the sampling approach is here the random phenomenon. One of the earliest examples of a random variable, which is in a way unrelated to our sampling discussion, is, for instance, the age of death. Indeed, when somebody will die is unknown and in many ways random.

Random variables are very convenient to help quantify particular probabilities. John Graunt published in 1662 a *life table* for people in London. He provided probabilities of mortality at different age groups. For instance, he indicated that from 100 births, 36 of them would not grow older than 6 years of age, only one of them would reach an age of 76 years, and none of them would become 80 years or older (Glass 1950). In terms of mathematics, we can write such mortality probabilities in the form of $Pr(X \leq x)$, where the Pr indicates probability, X is the random variable for age at death, and x is a specific age of interest.[7] For instance, in terms of the life table of John Graunt: $Pr(X \leq 6) = 0.36$ represents the probability that a new born person would die before or at the age of 6 years old and it is equal to $0.36 = 36/100$.

The probability function $Pr(X \leq x)$ is a general concept and can be used for any random variable. The random variable X can be the number of hours per week that a school child watches television and we may ask what is the probability that a school child watches less than or equal to two hours per week: $Pr(X \leq 2)$. The probability $Pr(X \leq x)$ is also referred to as the *distribution function* obtained in x and it is denoted by $F(x) = Pr(X \leq x)$. Thus every random variable X has a distribution function F through $F(x) = Pr(X \leq x)$, but also every distribution function F has a random variable X, namely the random variable X that makes $Pr(X \leq x) = F(x)$. Thus the two concepts are directly related to each other and we then typically say that X is distributed according to F, i.e., $X \sim F$.

Each distribution function typically satisfies three conditions:

1. When the value x increases to infinity, the distribution function becomes equal to one, i.e., $\lim_{x \to \infty} F(x) = 1$. In terms of the examples of death and television watching this makes sense. When x is large, say larger than 168, every body has died before this age or watches this number of hours of television per week or less. Thus, in these examples $F(x) = 1$ for any $x > 168$.
2. When the value x decreases to minus infinity the distribution function becomes equal to zero, i.e., $\lim_{x \to -\infty} F(x) = 0$. Again this makes sense for the two examples, because no newborn baby would die before the age of zero nor does anybody watch less than zero hours of television per week. Thus, in these examples $F(x) = 0$ for any value $x < 0$.
3. The distribution function is a non-decreasing function, i.e., $F(x_1) \leq F(x_2)$ when $x_1 \leq x_2$. Indeed, the probability of dying within the age of x_1 cannot be larger

---

[7] In the analysis of life tables it is much more common to calculate probabilities of surviving after a specific age x, i.e., $Pr(X > x)$, but this is of course equal to $Pr(X > x) = 1 - Pr(X \leq x)$, as we discussed in Chap. 3.

than the probability of dying at a higher age x_2. In theory, it is possible that the probability will stay at the same level between x_1 and x_2, indicating that in the interval $[x_1, x_2]$ no population values exist (e.g., no one would die between the ages of 19 and 20 years, say, or no one watches 6 to 7 hours per week television)

There is a direct relation between distribution functions and densities. If we start with a PDF, we can define a distribution function in the following way:

$$F(x) = \int_{-\infty}^{x} f(z)dz. \tag{4.5}$$

Clearly, this function F is a distribution function. When the value x increases to ∞ we obtain the full area under the density, which is by definition equal to one. The area under the density must become zero when x goes to $-\infty$ (there is no unit in the population with the value $-\infty$. And finally, when x_2 is larger than x_1, the area under the PDF from $-\infty$ up to x_2 is not smaller than the same area under the density up to x_1. The distribution function F is often referred to as the *cumulative density function* (CDF). It also shows that the distribution function F is defined as a function from the real line $\mathbb{R}$ to the interval $[0, 1]$. Note that a PDF now also defines a random variable, since it defines a distribution function and a distribution function defines a random variable.

For each of the PDFs we discussed in the previous subsection there exist an accompanied distribution function. This does not mean that we have a closed form expression of each distribution function, since we do not have this for the normal and lognormal distribution functions. However, for the uniform and exponential distribution function we do have an explicit form.

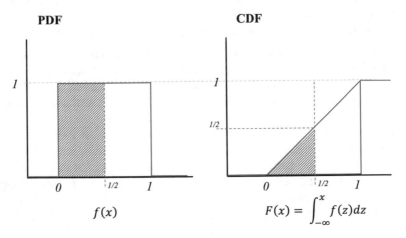

Fig. 4.4 Relationship between the PDF and CDF of a continuous random variable with a uniform distribution function

The uniform distribution function is given by

$$F_{\theta_0, \theta_1}(x) = \begin{cases} 0 & \text{for } x < \theta_0 \\ \frac{x}{\theta_1 - \theta_0} & \text{for } x \in [\theta_0, \theta_1] \\ 1 & \text{for } x > \theta_1 \end{cases}$$

This implies that the standard uniform distribution function is equal to $F_{0,1}(x) = x$ for $x \in [0, 1]$, $F_{0,1}(x) = 0$ for $x < 0$, and $F_{0,1}(x) = 1$ for $x > 1$. The relationship between the standard uniform PDF and CDF is illustrated in Fig. 4.4.

The exponential distribution function is given by

$$F_{\lambda}(x) = \begin{cases} 0 & \text{for } x \leq 0 \\ 1 - \exp\{-\lambda x\} & \text{for } x > 0. \end{cases}$$

The standard exponential distribution function is $F_1(x) = 1 - \exp\{-x\}$ for $x > 0$ and zero everywhere else.

There are a few additional characteristics that we need to mention. First of all, there are theoretical (or exotic) examples where we can define a distribution function without having a density. This has to do with distribution functions that are not differentiable. We do not treat these distribution functions in this book, thus we always assume that there is a PDF that defines the distribution function. Under this assumption, we can obtain that the probability that a random variable X has its outcome in an interval $(x_1, x_2]$ is equal to

$$\begin{aligned} \Pr(X \in (x_1, x_2]) &= \Pr(x_1 < X \leq x_2) \\ &= \Pr(X \leq x_2) - \Pr(X \leq x_1) \\ &= F(x_2) - F(x_1) \\ &= \int_{-\infty}^{x_2} f(z)dz - \int_{-\infty}^{x_1} f(z)dz \\ &= \int_{x_1}^{x_2} f(z)dz. \end{aligned}$$

This equality formalizes our earlier discussion in Sect. 4.2 that the integral from x_1 to x_2 represents the proportion of units in the population having its values in this interval. Note that we have used probability rule $\Pr(A \cup B) = \Pr(A) + \Pr(B) - \Pr(A \cap B)$ from Chap. 3 for the second equality sign. If we define the two events A and B by $A = (X \leq x_2)$ and $B = (X > x_1)$, we have $A \cap B = (x_1 < X \leq x_2)$ and $A \cup B = (X \in \mathbb{R})$. Thus we obtain now: $1 = \Pr(X \in \mathbb{R}) = \Pr(A \cup B) = \Pr(A) + \Pr(B) - \Pr(A \cap B) = \Pr(X \leq x_2) + \Pr(X > x_1) - \Pr(x_1 < X \leq x_2)$. This implies that $\Pr(x_1 < X \leq x_2) = \Pr(X \leq x_2) + \Pr(X > x_1) - 1 = \Pr(X \leq x_2) - \Pr(X \leq x_1)$.

Finally, as a consequence of the second characteristic, the density value $f(x)$ is **not** equal to $\Pr(X = x)$. The probability $\Pr(X = x)$ is equal to zero for continuous random variables, since there is no surface area under $f(x)$. It also implies that $\Pr(X < x) = \Pr(X \leq x)$ for continuous random variables. The fact that $f(x) \neq \Pr(X = x)$ is somewhat confusing and may be disappointing, since we started the introduction of a PDF as an approximation of the histogram for all values in a population. This emphasizes that the PDF is a mathematical abstractness or model for describing population values. The abstractness comes from the fact that densities are formulated for infinitely large populations. In terms of random measurement errors, it would make sense to assume that there is an infinite number of possible random errors that could influence the measurement.

4.4 Expected Values of Continuous Random Variables

In the subsection on probability densities we discussed the population mean and standard deviation. These population characteristics can now be more rigorously defined through the continuous random variables. The random variable represents a variable of the population without yet knowing its outcome. If we "average out" all the possible outcomes, where we weight each outcome with the PDF, we obtain a kind of "weighted average", similar to what we did in Chap. 2. However, we have many values to average out, in principle all values of $\mathbb{R}$, which we can not just average (there are far too many values). The averaging is then conducted by the use of integrals as a generalization of summation.

Let X be a continuous random variable with density f, then the *expected value of random variable X* is defined by

$$\mathbb{E}(X) = \int_{\mathbb{R}} x f(x) \, dx. \tag{4.6}$$

This expectation represent the population mean, which is typically denoted by the parameter μ as we used in Chap. 2. Thus, the population mean is $\mu = \mathbb{E}(X)$. Note that we have used the symbol $\mathbb{E}$ before in Chap. 2; at that point in the text we did not explain exactly what it meant, but now we know its formal definition. The population variance σ^2 can also be formulated in terms of an expected value of the random variable. The population variance σ^2 is now given by $\sigma^2 = \mathbb{E}(X - \mu)^2$, with

$$\mathbb{E}(X - \mu)^2 = \int_{\mathbb{R}} (x - \mu)^2 f(x) \, dx. \tag{4.7}$$

Clearly, we can generalize this concept. If we consider a (not necessarily continuous or differentiable) function $\psi : \mathbb{R} \to \mathbb{R}$, then the *expected value of the random variable $\psi(X)$* is defined by

$$\mathbb{E}\psi(X) = \int_{\mathbb{R}} \psi(x) f(x) \, dx. \tag{4.8}$$

The population mean is now obtained by taking $\psi(x) = x$ and the population variance is obtained by taking $\psi(x) = (x - \mu)^2$. Thus the function ψ may depend on population parameters.

The mean μ is also called the *first moment* of the random variable X and the variance is called the *second central moment* of the random variable X, since it squares the random variable after the mean is subtracted. We can also investigate other moments of the random variable X. The *pth moment* of random variable X is obtained by Eq. (4.8) with $\psi(x) = x^p$ and the *pth central moment* of random variable X is obtained by choosing $\psi(x) = (x - \mu)^p$ in Eq. (4.8). The third and fourth central moments are related to the skewness and kurtosis of the population values. The skewness is equal to $\gamma_1 = \mathbb{E}(X - \mu)^3/\sigma^3$ and the kurtosis is $\gamma_2 = \mathbb{E}(X - \mu)^4/\sigma^4 - 3$. Note that the moments of a random variable X may not always exist: this depends on the density f.

In the following table we provide the mean, variance, skewness, and kurtosis of the five parametric distributions we introduced in Sects. 4.2 and 4.3. Section 4.3 has already provided the mean and variance, but not the skewness and kurtosis. Note that we have used the following notation $\tau^2 = \exp\{\sigma^2\}$ in the table.

	Mean	Variance	Skewness	Kurtosis
Normal distribution (μ, σ)	μ	σ^2	0	0
Lognormal distribution (μ, σ)	$\exp\{\mu\}\tau$	$(\tau^2 - 1)\tau^2\exp\{2\mu\}$	$(\tau^2 + 2)\sqrt{\tau^2 - 1}$	$\tau^8 + 2\tau^6 + 3\tau^4 - 6$
Uniform distribution (θ_0, θ_1)	$[\theta_1 - \theta_0]/2$	$[\theta_1 - \theta_0]^2/12$	0	$-6/5$
Exponential distribution λ	λ^{-1}	λ^{-2}	2	6
Double exponential distribution λ	0	$2\lambda^{-2}$	0	3

In Eq. (4.8) we used the function ψ and therefore discussed the expected value of random variable $\psi(X)$. The random variable $\psi(X)$ can be seen as a mathematical transformation of the original random variable. Knowing the expected value of this transformed random variable provides us the mean of the population of transformed values. However, if we can establish the full distribution function of $\psi(X)$, this gives us much more information about the population of transformed values than just the mean.

The full distribution function of $\psi(X)$ can always be established, but it does not always have a simple workable form. To illustrate a case in which we can obtain the full distribution function of a transformed random variable in workable form, we will start with X being normally distributed with parameters μ and σ, and consider the function $\psi(x) = \exp\{x\}$. The distribution function of the normally distributed random variable X is given by

$$\Pr(X \leq x) = \int_{-\infty}^{x} \frac{1}{\sigma} \phi \left(\frac{z - \mu}{\sigma} \right) dz = \int_{-\infty}^{(x-\mu)/\sigma} \phi(z)dz,$$

and this normal distribution function is often denoted by $\Phi((x - \mu)/\sigma)$, which is defined as $\Phi(x) = \int_{-\infty}^{x} \phi(z)dz$. Then for any value $x > 0$, we can obtain the distribution function of $\exp\{X\}$ by

$$\Pr(\exp\{X\} \leq x) = \Pr(X \leq \log(x))$$
$$= \Phi \left(\frac{\log(x) - \mu}{\sigma} \right)$$
$$= \int_{-\infty}^{\log(x)} \frac{1}{\sigma} \phi \left(\frac{z - \mu}{\sigma} \right) dz$$
$$= \int_{0}^{x} \frac{1}{z\sigma} \phi \left(\frac{\log(z) - \mu}{\sigma} \right) dz. \tag{4.9}$$

Since the last integral contains the lognormal PDF, we have obtained that the distribution of $\exp\{X\}$ is lognormally distributed with parameters μ and σ. Note that the integral does not start from $-\infty$, but we know that the lognormal density is zero for $x \leq 0$, thus the integral from $-\infty$ to 0 does not contribute. Thus we see that the lognormal PDF is related to the random variable $\exp\{X\}$ when X is normally distributed. We have now learned that the logarithm of a lognormally distributed random variable is normally distributed.

Now we can generalize this for any random variable X and any monotone differentiable function ψ. Let F be the distribution function of X and f the PDF, we then have

$$\Pr(\psi^{-1}(X) \leq x) = \Pr(X \leq \psi(x))$$
$$= F(\psi(x))$$
$$= \int_{-\infty}^{\psi(x)} f(z)dz$$
$$= \int_{-\infty}^{x} \psi'(z) f(\psi(z))dz,$$

with ψ' the derivative of ψ. The calculations now show that the distribution function of random variable $\psi^{-1}(X)$ is equal to $F(\psi(x))$ and the PDF is $\psi'(x) f(\psi(x))$. An interesting consequence of this finding is that the distribution function of the random variable $F^{-1}(U)$, with U a standard uniform distributed random variable and F any invertible distribution function, is now given by F. Indeed, just applying the same procedure as above, we find that $\Pr(F^{-1}(U) \leq x) = \Pr(U \leq F(x)) = F(x)$. Thus the PDF of random variable $F^{-1}(U)$ must now be equal to f. This result for the standard uniform random variable U is very convenient if we want to simulate data from some kind of distribution function F, as we will explain in Sect. 4.8.3.

4.5 Distributions of Discrete Random Variables

Not all the data that we collect and observe are realizations of continuous random variables. Many applications provide us with discrete numerical data, e.g., the number of defective products, the number of microorganisms in a production environment, the presence or absence of a disease, the score on an intelligence test, etc. For these discrete numerical variables, we can also formulate random variables. A discrete random variable X is a random variable that takes its values in the set $\mathbb{N} = \{0, 1, 2, 3, ...\}$.[8]

For discrete random variables we can define $p_k = \Pr(X = k)$ as the probability of observing the outcome k. This is referred to as the *probability mass function* (PMF) if the probabilities p_k satisfy two conditions. First, all probabilities p_k should be nonnegative ($p_k \geq 0$, $\forall k$) and secondly, the probabilities need to add up to one, i.e., $\sum_{k=0}^{\infty} p_k = 1$. This second condition is related to the way we constructed probabilities: the probability that one of the events (in this case outcome k) happens—regardless of which one—is equal to 1.[9] Sometimes we would like to use the notation $f(x) = \Pr(X = x)$, with $x \in \mathbb{N}$, for the PMF, and we can then write the PMF out in full

$$
f(x) = \begin{cases}
0 & \text{if } x < 0 \\
p_0 & \text{if } x = 0 \\
p_1 & \text{if } x = 1 \\
p_2 & \text{if } x = 2 \\
\vdots & \vdots \vdots \\
p_k & \text{if } x = k \\
\vdots & \vdots \vdots
\end{cases}
$$

If only a few numbers of discrete values are possible, like the outcomes for throwing a die, then most of the probabilities p_k will be equal to zero. For throwing a fair die, we may have the random variable X taking its values in the set $\{1, 2, 3, 4, 5, 6\}$ and the probabilities $p_0 = 0$, $p_1 = p_2 = p_3 = p_4 = p_5 = p_6 = 1/6$, and $p_k = 0$ for $k > 6$. For binary events, like the occurrence of a disease, we can introduce the random variable X that takes its value in the set $\{0, 1\}$. The probabilities p_k may then be defined as $p_0 = 1 - p$, $p_1 = p$, and $p_k = 0$ for $k > 1$ and some value $p \in [0, 1]$.

The PMF for a discrete random variable is the equivalent of the PDF for a continuous random variable. This means that there is also a distribution function for a discrete random variable. The distribution function or cumulative density function (CDF) for a discrete random variable X is now given by $F(x) = \Pr(X \leq x) = \sum_{k=0}^{x} f(k)$.[10]

---

[8] Discrete does not always mean that we observe values in $\mathbb{N}$. For instance, grades on a data science test may take values in $\{1, 1.5, 2.0, 2.5, \ldots, 9.0, 9.5, 10\}$. Thus, it would be more rigorous to say that a discrete random variable X takes its values in the set $\{x_0, x_1, x_2, \ldots, x_k, \ldots\}$, with x_k an element of the real line ($x_k \in \mathbb{R}$) and with an ordering of the values $x_0 < x_1 < x_2 < \cdots$. However, in many practical settings we can map this set to a subset of $\mathbb{N}$ or to the whole set $\mathbb{N}$.

[9] In the more general setting, the probability can be defined as $P(X = x_k) = p_k$.

[10] If the set is $\{x_0, x_1, x_2, \ldots, x_k, \ldots\}$, with $x_0 < x_1 < x_2 < \cdots$, then the CDF is defined as $F(x) = \sum_{k=0}^{m_x} f(x_k)$, with m_x the largest value for k that satisfies $x_k \leq x$.

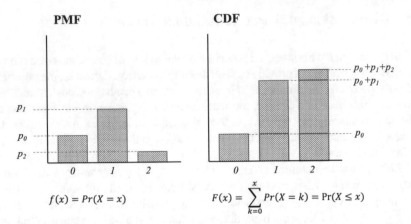

$$f(x) = Pr(X = x) \qquad\qquad F(x) = \sum_{k=0}^{x} Pr(X = k) = Pr(X \le x)$$

Fig. 4.5 Relationship between the PMF and CDF of a discrete random variable with three possible outcomes

The CDF represents the probability that the random variable X will have an outcome less than or equal to the value x, the same as for continuous random variables. We can write the CDF in full by stating:

$$F(x) = \begin{cases} 0 & \text{if } x < 0 \\ p_0 & \text{if } x \le 0 \\ p_0 + p_1 & \text{if } x \le 1 \\ p_0 + p_1 + p_2 & \text{if } x \le 2 \\ \vdots & \vdots \; \vdots \\ p_0 + p_1 + \cdots + p_k & \text{if } x \le k \\ \vdots & \vdots \; \vdots \end{cases}$$

Knowing either the PMF or the CDF of a discrete random variable suffices to describe the probabilities associated with the values that the discrete random variable can take. Clearly, the PMF and CDF are closely related: Fig. 4.5 demonstrates the relationship between the PMF and the CDF.

The PMF and CDF for a discrete random variable that we introduced based on our coin-tossing and dice throwing scenarios presented earlier are:

$$f(x) = \begin{cases} \frac{1}{2} & \text{if } x = 0 \\ \frac{1}{2} & \text{if } x = 1 \end{cases}$$

and

$$F(x) = \begin{cases} 0 & \text{if } x < 0 \\ \frac{1}{2} & \text{if } 0 \le x < 1 \\ 1 & \text{if } x \ge 1 \end{cases}$$

Note that we do not have to restrict the value x to values in $\mathbb{N}$ for the CDF. We can also use any x in $\mathbb{R}$. If we define $\lfloor x \rfloor$ as the highest integer value that is less than or equal to x, the CDF for any value $x \in \mathbb{R}$ is now given by $F(x) = F(\lfloor x \rfloor)$. This means that the CDF is a *step function*, see Fig. 4.5. For any value $x \in [\lfloor x \rfloor, \lfloor x + 1 \rfloor)$ the CDF is constant.

4.6 Expected Values of Discrete Random Variables

In this section we discuss expected values of a discrete random variable X, similar to what we did for continuous random variables. More specifically, we will discuss the expectation of the discrete random variable $\psi(X)$. The definition is similar to the definition for continuous random variables, but for discrete random variables we can use summation instead of using integrals. Thus, the expectation of a discrete random variable $\psi(X)$ is given by

$$\mathbb{E}(\psi(X)) = \sum_{k=0}^{\infty} \psi(k) p_k = \sum_{k=0}^{\infty} \psi(k) \Pr(X = k). \qquad (4.10)$$

This definition is closely related to the definition of the expected population parameter for an estimator T as discussed in Chap. 2. If we would collect many realizations of the discrete random variable X, say N realizations, we expect to see value k with frequency $N \cdot p_k$. Thus, the mean value of the random variable $\psi(X)$ would be calculated with Eq. (4.10) when the number of realizations N becomes really large. This was the same argument used in Chap. 2 for an estimator T that was used on a sample of data that was collected with probability sampling.

If we choose $\psi(x) = x$ we obtain the expected value of X and this is again referred to as the *mean* of the random variable or the mean of the population, the same as for continuous random variables. We also use the same notation μ for this mean, i.e., $\mu = \mathbb{E}(X)$. By choosing $\psi(x)$ equal to $\psi(x) = (x - \mu)^2$ and using this in Eq. (4.10) we obtain the variance of a discrete random variable X, and denote this by $\sigma^2 = \mathbb{E}(X - \mu)^2$.

Similar as for the continuous random variables, we can investigate other moments of the discrete random variable X. The pth moment of a discrete random variable X is obtained by Eq. (4.10) with $\psi(x) = x^p$ and the pth central moment of a discrete random variable X is obtained by choosing $\psi(x) = (x - \mu)^p$ in Eq. (4.10). The skewness and kurtosis of a discrete random variable X (or equivalently the skewness and kurtosis for a population with discrete values), are equal to $\gamma_1 = \mathbb{E}(X - \mu)^3/\sigma^3$ and $\gamma_2 = \mathbb{E}(X - \mu)^4/\sigma^4 - 3$, respectively, using Eq. (4.10) for the expectation $\mathbb{E}$. Note that the moments of a discrete random variable X may not always exist: this depends on the choice of probabilities p_k.

4.7 Well-Known Discrete Distributions

Similar to the case of PDFs, the PMFs may typically have a particular form that is known up to a set of one or more parameters $\boldsymbol{\theta} = (\theta_1, \theta_2, \dots, \theta_m)^T$. The PMF is then denoted by $f_{\boldsymbol{\theta}}(x)$, using the same notation as for continuous PDFs, where again we may use other symbols for the parameters $\boldsymbol{\theta}$, since this is more aligned with literature. People have studied different forms of the distribution of discrete random variables for a large number of applications. In this section we introduce four famous discrete distributions.[11] We do so by providing a story that motivates the random variable.

4.7.1 Bernoulli Probability Mass Function

The story of the Bernoulli random variable is simple: Bernoulli random variables are motivated by considering *binary* random variables: i.e., random variables that take on values 0 or 1. The simplest example of this is a single coin toss where we map tails to 0 and heads to 1. Now introduce the parameter p, with $0 \leq p \leq 1$, for the probability that the binary random variable X will be equal to 1. This gives rise to the following PMF:

$$\Pr(X = x) = f_p(x) = p^x(1 - p)^{1-x},$$

with $x \in \{0, 1\}$ and $f_p(x) = 0$ for any other value of x. A binary random variable with the above PMF is said to be Bernoulli distributed. Also, note that we often write $X \sim \mathscr{B}(p)$ to denote that X is Bernoulli distributed with parameter p. We leave it to the reader to specify the Bernoulli CDF.

The mean and variance of a Bernoulli random variable are easily determined by using Eq. (4.10). The mean μ is equal to $\mu = \mathbb{E}(X) = \sum_{k=0}^{1} k p^k (1 - p)^{1-k} = p$ and the variance σ^2 is equal to $\sigma^2 = \mathbb{E}(X - p)^2 = p^2(1 - p) + (1 - p)^2 p = p(1 - p)$. Thus the mean and variance are just functions of the parameter p. The mean represents the average number of "ones" (or events), which is obviously equal to the probability p of observing the value 1.

4.7.2 Binomial Probability Mass Function

The binomial distribution follows from the idea that we might be interested in the total number of heads if we toss a coin multiple times or the total number of airplane accidents or crashes per year. The quantity of interest is the total number of ones S_n (e.g., heads for the coin and crashes for the airplanes), when n represents the total number of tosses or flights per year. The random variable S_n is then given by

---

[11] Many more distribution functions are known and often used and studied; we present only a small selection.

$S_n = X_1 + X_2 + \cdots + X_n$, with X_k the binary random variable for toss or flight k. Obviously, the random variable S_n can attain the outcome values $0, 1, 2, \ldots, n$. The small letter s is in this case used to indicate a possible value that S_n can take. If the outcome is equal to zero ($s = 0$) we have not seen a single head in n throws or any accident in n flights, while if $s = n$ all throws were heads or all airplane flights resulted in an accident.

The PMF of the binomial is given by:

$$P\left(S_n = s\right) = f_{n,p}(s) = \binom{n}{s} p^s \left(1 - p\right)^{n-s},$$

where

$$\binom{n}{s} = \frac{n!}{s!\,(n-s)!},$$

and $s!$ is the total number of permutations of a set of s (different) values. While already briefly introduced in Chap. 2, we discuss permutations in more detail in the additional materials at the end of this chapter. We have visualized three binomial PMFs in Fig. 4.6.

The binomial PMF has two parameters: probability $p \in [0, 1]$ and the number of trials n. In many settings the number of trials will be known, and only p is unknown. For instance, the probability p of passing a data science test would be unknown, but the number of students taking the exam is known upfront. However, in some settings the number of trials is not known, while the probability p is assumed known. For instance, the estimation of the number of microorganisms in a container solution (e.g., milk container) based on a set of binary test scores of small sample volumes from the container (Cochran 1950; van den Heuvel 2011).

To obtain the mean and variance for a binomial random variable with formula (4.10) is somewhat more work than for the Bernoulli. However, there exist closed form expressions. The mean is equal to $\mu = \mathbb{E}(S_n) = np$ and the variance is

Fig. 4.6 PMFs for the binomial distribution

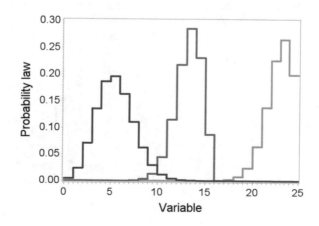

$\sigma^2 = \mathbb{E}(S_n - np)^2 = np(1 - p)$. Since the binomial random variable S_n is the sum of n independent Bernoulli variables, it may be expected that the mean is just n times the mean of a Bernoulli random variable. However, this rule also seems to hold for the variance, which may be less expected. In Sect. 4.10 we will show that these rules hold true in general, irrespective of the underlying PMF.

4.7.3 Poisson Probability Mass Function

A disadvantage of a random variable with a binomial distribution function is that it is bounded by the number of trials n. There are, however, many applications where the number of events or count is not necessary bounded by a fixed number (or at least it is difficult to formulate this bound). In such cases we can use the Poisson distribution, which is often used to express the probability of a given number of events occurring in a fixed interval of time when these events occur with a known constant rate and independently of the time since the last event.

Let X be a random variable with outcome set $\{0, 1, 2, 3, \ldots\}$; then X has a Poisson PMF with parameter $\lambda > 0$ when the probability of observing k events is given by

$$P(X = k) = f_\lambda(k) = \frac{\lambda^k}{k!} \exp(-\lambda)$$

Figure 4.7 shows three different choices of the Poisson PMF.

The mean of a Poisson random variable is equal to $\mu = \mathbb{E}(X) = \lambda$. Thus the parameter λ represents the average number of counts. When the mean is larger than 5, the shape of the PMF looks very much like the normal PDF (see the rightmost PMF in Fig. 4.7). The normal PDF is then viewed as a smooth version of the discrete Poisson PMF. The variance of a Poisson random variable is equal to the mean, i.e.,

Fig. 4.7 PMFs for the Poisson distribution

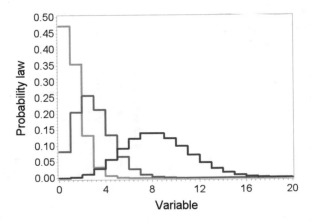

$\sigma^2 = \mathbb{E}(X - \lambda)^2 = \lambda$. It requires some in-depth calculations to obtain the mean and variance using Eq. (4.10).

4.7.4 Negative Binomial Probability Mass Function

The negative binomial PMF is often considered a Poisson PMF with an extra amount of variation, even though it originated from a different type of application. In this original application, a random variable X has a negative binomial PMF when it represents the number of trials needed to obtain a fixed known number of binary events. For instance, how many products (e.g., light bulbs) X should be tested before we observe, say r, defective products (e.g., not working light bulbs), when each product has the same probability p of being defective. Thus the negative binomial has two parameters p and r, with r typically known in this application. However, in this form the connection with the Poisson is less obvious.

The negative binomial PMF that we will introduce has two parameters λ and δ, where λ still represents the mean, the same as for the Poisson random variable, but the δ represents an *overdispersion* parameter, indicating the extra amount of variation on top of the Poisson variation. The PMF for the original application is the same as the PMF we will introduce, but it is just a different way of parameterizing the PMF.

A negative binomial random variable X has its outcomes in the set $\{0, 1, 2, 3, ...\}$, like the Poisson random variable. The PMF is defined by

$$\Pr(X = k) = f_{\lambda, \delta}(k) = \frac{\Gamma(k + \delta^{-1})}{\Gamma(k + 1)\Gamma(\delta^{-1})} \frac{(\delta\lambda)^k}{(1 + \delta\lambda)^{k + \delta^{-1}}},$$

with Γ the gamma function given by $\Gamma(z) = \int_0^\infty x^{z-1} \exp\{x\}dx$. Note that $\Gamma(k) = (k-1)!$, when k is an integer. A few choices for the negative binomial PMF are provided in Fig. 4.8. The mean and variance of the negative binomial random variable

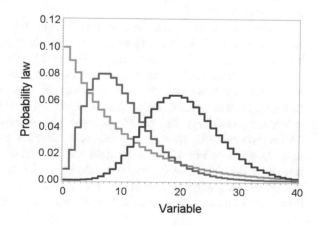

Fig. 4.8 PMFs for the negative binomial distribution

are given by $\mu = \mathbb{E}(X) = \lambda$ and $\sigma^2 = \mathbb{E}(X - \lambda)^2 = \lambda + \delta\lambda^2$. Clearly, in case the parameter δ converges to zero, the variance converges to the variance of a Poisson random variable. This is the reason that the parameter δ is called the overdispersion.

4.7.5   Overview of Moments for Well-Known Discrete Distributions

The following table provides the expected value, variance, skewness, and kurtosis for the four discrete distributions we introduced above. Note that we have already provided the means and variances.

	Mean	Variance	Skewness	Kurtosis
Bernoulli $f_p(x)$	p	$p(1-p)$	$\dfrac{1-2p}{\sqrt{p(1-p)}}$	$\dfrac{1-6p(1-p)}{p(1-p)}$
Binomial $f_{n,p}(x)$	np	$np(1-p)$	$\dfrac{1-2p}{\sqrt{np(1-p)}}$	$\dfrac{1-6p(1-p)}{np(1-p)}$
Poisson $f_\lambda(x)$	λ	λ	$1/\sqrt{\lambda}$	$1/\lambda$
Negative Binomial $f_{\lambda,\delta}(x)$	λ	$\lambda + \delta\lambda^2$	$\dfrac{1+2\delta\lambda}{\sqrt{\lambda(1+\delta\lambda)}}$	$6\delta + [\lambda(1+\delta\lambda)]^{-1}$

When the number of trials n for the binomial random variable is large, the skewness and kurtosis are close to zero. Actually, it is not the number of trials that is important, but either the number of events np or the number of non-events $n(1 - p)$ that should be large to make the skewness and kurtosis close to zero. In that case the Binomial PMF looks very much like the normal PDF. An example of this situation is given by the most right PMF in Fig. 4.6. Here the number of trials is equal to $n = 25$ and the probability of an event is $p = 0.20$. This gives a skewness of 0.3 and a kurtosis of 0.01. See also Sect. 4.9.

Something similar is also true for the Poisson and Negative Binomial random variables. For the Poisson the mean λ should be relatively large to have a shape that is similar to a normal PDF. We already indicated this. The most right PMF in Fig. 4.7 is close to a normal PDF. The mean of this Poisson PMF was equal to $\lambda = 8$, which makes the skewness equal to 0.35 and the kurtosis equal to 0.125. For the Negative Binomial random variable the mean λ should also be large, but the overdispersion should not be too large. Indeed, a large mean will put the skewness close to zero, but the kurtosis may still be away from zero when delta is too large, due to the term 6δ in the kurtosis. In Fig. 4.8 the most right PMF is closest to a normal PDF, although it is still a little bit skewed and has a little bit thicker tails than the normal density. This PMF has a mean of $\lambda = 20$ and an overdispersion of $\delta = 0.05$, which makes the skewness equal to 0.47 and the kurtosis equal to 0.325.

4.8 Working with Distributions in R

The functions and packages in R can support us when working with random variables, PDFs, PMFs, and CDFs. First of all, R can help us calculate particular probabilities. As we have seen, not all CDFs have closed-form expressions. Thus, to determine a CDF value requires either calculation of integrals or otherwise summations of many terms. Numerical approaches have been implemented in R to help us do this with the computer. Secondly, R can help us create population values or samples from populations which are described by a PDF or PMF. We will demonstrate how you can use R to—by means of Monte-Carlo (MC) simulation—compute summaries of random variables with complex distribution functions. Summaries are sometimes obtained exact, like the means, variance, skewness, and kurtosis reported earlier, but not every type of summary can always be determined exactly or it may be more time-consuming than just using a MC simulation. Finally, we will also demonstrate a method that allows you to obtain realizations (draws) of a random variable with a specific distribution function that you may have created your self or that exists in the literature but not in R.

4.8.1 R Built-In Functions

R offers a number of well-known distribution functions. It uses a standardized naming scheme to name the functions that relate to probability distributions. The name always consists of (an abbreviation of) the mathematical name of the distribution function—for example `norm` for the normal distribution function—with one of the following prefixes:

- `d-` A distribution function with the prefix `d-`, for example `dnorm`, allows you to evaluate the PDF or PMF at a specific value. Thus, a call to `dnorm(x, mu, sigma)` evaluates the PDF of the normal distribution function with mean `mu` and standard deviation `sigma` at `x`.
- `p-` A distribution function with the prefix `p-`, for example `pnorm`, allows you to evaluate the CDF at a specific value. Thus, a call to `pnorm(x, mu, sigma)` evaluates the CDF in x of the normal distribution function with mean `mu` and standard deviation `sigma` at `x`.
- `q-` A distribution function with the prefix `q-`, for example `qnorm`, allows you to evaluate the so-called quantile or inverse function of the distribution functions. The quantile function specifies the value of the random variable that gives you the specific probability of the variable being less than or equal to that value. Thus the quantile function gives $x = F^{-1}(u)$ for probability or value $u \in (0, 1)$, with F the CDF.
- `r-` A distribution function with the prefix `r-`, for example `rnorm`, allows you to generate draws of a random variable with a specific distribution function. Thus `rnorm(1000, mu, sigma)` returns a vector of length 1,000 containing draws from a normal random variable with mean `mu` and standard deviation `sigma`.

The normal, uniform, and exponential distributions we covered in this chapter are called -norm, -unif and -exp. The lognormal distribution function does not exist in R, but it can be created through the normal distribution function. The binomial, Poisson, and negative binomial distributions are given by -binom, -pois, and -nbinom. The Bernoulli is just a special case of the binomial with a size (or number of trials) of one. It should be mentioned that the parametrization of the negative binomial in R is different from our formulation, which means that we need to study the difference between the parametrization in R and our formulation (when you investigate -nbinom, R gives info on this difference). A full overview of the built-in distribution functions can be found at https://stat.ethz.ch/R-manual/R-devel/library/stats/html/Distributions.html.

The code below gives an example in which we first evaluate the PDF and the CDF of the normal for a given mean and standard deviation. Next, we demonstrate the equality concerning the quantile function we highlighted above, and finally we obtain 10 draws from the same normal distribution function.

```
> mu <- 0          # Mean
> s2 <- 1          # Variance
> s <- sqrt(s2)    # Standard deviation
>
> x <- 1
> dnorm(x, mean=mu, sd=s) # PDF of the normal distribution
    evaluated at x
[1] 0.2419707
> pnorm(x, mean=mu, sd=s) # CDF of the normal distribution
    evaluated at x
[1] 0.8413447
>
> p <- pnorm(x, mean=mu, sd=s)
> qnorm(p, mean=mu, sd=s) # The so-called quantile function Q(p)
    = x if and only if p = F(x)
[1] 1
>
> set.seed(982749)
> n <- 10
> rnorm(n, mean=mu, sd=s) # Get 10 draws / realizations from the
    distribution
 [1]  0.15958190 0.60671592 1.10638675 -1.03021164 0.14672386
      0.37733998
 [7]  0.55563879 0.77358142 0.61140111 -0.09188106
```

4.8.2 Using Monte-Carlo Methods

The ability to easily obtain draws from a distribution function allows us to approximate the properties of distribution functions by computing summaries of the draws obtained. This method is called Monte Carlo (or MC) simulation, and we can use it to check our analytical results. For example, we can approximate the expected value of a random variable $X \sim \mathcal{N}(2, 9)$ using the following code[12]:

```
> draws <- rnorm(10^6, mean=2, sd=3)
> mean(draws)
[1] 1.996397
```

In this case we already knew that the mean was equal to 2. The simulation shows that we obtain a value very close to 2 and confirms our knowledge. This MC approach is closely related to simple random sampling in Chap. 2, but now we sample from an infinitely large population that is described by the normal distribution function. The example is somewhat simple, but it shows how simulation works. MC becomes more relevant when more complicated random variables are being studied. For instance, the expected value of $\exp\{\sqrt{X}\}$, with $X \sim \mathcal{N}(\mu, \sigma^2)$, is less easy to determine mathematically. You may think that this may be an exotic random variable to study, but practice often studies very interesting and complex random variables, often a combination of several random variables. Instead of evaluating the mean of the random variables, we could also study the variance and other moments (like skewness and kurtosis), which will be even more difficult to obtain mathematically.

To illustrate MC with multiple random variables, we can easily imagine a random variable Z whose distribution function is a combination of two normal distribution functions with different means and variances, which is something we call a *mixture* distribution: for instance, the distribution of body weight of women and men or the tensile strengths of plastic tubes produced from two production lines. One way of constructing such a variable is by imagining that we first throw a coin, $Y \sim \mathcal{B}(1/3)$, and subsequently we obtain a draw from one of two different normals: $Z_0 \sim \mathcal{N}(10, 1)$ if $Y = 0$ and $Z_1 \sim \mathcal{N}(20, 5)$ if $Y = 1$. Or, more generally, the random variable of interest Z is constructed as $Z = Y Z_0 + (1 - Y)Z_1$ where $Y \sim \mathcal{B}(p)$, $Z_0 \sim \mathcal{N}(\mu_0, \sigma_0^2)$, and $Z_1 \sim \mathcal{N}(\mu_1, \sigma_1^2)$ with $p = 1/3$, $\mu_0 = 10$, $\sigma_0^2 = 1$, $\mu_1 = 20$, and $\sigma_1^2 = 5$.

Since Z is a function of random variables, it is itself a random variable. However, it is not immediately clear (yet) how we could compute summaries such as its expectation, variance, moments, or even the percentage of values above a certain level (say 15). In Sect. 4.9 we will look at this variable more mathematically, but here we will study the variable through MC simulation:

```
> # Set number of draws, probability of coin, and mean and
      standard deviation of first normal distribution
> n <- 10^6
> p <- 1/3
> mu_1 <- 10
> s_1 <- 1
>
> # Flip a coin with probability p
> Y <- rbinom(n, size=1, prob=p)
>
```

---

[12] Note that -norm uses standard deviations instead of variances. You can always type `?rnorm` to see the exact arguments.

```
> # Generate the mixture draws (note that 20 = 1*10+10 and 5 = 1*
    4+1)
> Z <- rnorm(n, mean=(Y*10)+mu_1, sd=(Y*4)+s_1)
>
> # Plot the draws in a histogram
> hist(Z, freq=FALSE, breaks=50)
>
> # Compute mean and variance (increase the power to compute
    higher central moments)
> mean_Z <- mean(Z)
> var_Z <- mean((Z-mean(Z))^2)
> mean_Z
[1] 13.31672
> var_Z
[1] 31.09315

> # Compute percentage of values above 15
> P <- (Z>15)
> mean_P <- mean(P)
> mean_P
[1] 0.28147
```

Figure 4.9 shows the histogram of the draws generated using this code. The mixing of the two normals is clearly visible. Note that we use `freq=FALSE` to print a histogram that has probabilities on the *y*-axis instead of frequencies (which is the default we saw in Chap. 1).

The example above shows how we can better understand the properties of a distribution or a random variable, but it can also be used to evaluate estimators that are being used on samples from a population, as we discussed in Chap. 2. The only difference with Chap. 2 is that we are making certain assumptions on the population values in this simulation. Indeed, we have assumed that the population is described by a particular PDF or PMF or a particular set of random variables. If we study one large simulation of many draws (like 10^6 in the example above) we obtain a population that can give us better insight in the population characteristics, but if we

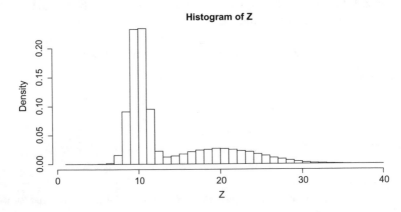

Fig. 4.9 The histogram of a mixture distribution

simulate repeatedly (say 10^3) a smaller number of draws (equal to an anticipated sample size n) and calculate our estimator on each simulation sample, we can evaluate the performance of the estimator. Using the (10^3) repeated simulations, we can approximate the mean, variance and other moments of the estimator, like we did in Chap. 2.

4.8.3 Obtaining Draws from Distributions: Inverse Transform Sampling

It is clear that R is very useful for working with probability distributions; we can evaluate PMFs, PDFs, and CDFs, and we can use MC simulation to compute expectations, variances, and moments of random variables and estimators—this is even possible when we might not be able to do so analytically. However, in the examples above we are inherently limited by Rs default functions; hence, we can only work with the well-known distributions that R supports. Although there are packages that implement more distributions, sometimes we might want to obtain draws from a distribution that is not well-known and implemented by others. If this is the case, we can sometimes use *inverse transform sampling* as we discussed in Sect. 4.4; we demonstrate it here in more detail to generate draws of an exponential distribution function—which actually *is* implemented in R—and than check our results.

Figure 4.10 shows the idea behind inverse transform sampling. As long as we know the CDF of a random variable, we can use draws from a uniform distribution function to generate draws from the variable of interest by evaluating the inverse of the CDF (obviously this is something we need to derive ourselves). For example, consider the exponential distribution function: the exponential distribution function with parameter λ has the following PDF and CDF:

$$f_\lambda(x) = \lambda \exp\{-\lambda x\} \qquad F_\lambda(x) = 1 - \exp\{-\lambda x\},$$

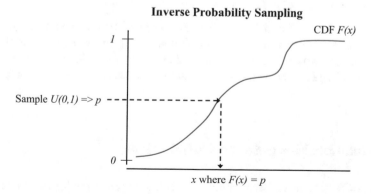

Inverse Probability Sampling

Fig. 4.10 The CDF of a complex continuous distribution function

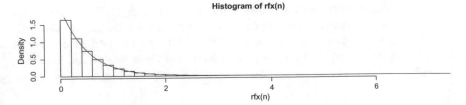

Fig. 4.11 Exponential approximated using draws obtained by inverse transform sampling

with $x > 0$. The inverse of the CDF is now equal to

$$F_\lambda^{-1}(u) = -\log(1-u)/\lambda,$$

with $u \in (0, 1)$. Now, we can implement the inverse CDF, $F^{-1}(x)$ in R, and use the inverse transform sampling trick:

```
> # lambda = 2
> # f(x)    = 2*exp(-2*x)
> # F(x)    = 1-exp(-2*x)
> # F(u)^-1 = -log(1-u)/2
>
> cdf_inverse <- function(u) {
+     -log(1-u)/2
+ }
>
> rfx <- function(n) {
+     u <- runif(n, min=0, max=1)
+     cdf_inverse(u)
+ }
>
> n <- 10^6
> draws <- rfx(n)
> hist(draws, freq=FALSE, breaks=40)
> curve(dexp(u, rate=2), col="red", add=TRUE)
```

The last two lines plot a histogram (see Fig. 4.11) of the obtained draws using our trick, and superimpose a curve using the built-in `dexp` function in R; it is clear that our sampling approach works very well! Thus using the uniform random variable we can in principle simulate any other random variable if we know the CDF and its inverse.

4.9   Relationships Between Distributions

We have only introduced a small number of well-known distribution functions above; many probability textbooks will provide many more examples of well-known distribution functions. However, our aim was just to introduce the main concepts; it's

Table 4.1 Comparisons of the probabilities of Poisson and binomial distribution functions

X	Binomial	Poisson	X	Binomial	Poisson	X	Binomial	Poisson
0	0.00000	0.00001	7	0.01456	0.04368	14	0.12441	0.09049
1	0.00000	0.00007	8	0.03550	0.06552	15	0.07465	0.07239
2	0.00000	0.00044	9	0.07099	0.08736	16	0.03499	0.05492
3	0.00004	0.00177	10	0.11714	0.10484	17	0.01235	0.03832
4	0.00027	0.00531	11	0.15974	0.11437	18	0.00309	0.02555
5	0.00129	0.01274	12	0.17971	0.11437	19	0.00049	0.01614
6	0.00485	0.02548	13	0.16588	0.10557	20	0.00004	0.00968

easy to look up the PDFs, CDFs, expectations and moments of specific distributions online (Wikipedia is actually a good source in this regard). Here we highlight a few well-known relationships between distribution functions. We have already seen the relationship between the Bernoulli and binomial distribution functions; we now discuss two more.

4.9.1   Binomial—Poisson

Although the Poisson and binomial distribution functions are different, as they have different supports, they can be close to each other. To demonstrate, Table 4.1 shows the Poisson probabilities next to the binomial probabilities when $\lambda = np$ is equal to 12 and $n = 20$. The probabilities are reasonably close although not extremely close.

The Poisson and binomial distribution functions are quite close whenever λ is equal to np and n is relatively large. It can be shown that the Poisson probabilities are the limit of the binomial probabilities when n converges to infinity under the condition that np converges to λ.

4.9.2   Binomial—Normal

Although the normal distribution function provides probabilities for continuous outcomes and the binomial distribution function provides probabilities for discrete outcomes, the normal distribution function may approximate the binomial distribution function (as we have already discussed in Sect. 4.7.5). The approximation is quite good under certain conditions, in particular when the skewness and kurtosis of the binomial distribution are close to zero.

Let S_n be a binomial random variable with parameters n and p. Probability calculation with the binomial distribution function can be adequately approximated with a normal distribution function when the mean np and the value $n(1 - p)$ are both larger than 5 and the sample size is at least 20 ($n \geq 20$). The binomial probability is then approximated by a normal probability as follows:

$$P\left(S_n \leq k\right) \approx P\left(Z \leq (k + 0.5 - np) / \sqrt{np\left(1 - p\right)}\right),$$

with Z the random variable from a standard normal distribution function.

For large sample sizes this approximation can be very useful, since binomial probabilities may not be easily calculated with a computer, due to the complexity of calculating the number of permutations $n!$ when n is large.

4.10   Calculation Rules for Random Variables

In our discussion of the binomial random variable we saw that it can be formulated as the sum of n binary random variables. In the Monte Carlo simulation we created $Z = YZ_1 + (1 - Y)Z_2$, with Y a binary variable and Z_k a normally distributed random variable having mean μ_k and variance σ_k^2. This shows that we are often interested in functions of random variables. In some cases we are able to determine the properties of these constructed random variables that are functions of the properties of the random variables that were used in the construction. This section will provide some calculation rules that apply in all cases (when certain independence conditions are satisfied), whatever the underlying PDF or PMF is used. Thus they are very generic rules. We will state the rules without giving proofs.

4.10.1   Rules for Single Random Variables

Here we assume that we have a random variable X, either discrete or continuous, and a constant c. The following rules hold true.

$$\mathbb{E}(c) = c$$
$$\mathbb{E}(cX) = c\mathbb{E}(X)$$
$$\mathrm{Var}(X) \geq 0$$
$$\mathrm{Var}(X + c) = \mathrm{Var}(X)$$
$$\mathrm{Var}(cX) = c^2 \mathrm{Var}(X)$$
$$\mathrm{Var}(X) = \mathbb{E}(X^2) - [\mathbb{E}(X)]^2$$

4.10.2 Rules for Two Random Variables

Here we assume that we have a random variable X and a random variable Y. The following rules always hold true:

$$\mathbb{E}(X) = \mathbb{E}(Y), \quad \text{when } X = Y$$
$$\mathbb{E}(X + Y) = \mathbb{E}(X) + \mathbb{E}(Y)$$

If we assume that the random variables X and Y are independent of each other (see also Chap. 6), we can provide a few other simple rules. Independence means that the outcome of X has nothing to do with the outcome of Y. For instance, when X represents the body weight of women and Y the body weight of men, and we draw randomly one woman and one man from the population, the weight of the woman will be unrelated to, or independent of, the weight of the man.[13]

More mathematically, independence can be defined through our definition of independent events in Chap. 3. If we introduce the events $A = \{X \le x\}$ and $B = \{Y \le y\}$, then independence of the two events is given by $\Pr(X \le x, Y \le y) = \Pr(A \cap B) = \Pr(A)\Pr(B) = \Pr(X \le x)\Pr(Y \le y) = F_X(x)F_Y(y)$, with F_X and F_Y the CDFs for X and Y, respectively. In Chap. 6 we will see that $\Pr(X \le x, Y \le y)$ is the joint CDF of X and Y, denoted by $F_{XY}(x, y)$. The two random variables X and Y are now considered independent, when this product of probabilities occurs for every x and y, i.e., when $F_{XY}(x, y) = F_X(x)F_Y(y)$ for all x and y. Note that, when X and Y are independent, the random variables $\varphi(X)$ and $\psi(Y)$ are independent, whatever functions φ and ψ are chosen.

The following rules will hold true when X and Y are independent.

$$\mathbb{E}(XY) = \mathbb{E}(X)\mathbb{E}(Y)$$
$$\mathrm{Var}(X + Y) = \mathrm{Var}(X) + \mathrm{Var}(Y)$$
$$\mathrm{Var}(X - Y) = \mathrm{Var}(X) + \mathrm{Var}(Y)$$
$$\mathrm{Var}(XY) = \mathrm{Var}(X)\mathrm{Var}(Y) + \mathrm{Var}(X)(\mathbb{E}(Y))^2 + \mathrm{Var}(Y)(\mathbb{E}(X))^2$$

The second rule shows why the variance of a binomial random variable is n times the variance of a Bernoulli random variable. This rule is applied (sequentially) to the random variable $S_n = X_1 + X_2 + \cdots + X_n$, with $X_1, X_2, \ldots, X_n$ independent random variables. Furthermore, the second and thrid rule tell us that variances of independent random variables always add up, even if we subtract random variables from each other.

---

[13] Yes, you are correct, practice is more complicated since a man and a woman may share a household and therefore their weights may be related.

Note that the first rule above shows that we could have calculated the mean of the mixture distribution discussed in Sect. 4.8.2 analytically:

$$
\begin{aligned}
\mathbb{E}(YZ_0 + (1 - Y)Z_1) &= \mathbb{E}(YZ_0) + \mathbb{E}((1 - Y)Z_1) \\
&= \mathbb{E}Y\mathbb{E}Z_0 + \mathbb{E}(1 - Y)\mathbb{E}Z_1 \\
&= p\mu_0 + (1 - p)\mu_1.
\end{aligned}
$$

The fourth rule helps us calculate the variance of YZ_0, which is $p(1 - p)\sigma_0^2 + p(1 - p)\mu_0^2 + \sigma_0^2 p^2 = p\sigma_0^2 + p(1 - p)\mu_0^2$. However, computing the variance of $YZ_0 + (1 - Y)Z_1$ is more difficult, since YZ_0 and $(1 - Y)Z_1$ are not independent, due to the common random variable Y.

4.11   Conclusion

In this chapter we have introduced one more snippet of theory that we need to advance our analysis of data: we introduced random variables and distributions of random variables. In the next chapter we will, using all the theory that we have now developed, relate back to sample data. We will first discuss distribution functions of sample statistics over repeated random sampling, and we will find that these depend on the parameters of the population distributions that we assume. We will than discuss two methods of estimating these population parameters.

Problems

4.1 In a trial the patients ($n = 20$) are randomly assigned to the groups A and B. The randomization is done by throwing an unbiased die. When the number of dots is even, the patient will be in group A, otherwise in group B.

1. What is the probability that exactly 10 patients will be in group A?
2. What is the probability that at most 9 patients will be allocated to group A?

4.2 In Sect. 4.5 we discussed multiple discrete distribution functions by providing the PMF ($f(x)$) and discussing their means, variances, and central moments.

1. Derive the CDF of the Bernoulli distribution.
2. Determine mathematically that the mean of a binomially distributed random variable X with parameters p and n is equal to $\mathbb{E}X = np$.
3. Determine mathematically that the variance of a binomially distributed random variable X with parameters p and n is equal to $\mathbb{E}(X - np)^2 = np(1 - p)$.
4. Determine mathematically that the mean of a Poisson distributed random variable X with parameter λ is equal to $\mathbb{E}X = \lambda$.

5. Determine mathematically that the variance of a Poisson distributed random variable X with parameter λ is equal to $\mathbb{E}(X - \lambda)^2 = \lambda$.
6. Use R to make a figure displaying the CDF of the Poisson distribution with $\lambda = 5$.
7. Determine mathematically that the mean of a uniform distributed random variable X with parameters a and b, with $a < b$, is equal to $\mathbb{E}X = (a + b)/2$.
8. Determine mathematically that the variance of a uniform distributed random variable X with parameters a and b, with $a < b$, is equal to $\mathbb{E}(X - (a + b)/2)^2 = \frac{1}{12}(b - a)^2$.

4.3 Let us assume that the probability of a person in the Netherlands being left-handed is 0.10. What is the probability that in a random group of 20 persons from the Netherlands you will find at least three left-handed persons?

4.4 A specific diagnostic test has a known sensitivity of 0.9 for the related disease. Five patients, all carriers of the disease, do the diagnostic test. Give the probability distribution function of the number of positive tests. This means that you need to calculate $P(S_5 = 0), P(S_5 = 1), \ldots, P(S_5 = 5)$, with S_5 the random variable that indicates the number of positive tests.

4.5 Consider the exponential CDF $F(x) = 1 - \exp(-\lambda x)$, for $x > 0$ and otherwise equal to zero. Now let X be distributed according to this exponential distribution.

1. Determine the mean and variance of X.
2. What is the median value of the exponential distribution function? Use the definition of the median we discussed in Chap. 1.

4.6 Consider the PDF $f(x) = 3x^2$ on the interval $(0, 1]$.

1. Demonstrate that the function is indeed a density.
2. What are the mean, variance and standard deviation?
3. How likely is it that the outcome will be in between 0.25 and 0.75?

4.7 The following questions concern the use of R to work with random variables

1. Use R to make a figure of both the PDF and the CDF of the normal distribution with parameters $\mu = 10$ and $\sigma^2 = 3$.
2. Compute the expected value and variance for the $\mathcal{N}(\mu = 10, \sigma^2 = 3)$ distribution using Monte Carlo simulation.

4.8 Implement inverse transform sampling for the PDF $f(x) = 1/2x$ defined from 0 to 2.

Additional Materials I: From Bernoulli to Binomial

In Chap. 2 we have already discussed permutations. Here we will repeat this for binary values and then discuss how the binomial distribution is generated from Bernoulli distributed random variables. Thus we will consider units (or subjects) i that have a Bernoulli random variable X_i with parameter p.

Consider a sample of three subjects ($n = 3$), and let x_1, x_2, and x_3 be the outcomes or realizations. These values can be ordered in six ($3! = 3 \cdot 2 \cdot 1 = 6$) different ways, or in other words there are six permutations, namely (x_1, x_2, x_3), (x_1, x_3, x_2), (x_2, x_1, x_3), (x_2, x_3, x_1), (x_3, x_1, x_2), and (x_3, x_2, x_1). To see this, we can see that for the first position there are three possibilities to choose from (x_1, x_2, or x_3), then for the second position there are only $2 = (3 - 1)$ possibilities left because the first position is already taken by one of the outcomes. Then for the third position there is only $1 = (3 - 2)$ possibility left, since the previous two positions are taken. Clearly, this can be generalized to k different values, leading to $k! = k \cdot (k - 1) \cdot (k - 2) \cdots 2 \cdot 1$ permutations.

For the binomial distribution function, the values or outcomes from the subjects are not all different, as they are either equal to zero or equal to one (they come from a binary random variable). For instance, when we consider again the three values x_1, x_2, and x_3, with the assumption that $x_1 = 1$, $x_2 = 0$, and $x_3 = 0$, there are still six permutations, but these permutations are not all unique in the sense that the sum over all the values is still the same. The permutation (x_1, x_2, x_3) is exactly the same as permutation (x_1, x_3, x_2), since they are both equal to $(1, 0, 0)$.

The number of unique permutations (also referred to as the number of combinations) is in this case thus three, since they are $(1, 0, 0)$, $(0, 1, 0)$, and $(0, 0, 1)$. Clearly, for a given permutation we could permute all the zero's and all the ones without affecting the result. Thus the number of unique permutations is determined by the total number of permutations, divided by the number of permutations that can be made with the zero's and with the ones. Thus in the example we find $3! / (2! \cdot 1!) = 6 / (2 \cdot 1) = 3$. More generally, when the outcomes consist of zero's and ones and the number of ones is for instance k, then the number of unique permutations is given by the binomial coefficient:

$$\binom{n}{k} = \frac{n!}{k! \, (n - k)!}$$

This so-called *binomial coefficient* is pronounced n over k or n choose k.

Each of the $n!/(k!(n - k)!)$ outcomes result in the exact same probability of occurrence, namely $p^k (1 - p)^{n-k}$. To illustrate this with the three outcomes x_1, x_2, and x_3, the probability that $(1, 0, 0)$ occurs is $p(1 - p)(1 - p)$, that $(0, 1, 0)$ occurs is $(1 - p)p(1 - p)$, and that $(0, 0, 1)$ occurs is $(1 - p)(1 - p)p$. Thus all three outcomes have a probability of $p(1 - p)^2$ of occurrence. Thus the probability that we see k events (or ones) is now equal to $[n!/(k!(n - k)!)]p^k(1 - p)^{n-k}$, which results into the binomial distribution function.

Additional Materials II: The Log Normal Distribution

Determining the moments is a little bit of work, but can be determined by standard calculus methods and the knowledge that $\int_{\mathbb{R}} \sigma^{-1} \phi((x - \mu)/\sigma) dx = 1$. For instance, the first moment of a lognormal distributed random variable $X \sim \mathscr{LN}(\mu, \sigma^2)$ is given by

$$
\begin{aligned}
\mathbb{E}(X) &= \int_{\mathbb{R}} x f_{\mu,\sigma}(x) dx \\
&= \int_0^\infty x \frac{1}{\sigma x} \phi \left(\frac{\log(x) - \mu}{\sigma} \right) dx \\
&= \int_{\mathbb{R}} \exp(z) \frac{1}{\sigma} \phi \left(\frac{z - \mu}{\sigma} \right) dz \\
&= \int_{\mathbb{R}} \frac{1}{\sigma \sqrt{2\pi}} \exp \left(-\frac{(z-\mu)^2}{2\sigma^2} + z \right) dz \\
&= \int_{\mathbb{R}} \frac{1}{\sigma \sqrt{2\pi}} \exp \left(-\frac{z^2 - 2z(\mu + \sigma^2) + \mu^2}{2\sigma^2} \right) dz \\
&= \exp \left(\frac{(\mu + \sigma^2)^2 - \mu^2}{2\sigma^2} \right) \int_{\mathbb{R}} \frac{1}{\sigma \sqrt{2\pi}} \exp \left(-\frac{(z - \mu - \sigma^2)^2}{2\sigma^2} \right) dz \\
&= \exp \left(\mu + 0.5\sigma^2 \right) \int_{\mathbb{R}} \frac{1}{\sigma} \phi \left(\frac{z - \mu - \sigma^2}{\sigma} \right) dz \\
&= \exp \left(\mu + 0.5\sigma^2 \right)
\end{aligned}
$$

Note that the population mean $\mathbb{E}(X)$ is a function of the density parameters μ and σ^2, which is typically different from the normal distribution. It also implies that the parameters μ and σ do not represent the mean and standard deviation of the random variable $X \sim \mathscr{LN}(\mu, \sigma^2)$.

Using similar calculus techniques, the variance is given by

$$
\mathbb{E}(X - \exp(\mu + 0.5\sigma^2))^2 = \exp \left(2\mu + \sigma^2 \right) \left(\exp \left(\sigma^2 \right) - 1 \right),
$$

which again is a function of the density parameters μ and σ. This implies that the relative standard deviation is now equal to $RSD = 100\% \sqrt{\exp (\sigma^2) - 1}$. Thus the relative standard deviation is now only a function of σ^2 and does not depend on μ. The skewness and excess kurtosis are a little more elaborate. They are functions of just the parameter σ and do not depend on μ. To get some feeling about the values of the skewness and kurtosis, we visualized them as function of σ in Fig. 4.12. This figure also plots the relative standard deviation (not expressed as percentage).

The figure suggest that for larger values of σ, the skewness and kurtosis are deviating from the value zero. Thus for larger values of σ, the lognormal distribution function really deviates from the normal distribution function. Additionally, the relative standard deviation is also increasing with σ. For instance, a value of $\sigma = 0.5$ gives an $RSD = 53.29\%$.

The quantiles of the lognormal distribution function can be obtained by the quantiles z_p of the standard normal distribution function. Indeed, let x_p be the pth quantile of the lognormal distribution function: then we know that $F_{\mu,\sigma}(x_p) = p$, with $F_{\mu,\sigma}$ the CDF for $f_{\mu,\sigma}$ in Eq. (4.2) and hence, using relationship Eq. (4.9), we obtain that $x_p = \exp \left(\mu + \sigma z_p \right)$. It follows immediately that the median of the lognormal

Fig. 4.12 RSD, skewness, and excess kurtosis of lognormal distribution: light gray curve: RSD; gray curve: γ_1, dark gray curve: γ_2

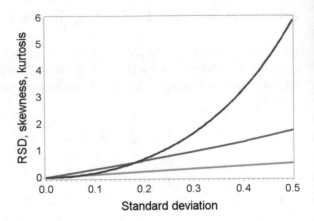

distribution function is equal to $\exp(\mu)$, since $z_{0.5} = 0$. The first and third quartiles are equal to $\exp(\mu - 0.67449\sigma)$ and $\exp(\mu + 0.67449\sigma)$, respectively, since $\Phi(-0.67449) = 1 - \Phi(0.67449) = 0.25$.

References

R.E. Barlow, Mathematical theory of reliability: a historical perspective. IEEE Trans. Reliab. **33**(1), 16–20 (1984)

W.G. Cochran, Estimation of bacterial densities by means of the "most probable number". Biometrics **6**(2), 105–116 (1950)

D. Glass, Graunt's life table. J. Inst. Actuar. **76**(1), 60–64 (1950)

A. Hald, *A History of Parametric Statistical Inference from Bernoulli to Fisher, 1713–1935*. (Springer Science & Business Media, 2008)

S. Kotz, N. Balakrishnan, N.L. Johnson, *Continuous Multivariate Distributions, Volume 1: Models and Applications* (Wiley, Hoboken, 2004)

C. Liu, D. Zheng, C. Griffiths, A. Murray, Comparison of repeatability of blood pressure measurements between oscillometric and auscultatory methods, in *2015 Computing in Cardiology Conference (CinC)* (IEEE, 2015), pp. 1073–1076

O.B. Sheynin, Cf gauss and the theory of errors. Arch. Hist. Exact Sci. **20**(1), 21–72 (1979)

O. Sheynin, Density curves in the theory of errors. Arch. Hist. Exact Sci. **49**(2), 163–196 (1995)

E. van den Heuvel, Estimation of the limit of detection for quantal response bioassays. Pharm. Stat. **10**(3), 203–212 (2011)

Chapter 5
Estimation

5.1 Introduction

The field of *inferential statistics* tries to use the information from a sample to make statements or decisions about the population of interest. It takes into account the uncertainty that the information is coming from sampling and does not perfectly represent the population, since another sample would give different outcomes. An important aspect of inferential statistics is estimation of the population parameters of interest. We have discussed the step from descriptions of a sample to those of a population already in Chap. 2; however, now that we have the theory of random variables at our disposal we can do much more than we did before. This is what we explore in this chapter.

This chapter can be split up into two parts: in Sects. 5.2–5.4 we consider the distributions of sample statistics or *estimators* given assumptions regarding the distribution of the variables of interest in the population. Sample statistics themselves are random variables, and hence we can study their distribution functions, expectations, and higher moments. We first study the distribution functions of sample statistics in general, assuming that the variable of interest has some distribution in the population but without further specifying the shape of this distribution function. Next, we study the distributions of sample statistics when we assume the variable of interest to be either normally or log normally distributed in the population. We devote more attention to so-called normal populations because of their prominence in statistical theory.

The second part of this chapter is Sect. 5.5, where we change our focus to estimation: in the subsections we discuss two different methods to obtain estimates $\hat{\theta}$ of the parameters of a population distribution $F_\theta(x)$ given sample data. The methods we discuss are the *method of moments* and the *maximum likelihood method*. In these sections, to provide a concrete example, we study the log normal distribution function, as this is one of the distribution functions for which the estimates originating from the two estimation methods differ.

© Springer Nature Switzerland AG 2022
M. Kaptein and E. van den Heuvel, *Statistics for Data Scientists*, Undergraduate Topics in Computer Science, https://doi.org/10.1007/978-3-030-10531-0_5

Hence, to summarize, in this chapter we will discuss:

- The definition of sample statistics
- Distributions of sample statistics
- The central limit theorem
- Confidence intervals
- Estimation procedures: method of moments and maximum likelihood

We will obviously discuss how to use R to do the computations involved.

5.2   From Population Characteristics to Sample Statistics

Now that we have introduced PDFs and PMFs in Chap. 4 to describe or approximate the population values, we can visualize the concept of statistical inference for the analysis of data. Figure 5.1a shows our approach in Chap. 2: we considered random samples directly from a population without using the notion of random variables or density functions. In this chapter, as depicted in Fig. 5.1b, the population (visualized as a histogram) may still be seen as a finite set of values $x_1, x_2, \ldots, x_N$, as we discussed in Chap. 2. However, contrary to our earlier approach, PDFs and PMFs, visualized on the right-hand side of Fig. 5.1b, are considered *statistical models* for the population values, as we have already indicated in Chap. 4. While we still use probability sampling to obtain our actual data, our process of statistical inference changes due to the introduction of our statistical models. Before collecting our sample, we consider the data to be represented by random variables (our model), where the CDFs connect the random variables to their PDFs or PMFs. When the data are truly collected we obtain the observed sample data, which we consider to be realizations from these random variables. Thus, the difference from Chap. 2 is that we now use the statistical models for describing populations using the theory introduced in Chap. 4 to get from the population to the sample data.

The introduction of PDFs and PMFs changes the way we make inference back from the sample data to the population. In Chap. 2 we discussed how we could use descriptive statistics from Chap. 1 to directly estimate population characteristics like population means, variances, and proportions. We have already called this statistical inference, but in this chapter we will use the sample data to estimate aspects or characteristics of the assumed PDF or PMF. More precisely, we will estimate the *parameters* of the PDF or PMF. These estimates will be considered realizations of corresponding estimators that are created from the random variable(s) of interest $X_1, X_2, \ldots, X_n$. Based on our understanding of the assumed PDF or PMF through the estimated parameters, we can draw conclusions regarding the population if the PDF or PMF describes the population appropriately. Going from the data to the PDF or PMF to the population is also called statistical inference, but now we are using (parametric) statistical models. Introducing these statistical models allows us to make more detailed statements about the population than we were able to without the introduction of these models.

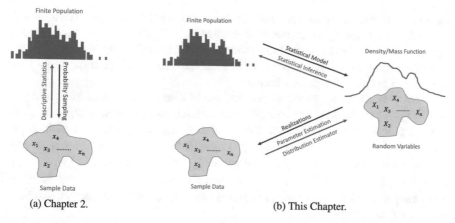

(a) Chapter 2. (b) This Chapter.

Fig. 5.1 Overview of the changes in our estimation procedure in this chapter compared to our approach in Chap. 2: in this chapter we use the theory of random variables and distribution functions as developed in the previous two chapters to allow us to make more detailed statements about the population of interest

Below we study the properties and distribution functions of different sample statistics or estimators—which we define in Sect. 5.2.2 below—given different choices or *assumptions* about the distribution function F_θ in the population. However, we first repeat some of our earlier defined population characteristics that might be of interest.

5.2.1 Population Characteristics

If the population is described by a PDF or PMF, say f, the population mean and population variance are determined by

$$\mu(f) = \mathbb{E}(X) = \int_\mathbb{R} x f(x) dx \tag{5.1}$$

$$\sigma^2(f) = \mathbb{E}(X - \mu(f))^2 = \int_\mathbb{R} (x - \mu(f))^2 f(x) dx \tag{5.2}$$

The population standard deviation is denoted by $\sigma(f)$. Note that we are now using the notation $\mu(f)$ to make explicit that the population mean μ will depend on our choice of f.

Other moments than the first two moments may help characterize the population as well, although they may not be used frequently in practice.[1] Two moments that are

---

[1] Higher moments of the population density f are difficult to determine or to estimate from a sample with small sample sizes. Thus, whenever data are sparse, higher moments will be considered less relevant, but in cases with big data, we anticipate that higher moments may become more important since the large sample size would make it possible to estimate these population characteristics better.

not uncommon to study are the third and fourth moment of the standardized variable $Z = (X - \mu(f))/\sigma(f)$. As we have already seen in Chap. 4, the third moment $\gamma_1(f) = \mathbb{E}(Z)^3$ of the standardized random variable is called the *skewness* and the fourth moment minus the value 3 of the standardized random variable $\gamma_2(f) = \mathbb{E}(Z)^4 - 3$ is called the excess kurtosis.

Other characteristics of the population might also be of interest, such as a percentile x_p. A percentile x_p separates the population in two pieces such that $p \times 100\%$ of the population has a value less than or equal to x_p and $(1 - p) \times 100\%$ of the population has a value larger than x_p. In Chap. 4 we called such values *quantiles*. If the population is described by a PDF f, the percentile $x_p(f)$ is given by $x_p(f) = F^{-1}(p)$, since $F(x_p(f)) = \int_{-\infty}^{x_p(f)} f(x)\,dx = p$, with F^{-1} the inverse distribution function. Thus the percentiles $x_{0.25}(f)$, $x_{0.5}(f)$, and $x_{0.75}(f)$ are the first, second, and third quartiles of the population.[2]

5.2.2 *Sample Statistics Under Simple Random Sampling*

We can view a sample X_1, X_2, ..., X_n of size n from a population as a set of random variables all coming from the same distribution function F. Although we have seen that the distribution of the finite population would change if we sample without replacement (due to the fact that the population is finite), it does not change if we sample with replacement or the change is really small if the sample is small with respect to the population size. Alternatively, in settings of repeated experiments (e.g., tossing a coin, producing products, measuring blood pressure) the population of values can be viewed as an infinite population and we may view the sample X_1, X_2, ..., X_n as random variables from some kind of mechanism F that would create random outcomes. Thus we will consider the sample X_1, X_2, ..., X_n as a set of random variables and $x_1, x_2, \ldots, x_n$ as the set of realizations that we would see or observe if we have conducted the (sampling) experiment.

In the simplest setting possible, we will assume that the random variables X_1, X_2, ..., X_n are all distributed according to F and they are all independent of each other, i.e. *mutually independent*, see Chaps. 4 and 6 for definitions of independence. This setting is often denoted by stating that "X_1, X_2,...., X_n *are independent and identically distributed (i.i.d) with distribution function F*". Thus in many situations we would start with something like "Let $X_1, X_2, \ldots, X_n$ be i.i.d. with $X_i \sim F$". Note that when doing so we have essentially assumed that we collected a simple random sample.

A sample statistic $T_n \in \mathbb{R}$ is now defined as any function $T_n \equiv T(X_1, X_2, \ldots, X_n)$ that is applied to the sample $X_1, X_2, \ldots, X_n$. As T_n is a function of random variables, it is itself a random variable. If we observe a realization $x_1, x_2, \ldots, x_n$ for the random

---

[2] We used the notation $\mu(f), \sigma(f), \gamma_1(f), \gamma_2(f)$, and $x_p(f)$ to indicate that these population characteristics are dependent on the density function f. In many texts however, the explicit dependence on (f) is omitted for convenience, as we already did in Chap. 4.

sample $X_1, X_2, \ldots, X_n$, we immediately observe a realization for the sample statistic T_n, which we will denote by t_n, i.e. $t_n = T(x_1, x_2, \ldots, x_n)$. Note that this latter expression aligns with our definition of sample statistics given in Chap. 1.

In Chap. 1 we listed multiple sample statistics. Here we list a number of often used sample statistics as random variables (instead of realizations):

- Sample average: $T_n = \bar{X} = \sum_{i=1}^{n} X_i/n$
- Sample variance: $T_n = S^2 = \sum_{i=1}^{n} (X_i - \bar{X})^2 / (n - 1)$
- Sample standard deviation: $T_n = S = \sqrt{\sum_{i=1}^{n} (X_i - \bar{X})^2 / (n - 1)}$
- Sample skewness: $T_n = b_1 = \frac{1}{n} \sum_{i=1}^{n} (X_i - \bar{X})^3 / S^3$
- Sample excess kurtosis: $T_n = b_2 = \frac{1}{n} \sum_{i=1}^{n} (X_i - \bar{X})^4 / S^4 - 3$
- Sample minimum: $T_n = X_{(1)} = \min\{X_1, X_2, \ldots, X_n\}$
- Sample maximum: $T_n = X_{(n)} = \max\{X_1, X_2, \ldots, X_n\}$

The minimum and maximum are considered the first and last random variables from the ordered set of random variables: $X_{(1)}, X_{(2)}, \ldots, X_{(n)}$. The ordered set of random variables puts the variables in order from small to large, i.e. $X_{(1)} < X_{(2)} < \cdots < X_{(n)}$.[3]

The order statistics $X_{(1)}, X_{(2)}, \ldots, X_{(n)}$ can be used to determine the quantiles x_p of the population. To determine the pth quantile x_p, we need to calculate np first. If $np \in \mathbb{N}$ is an integer, the quantile x_p is estimated by the average of two sequential order statistics: $q_p = [X_{(np)} + X_{(1+np)}]/2$. If $np \notin \mathbb{N}$ is not an integer, we take the smallest integer value that is larger than or equal to np, which is denoted by $\lceil np \rceil$. For example, assume $\lceil 7.29 \rceil = 8$ and $\lceil 7 \rceil = 7$, then the quantile x_p is estimated by the order statistic $q_p = X_{(\lceil np \rceil)}$.

In Chap. 1 we have already seen how to compute a number of the realized sample statistics using R.

5.3 Distributions of Sample Statistic T_n

Using the theory of random variables we developed in the previous chapter, we can now examine the distributions of the random variables T_n in more detail. The random variable T_n would itself have a CDF, which we denote by F_{T_n}. It is defined by $F_{T_n}(x) = P(T_n \leq x)$. This CDF is often referred to as the *sample distribution function* of statistic T_n, as it describes how the sample statistics vary if we draw (new) samples from the population. It should not be confused with the population distribution function F, which describes how the units from the population vary. The

---

[3] Note that the random variables $X_1, X_2, \ldots, X_n$ would indicate the order of sampling. In the order of sampling there is no guarantee that they also represent the order of size. Furthermore, it should be noted that ordering the random variables (like we do for the minimum and maximum) is only unique when the random variables are continuous, since each realization will produce different values. In practice though, rounding may violate uniqueness.

sample distribution function F_{T_n} is determined by the choice of T_n and the underlying distribution function F for the random sample $X_1, X_2, \ldots, X_n$.[4]

In many settings, the sample distribution function has a sample PMF or PDF f_{T_n}, often referred to as the *sample density function* of T_n. Unfortunately, it is not always easy to determine the sample distribution or sample density function in general. Only in special cases is it possible to determine closed-form functions. In the sections below we examine a number of these cases.

Note that we study the distributions of T_n, as these distribution functions effectively allow us to examine the *quality* of our estimators: the distribution function of the random variable T_n provides a measure of how well an estimator approximates a population value. For example:

- The expected value of an estimator $\mathbb{E}(T_n)$ is a measure for the central tendency of a sample statistic, and,
- The standard deviation of T_n provides a measure for the variability of a sample statistic.

Note that we have already studied these concepts in Chap. 2 when we studied the bias, MSE, and standard error (SE). In fact, the standard deviation of a sample statistic T_n is the *standard error* of that sample statistic; the standard error of a sample statistic of interest is commonly reported in scientific texts to quantify the uncertainty associated with the sample statistic.

5.3.1 Distribution of the Sample Maximum or Minimum

The sampling distribution of the minimum or maximum of $X_1, X_2, \ldots, X_n$ are easily determined in generality. Let T_n be the maximum of $X_1, X_2, \ldots, X_n$, the distribution function of T_n is given by

$$
\begin{aligned}
F_{X_{(n)}}(x) &= P\left(X_{(n)} \leq x\right) = P\left(\max\{X_1, X_2, \ldots, X_n\} \leq x\right) \\
&= P\left(X_1 \leq x, X_2 \leq x, \ldots, X_n \leq x\right) = \prod_{i=1}^{n} P\left(X_i \leq x\right) \\
&= [F(x)]^n
\end{aligned}
$$

Note that we have used both assumptions of i.i.d of the random sample X_1, $X_2, \ldots, X_n$ in the derivation of the sample distribution of the maximum. The independence is used to obtain a product of probabilities and the assumption that $X_i \sim F$ is used to obtain $[F(x)]^n$.

The distribution function for the minimum $X_{(1)} = \min\{X_1, X_2, \ldots, X_n\}$ can be determined in a similar way and is equal to $F_{X_{(1)}}(x) = 1 - [1 - F(x)]^n$.

The sample density functions of the maximum and minimum can be determined by taking the derivative of the sample distribution functions with respect

---

[4] Note that we are assuming simple random sampling here; when assuming other types of sampling procedures F_{T_n} might change. Thus F_{T_n} depends also on the sampling plan.

to x. Thus the sample densities for the maximum and minimum are given by $f_{X_{(n)}}(x) = n[F(x)]^{n-1} f(x)$ and $f_{X_{(1)}}(x) = n[1 - F(x)]^{n-1} f(x)$, respectively.

An interesting example of the sample minimum $X_{(1)}$ is when the random variables $X_1, X_2, \ldots, X_n$ are i.i.d. exponentially distributed, i.e. $X_i \sim \exp(\lambda)$. The exponential distribution function is given by $F(x) = 1 - \exp(-\lambda x)$ for $x > 0$ and zero otherwise. If we would substitute this distribution function in the distribution function $F_{X_{(1)}}$ of the sample minimum $X_{(1)}$ we would obtain $F_{X_{(1)}}(x) = 1 - \exp(-n\lambda x)$ for $x > 0$ and zero otherwise. Thus this implies that the sample distribution of the minimum $X_{(1)}$ is exponentially distributed, $X_{(1)} \sim \exp(n\lambda)$, but now with parameter $n\lambda$, when $X_1, X_2, \ldots, X_n$ are i.i.d. $\exp(\lambda)$ distributed.

5.3.2 Distribution of the Sample Average $\bar{X}$

The distribution function of the sample average $\bar{X} = \sum_{i=1}^{n} X_i / n$ is not so easy to determine in general. Only in special cases are we able to describe it. For example, we can describe it in the i.i.d. Bernoulli case. We have already seen in the previous chapter that if the random variables $X_1, X_2, \ldots, X_n$ are i.i.d. Bernoulli distributed, $X_i \sim \mathscr{B}(p)$, the sum of the random variables is binomial Bin (n, p) distributed. This implies that the distribution function of the sample average $\bar{X}$ of Bernoulli distributed random variables in $x \in [0, 1]$ is given by

$$F_{\bar{X}}(x) = P(\bar{X} \le x) = P(X_1 + X_2 + \cdots + X_n \le nx) = \sum_{k=0}^{\lfloor nx \rfloor} \binom{n}{k} p^k (1-p)^{n-k}$$

Although it is not easy to describe the sample distribution of the sample average in general, there are a few things that we can say about the moments of the distribution of the sample average.

The pth moment of a general sample statistic T_n is given by $\mathbb{E}(T_n^p) = \int_{\mathbb{R}} t^p f_{T_n}(t) \, dt$, if the sample density f_{T_n} exists, using the definition of moments for any random variable. It can, however, also be calculated in a different way, using the population density f and the fact that $X_1, X_2, \ldots, X_n$ are i.i.d. F. The pth moment of T_n is

$$\begin{aligned} \mathbb{E}(T_n^p) &= \int_{\mathbb{R}} t^p f_{T_n}(t) \, dt = \mathbb{E}(T_n^p(X_1, X_2, \ldots, X_n)) \\ &= \int_{\mathbb{R}^n} T_n^p(x_1, x_2, \ldots, x_n) f(x_1) f(x_2) \cdots f(x_n) \, dx_1 dx_2 \cdots dx_n \end{aligned} \quad (5.3)$$

A special case is the first moment $\mu(f_{T_n}) = \mathbb{E}(T_n)$. The advantage of Eq. (5.3) is that the moments of the sample statistic $T_n = \bar{X}$ can be expressed in moments of the random variables $X_1, X_2, \ldots, X_n$. Indeed, the first moment of $\bar{X}$ is now given by

$$
\begin{aligned}
\mu\left(f_{\bar{X}}\right) &= \mathbb{E}\left(\bar{X}\right) \\
&= \int_{\mathbb{R}^n} \left[\tfrac{1}{n}\sum_{i=1}^n x_i\right] f\left(x_1\right) f\left(x_2\right)\cdots f\left(x_n\right) dx_1 dx_2\cdots dx_n \\
&= \tfrac{1}{n}\sum_{i=1}^n \int_{\mathbb{R}^n} x_i f\left(x_1\right) f\left(x_2\right)\cdots f\left(x_n\right) dx_1 dx_2\cdots dx_n \\
&= \tfrac{1}{n}\sum_{i=1}^n \int_{\mathbb{R}} x_i f\left(x_i\right) dx_i \\
&= \tfrac{1}{n}\sum_{i=1}^n \mathbb{E}\left(X_i\right) = \mu\left(f\right)
\end{aligned}
$$

Note that this relationship essentially proves why the rule $\mathbb{E}\left(X+Y\right) = \mathbb{E}\left(X\right) + \mathbb{E}\left(Y\right)$ in Chap. 4 is true. The pth central moment of T_n is now given by $\mathbb{E}(T_n - \mu(f_{T_n}))^p$ and can be written in a similar way as in Eq. (5.3). The second central moment is the variance of the sample statistic T_n. Taking the square root, we obtain the standard deviation of T_n.

For the sample average $\bar{X}$ we would obtain a variance $\mathbb{E}(\bar{X}-\mu(f))^2 = \sigma^2(f)/n$. To see this note that the square of $\bar{X} - \mu(f)$ is given by

$$
\begin{aligned}
\left(\bar{X}-\mu(f)\right)^2 &= \left(\tfrac{1}{n}\sum_{i=1}^n \left(X_i - \mu(f)\right)\right)^2 \\
&= \tfrac{1}{n^2}\sum_{i=1}^n \left(X_i - \mu(f)\right)^2 + \tfrac{2}{n^2}\sum_{i=1}^{n-1}\sum_{j=i+1}^n \left(X_i - \mu(f)\right)\left(X_j - \mu(f)\right)
\end{aligned}
$$

Using this relation and the rules on random variables from Chap. 4, we obtain

$$
\begin{aligned}
\sigma^2\left(f_{\bar{X}}\right) &= \mathbb{E}\left(\bar{X}-\mu(f)\right)^2 \\
&= \mathbb{E}\left[\tfrac{1}{n^2}\sum_{i=1}^n \left(X_i - \mu(f)\right)^2 + \tfrac{2}{n^2}\sum_{i=1}^{n-1}\sum_{j=i+1}^n \left(X_i - \mu(f)\right)\left(X_j - \mu(f)\right)\right] \\
&= \tfrac{1}{n^2}\sum_{i=1}^n \mathbb{E}\left(X_i - \mu(f)\right)^2 + \tfrac{2}{n^2}\sum_{i=1}^{n-1}\sum_{j=i+1}^n \mathbb{E}\left(X_i - \mu(f)\right)\left(X_j - \mu(f)\right) \\
&= \sigma^2\left(f\right)/n + \tfrac{2}{n^2}\sum_{i=1}^{n-1}\sum_{j=i+1}^n \mathbb{E}\left(X_i - \mu(f)\right)\mathbb{E}\left(X_j - \mu(f)\right) \\
&= \sigma^2\left(f\right)/n
\end{aligned}
$$

Thus the sample average has an expectation of $\mu\left(f\right)$ and a variance of $\sigma^2\left(f\right)/n$, irrespective of the population density f. In other words, the sample average $\bar{X}$ is an appropriate estimator for the population mean $\mu\left(f\right)$, and it has a standard error (SE) that is a factor $\sqrt{n}$ smaller than the standard deviation of the population, i.e. the standard error is $SE(\bar{X}) = \sigma\left(f\right)/\sqrt{n}$. Note that this standard error is typically unknown, since it depends on the unknown population standard deviation $\sigma\left(f\right)$ (see also Chap. 2). This unknown parameter can also be estimated from the sample data, as we will discuss in next subsection.

The skewness and kurtosis of the sample density of the sample average can also be expressed in terms of the skewness and kurtosis of the population density. The skewness is given by $\gamma_1(f_{\bar{X}}) = \gamma_1(f)/\sqrt{n}$ and the excess kurtosis is given by $\gamma_2(f_{\bar{X}}) = \gamma_2(f)/n$. Thus the skewness and excess kurtosis are close to zero when the sample size is getting large.

5.3.3 Distribution of the Sample Variance S^2

The distribution function of the sample variance $T_n = S^2 \equiv \sum_{i=1}^{n}(X_i - \bar{X})^2/(n-1)$ is in general unknown, but similar to the sample average, we are able to determine a few moments. To do so, it is easier to rewrite the variance first into $S^2 = \sum_{i=1}^{n}(X_i - \mu(f))^2/(n-1) - n(\bar{X} - \mu(f))^2/(n-1)$. The first moment is now easily determined as

$$
\begin{aligned}
\mu(f_{S^2}) &= \mathbb{E}(S^2) \\
&= \tfrac{1}{n-1}\sum_{i=1}^{n}\mathbb{E}(X_i - \mu(f))^2 - \tfrac{n}{n-1}\mathbb{E}\left(\bar{X} - \mu(f)\right)^2 \\
&= \tfrac{n}{n-1}\sigma^2(f) - \tfrac{1}{n-1}\sigma^2(f) \\
&= \sigma^2(f),
\end{aligned}
$$

using the rules on random variables from Chap. 4. Thus the sample variance S^2 is an unbiased estimator of the population variance $\sigma^2(f)$. With this information it also becomes possible to estimate the standard error of the sample average with $\hat{SE}(\bar{X}) = S/\sqrt{n}$, which we already knew from Chap. 2.

The second moment of the sample variance is more difficult to determine, but it is possible (as we also have seen in Chap. 2). It is given by

$$
\sigma^2(f_{S^2}) = \mathbb{E}\left(S^2 - \mu(f_{S^2})\right)^2 = \mathbb{E}\left(S^2 - \sigma^2(f)\right)^2 = \left[\frac{1}{n}\gamma_2(f) + \frac{2}{n-1}\right]\sigma^4(f)
$$

Thus the second moment of the sample variance depends on the excess kurtosis of the population density and the squared population variance.[5]

The standard deviation $\sigma(f_{S^2}) = \sigma^2(f)\sqrt{[(n-1)\gamma_2(f) + 2n]/[n(n-1)]}$ is also referred to as the standard error of the sample variance S^2. This standard error can be estimated by substituting S for $\sigma(f)$ and b_2 for $\gamma_2(f)$.

5.3.4 The Central Limit Theorem

Above, we discussed the moments of the sample average for any population density f (assuming that these moments existed). The mean of the sample average was determined as $\mu(f)$ and the standard deviation (or standard error) was determined as $\sigma(f)/\sqrt{n}$. If the sample size increases, the standard deviation vanishes ($\sigma(f)/\sqrt{n} \to 0$ if $n \to \infty$). This implies that the sample average converges to the population mean $\mu(f)$. This seems reasonable, since the increase in information ($n \to \infty$) would lead to a more precise understanding of the population mean. In other words, if we

---

[5] As we mentioned in Chap. 2, for normal population densities f the second moment of the sample variance is $2\sigma^4(f)/(n-1)$, which is just a function of the sample variance. Indeed, under the assumption of normality, the excess kurtosis is equal to zero.

measure almost all units from the population we will know the population mean almost exactly.[6]

If we study the standardized sample average, i.e. $Z_n = (\bar{X} - \mu(f))/(\sigma(f)/\sqrt{n}) = \sqrt{n}(\bar{X} - \mu(f))/\sigma(f)$, the mean would be equal to zero $(\mathbb{E}(Z_n) = 0)$ and the variance would be equal to one $(\mathbb{E}(Z_n^2) = 1)$. This is true irrespective of the sample size n. Thus if the sample size increases, the random variable Z_n does not change its mean or its standard deviation. It does not converge to zero either when the sample size n increases to infinity, as the mean and variance of Z_n remain zero and one, respectively. Note that the distribution function of Z_n may still depend on n, since the skewness and kurtosis of Z_n are given by $\gamma_1(f)/\sqrt{n}$ and $\gamma_2(f)/n$, respectively, and are different for different n.

What can we then say about the distribution function $\Pr(Z_n \leq z)$ of Z_n? Well, *the central limit theorem* tells us that this distribution function converges to the standard normal distribution function $\Phi(z) = \int_{-\infty}^{z} \phi(x)dx$. Thus in other words, if the sample size becomes large,

$$Z_n = \sqrt{n}(\bar{X} - \mu(f))/\sigma(f) \sim \mathcal{N}(0, 1), \tag{5.4}$$

becomes almost normal. Note that we did not imply anything about the shape of the population density f, just the existence of $\mu(f)$ and $\sigma^2(f)$ (and of course the assumption that $X_1, X_2, \ldots, X_n$ are i.i.d. with density f).

The central limit theory is formulated as follows (Patrick 1995).[7] Let X_1, $X_2, \ldots, X_n$ be i.i.d. with distribution function F and with mean $\mu(f) = \mathbb{E}(X_k)$ and with finite variance $\sigma^2(f) = \mathbb{E}(X_k - \mu(f))^2 < \infty$. The distribution function of $\sqrt{n}(\bar{X} - \mu(f))$ converges to the normal distribution function with mean zero and variance $\sigma^2(f)$. In other words, the distribution function of $\sqrt{n}(\bar{X} - \mu(f))/\sigma(f)$ converges to the standard normal distribution function.

A consequence of this formulation is that the distribution function of any statistic of the form $S_n = \sum_{i=1}^{n} \psi(X_i)/n$ would also converge to a normal distribution function when the mean $\mu_\psi(f) = \mathbb{E}(\psi(X_k))$ and variance $\sigma_\psi^2(f) = \mathbb{E}(\psi(X_k) - \mu_\psi(f))^2$ are finite. Using the central limit theorem, $\psi(X_1), \psi(X_2), \ldots, \psi(X_n)$ are i.i.d. and have a finite variance; thus, the statistic $\sqrt{n}(S_n - \mu_\psi(f))$ converges to a normal distribution with mean zero and variance $\sigma_\psi^2(f)$.

As we can see from the formulation of the central limit theorem the underlying distribution function F is irrelevant. Thus the central limit theorem can also be applied to the sample average of Bernoulli distributed random variables $X_1, X_2, \ldots, X_n$, with $X_k \sim \mathcal{B}(p)$. Thus the sample average of zero's and one's is also related to a normal distribution when the sample size is large enough. In this case the mean is $\mu(f) = p$ and the variance is $\sigma^2(f) = p(1 - p)$. Thus the distribution function of $Z_n = \sqrt{n}(\bar{X} - p)/\sqrt{p(1 - p)}$ converges to a standard normal distribution function

---

[6] It is interesting to think about what would happen if this was not true; how would we then go about stating something regarding a population based on a sample?

[7] There exist several formulations of the central limit theorem. We chose the classical theorem which is the Lindeberg-Levy formulation.

Φ. This may not be surprising, since we have already discussed in Chap. 4 that the binomial and normal distribution functions are close to each other whenever n is larger than 20 and the number np and $n(1-p)$ are larger than 5. The approximation comes from the central limit theorem.

5.3.4.1 Central Limit Theorem Applied to Variances

The central limit theorem can also be applied to the sample variance S^2, when the fourth central moment $\mathbb{E}(X_k - \mu(f))^4$ exists. The sample variance can be rewritten as

$$S^2 = \frac{1}{n-1} \sum_{i=1}^{n} (X_i - \mu(f))^2 - \frac{n}{n-1} (\bar{X} - \mu(f))^2.$$

We may apply the central limit theorem first to $\frac{1}{n} \sum_{i=1}^{n} (X_i - \mu(f))^2$, where $\psi(x) = (x - \mu(f))^2$. The mean of $\psi(X_i) = (X_i - \mu(f))^2$ is given by $\mu_\psi(f) = \mathbb{E}(X_i - \mu(f))^2 = \sigma^2(f)$ and the variance is given by

$$\sigma_\psi^2(f) = \mathbb{E}((X_i - \mu(f))^2 - \sigma^2(f))^2 = \mathbb{E}(X_i - \mu(f))^4 - \sigma^4(f) = [\gamma_2(f) + 2]\sigma^4(f),$$

with $\gamma_2(f)$ the excess kurtosis of population density f. The variance of $\frac{1}{n} \sum_{i=1}^{n} (X_i - \mu(f))^2$ is then given by $[\gamma_2(f) + 2]\sigma^4(f)/n$. Based on the central limit theorem, we obtain that the large sample distribution of $\sqrt{n}[\frac{1}{n} \sum_{i=1}^{n} (X_i - \mu(f))^2 - \sigma^2(f)]/(\sigma^2(f)\sqrt{\gamma_2(f) + 2})$ is equal to the standard normal distribution. In other words,

$$\frac{1}{\sqrt{n}} \sum_{i=1}^{n} [(X_i - \mu(f))^2 - \sigma^2(f)] \overset{n \to \infty}{\longrightarrow} \mathcal{N}\left(0, [\gamma_2(f) + 2]\sigma^4(f)\right),$$

Thus $\frac{1}{n} \sum_{i=1}^{n} (X_i - \mu(f))^2$ is approximately normally distributed with $\mathcal{N}\left(\sigma^2(f), [\gamma_2(f) + 2]\sigma^4(f)/\mathcal{N}\right)$, which implies that $\frac{1}{n-1} \sum_{i=1}^{n} (X_i - \mu(f))^2$ is approximately normally distributed with $\mathcal{N}\left(\sigma^2(f), [\gamma_2(f) + 2]\sigma^4(f)/n\right)$.

From the central limit theorem, we know that the distribution of $\sqrt{n}(\bar{X} - \mu(f))$ converges to the normal distribution $\mathcal{N}\left(0, \sigma^2(f)\right)$. This implies that

$$(\bar{X} - \mu(f))^2 = [\sqrt{n}(\bar{X} - \mu(f))]^2/n \to 0$$

when n converges to ∞. Since $n/(n-1)$ converges to 1 for $n \to \infty$, we now obtain that $\frac{n}{n-1}(\bar{X} - \mu(f))^2$ converges to zero. Combining the individual results above, we obtain that $\sqrt{n}(S^2 - \sigma^2(f))$ converges to a normal distribution

$\mathcal{N}\left(0,[\gamma_2(f)+2]\sigma^4(f)\right)$, with $\sigma^2(f)$ the population variance and $\gamma_2(f)$ the population excess kurtosis.[8]

5.3.5   Asymptotic Confidence Intervals

The sample distribution function F_{T_n} can help quantify how much the statistic is varying around the population characteristic it is trying to approach. From the sample distribution function F_{T_n} we may determine certain quantiles, say $x_p(f_{T_n})$ and $x_{1-p}(f_{T_n})$, for $p < 0.5$. Based on the definition of quantiles and the sampling distribution function F_{T_n}, the sample statistic will fall in the interval $(x_p(f_{T_n}), x_{1-p}(f_{T_n})]$ with probability $1 - 2p$. Indeed, the probability is equal to

$$\begin{aligned}
\Pr(T_n \in (x_p(f_{T_n}), x_{1-p}(f_{T_n})]) &= \Pr(T_n \leq x_{1-p}(f_{T_n})) - \Pr(T_n \leq x_p(f_{T_n})) \\
&= F_{T_n}(x_{1-p}(f_{T_n})) - F_{T_n}(x_p(f_{T_n})) \\
&= 1 - p - p = 1 - 2p
\end{aligned}$$

Now if we make a few assumptions about the large sample distribution of T_n, we would be able to quantify how close the sample statistic is to the population characteristic θ, with θ equal to for instance $\mu(f)$, $\sigma(f)$, or $x_p(f)$. Thus let's assume that T_n is trying to estimate the population characteristic θ and that the asymptotic sample distribution of $(T_n - \theta)/\tau_n$ is given by the standard normal distribution function Φ. Here τ_n is the standard error of the sample statistic T_n and we will assume that we can estimate it from the sample data. If $z_p = x_p(\phi)$ is the quantile of the standard normal distribution, the sample statistic T_n falls in the interval $(\theta + z_p\tau_n, \theta + z_{1-p}\tau_n] = (\theta - z_{1-p}\tau_n, \theta + z_{1-p}\tau_n]$ with probability approximately equal to $1 - 2p$. Indeed, if the sample size is large enough we obtain

$$\begin{aligned}
\Pr(T_n \in (\theta - z_{1-p}\tau_n, \theta + z_{1-p}\tau_n]) &= \Pr((T_n - \theta)/\tau_n \in (-z_{1-p}, z_{1-p}]) \\
&\approx \Phi(z_{1-p}) - \Phi(-z_{1-p}) \\
&= 1 - p - p = 1 - 2p
\end{aligned}$$

Alternatively, we can rewrite the probability $\Pr(T_n \in (\theta - z_{1-p}\tau_n, \theta + z_{1-p}\tau_n])$ into $\Pr(\theta \in (T_n - z_{1-p}\tau_n, T_n + z_{1-p}\tau_n])$, which means that the population characteristic is contained within limits $T_n - z_{1-p}\tau_n$ and $T_n + z_{1-p}\tau_n$ with probability equal to $1 - 2p$. The interval $(T_n - z_{1-p}\tau_n, T_n + z_{1-p}\tau_n]$ is now called an *asymptotic confidence interval* for θ with *confidence level* $1 - 2p$. It is common to choose the

---

[8] The central limit theorem holds true for sums of random variables, as we just indicated, but there are other examples that demonstrate that the large sample distribution of T_n can be normal. For instance, it is shown that the large sample distribution of $\sqrt{n}\left(X_{(\lceil np \rceil)} - x_p\right)$ converges to a normal distribution $N\left(0, p(1-p)/(f(x_p))^2\right)$, with f the population density, x_p the pth quantile, and $X_{(k)}$ the kth-order statistic. Thus the sample distribution function of the sample statistic T_n can sometimes be approximated by a normal distribution function, even if it is not always the sum of independent random variables.

confidence level equal to 95%, which means that $p = 0.025$. The confidence interval thus quantifies that, if the same population is sampled on numerous occasions and interval estimates are made on each occasion, the resulting intervals would include the true population parameter in (approximately) 95% of the cases.

In practice we still need to estimate the standard error τ_n to be able to calculate the confidence interval, as τ_n would be a function of the density parameters and is unknown in the calculation of the interval $(T_n - z_{1-p}\tau_n, T_n + z_{1-p}\tau_n]$. It is then common to replace τ_n by its estimator $\hat{\tau}_n$. In some cases, we would also change the normal quantile z_{1-p} by a quantile of the t-distribution if we could formulate a degrees of freedom for the estimator $\hat{\tau}_n$ (see Sect. 5.4).

5.3.5.1 Illustrating the Asymptotic Confidence Interval

To illustrate the asymptotic confidence interval, let's consider the sample average $\bar{X} = \sum_{i=1}^n X_i/n$ that tries to estimate the population mean $\mu(f)$. From the central limit theorem, we know that $\bar{X} \sim \mathcal{N}(\mu(f), \sigma^2(f)/n)$ is approximately normally distributed. The 97.5% quantile of the standard normal distribution function is equal to $z_{0.975} = 1.96$; see Chap. 4. Applying the 95% asymptotic confidence interval for $\mu(f)$ using the estimator $\bar{X}$, results in

$$\left(\bar{X} - 1.96\sigma(f)/\sqrt{n}, \bar{X} + 1.96\sigma(f)/\sqrt{n}\right]$$

Since the standard deviation $\sigma(f)$ is unknown, we may replace $\sigma(f)$ by an estimator. The most commonly used estimator is to use the sample standard deviation $S = \sqrt{\sum_{i=1}^n (X_i - \bar{X})^2/(n-1)}$. The 95% confidence interval on $\mu(f)$ that can be calculated from the data is then equal to

$$\left(\bar{X} - 1.96\, S/\sqrt{n}, \bar{X} + 1.96\, S/\sqrt{n}\right] \tag{5.5}$$

Applying this to the estimation of the mean number of hours per week that children watch television using the school-children data gives us the following statistics: sample size $n = 50{,}069$, sample average $\bar{x} = 14.22914$ and sample variance $s^2 = 108.9057$. The 95% confidence interval for the mean number of hours of television watching per week is now determined by $(14.14, 14.32]$ using the confidence interval in Eq. (5.5). After reading the school data `high-school.csv` into R, and calling this dataset `schooldata`, the following R code gives us the required results

```
> schooldata <- read.csv("high-school.csv")
> n <- dim(schooldata)
> n
[1] 50069    13
> mu <- mean(schooldata$TV)
> mu
[1] 14.22914
> v <- var(schooldata$TV)
```

```
> v
[1] 108.9057
> lcl <- mu - 1.96*sqrt(v/n[1])
> lcl
[1] 14.13773
> ucl <- mu + 1.96*sqrt(v/n[1])
> ucl
[1] 14.32055
```

We can also determine the 95% asymptotic confidence interval on the population variance of the number of hours per week watching television. This requires an R package that can estimate the excess kurtosis of a variable (like television watching). The following R code gives all the result:

```
> n <- dim(schooldata)
> n
[1] 50069    13
> v <- var(schooldata$TV)
> v
[1] 108.9057
> library(e1071) # included to have access to the kurtosis()
    function
> b2 <- kurtosis(schooldata$TV, type=3)
> b2
[1] 3.653064
> lcl <- v - 1.96*sqrt((b2+2)*(v^2)/n[1])
> lcl
[1] 106.6376
> ucl <- v + 1.96*sqrt((b2+2)*(v^2)/n[1])
> ucl
[1] 111.1738
```

Note that we have made use of the asymptotic distribution of the sample variance, which is certainly appropriate with a sample size of more than 50,000.

5.4 Normally Distributed Populations

In cases in which we assume that the random variables $X_1, X_2, \ldots, X_n$ are i.i.d. normally distributed, $X_i \sim \mathcal{N}(\mu, \sigma^2)$, we are able to make finite sample statements for a few of the sample statistics, due to a few nice properties of the normal distribution.

1. **Property 1:** The sum of the random variables $\sum_{i=1}^{n} X_i$ is again normally distributed, but now with mean $n\mu$ and variance $n\sigma^2$. Thus this implies that the sample average $\bar{X}$ has a normal distribution with mean μ and variance σ^2/n and thus has $(\bar{X} - \mu)/(\sigma/\sqrt{n})$, a standard normal distribution function.
2. **Property 2:** The sum of the squared standardized random variables $\sum_{i=1}^{n} (X_i - \mu)^2/\sigma^2$ is known to be chi-square distributed with n degrees of freedom. The chi-square PDF is given by

$$f_{\chi^2}(x) = \frac{1}{\Gamma(n/2)2^{n/2}}\, x^{(n-2)/2} \exp(-x/2),$$

for $x > 0$ and zero otherwise. The function Γ is the gamma function defined implicitly through an integral: $\Gamma(x) = \int_0^\infty z^{x-1} \exp(-z)\, dz$; see also the negative binomial PMF in Chap. 4. A graphical representation of the chi-square PDF is given in Fig. 5.2. Additionally, the sum $\sum_{i=1}^n (X_i - \bar{X})^2 / \sigma^2$ is chi-square distributed with $n-1$ degrees of freedom. We will see the use of the chi-square distribution when we calculate confidence intervals for the standard deviation of normally distributed random variables.

The first four moments of the chi-square distribution with n degrees of freedom can be determined. They are given by

$$\mu\left(f_{\chi^2}\right) = n, \quad \sigma^2\left(f_{\chi^2}\right) = 2n, \quad \gamma_1\left(f_{\chi^2}\right) = \sqrt{\frac{8}{n}}, \quad \gamma_2\left(f_{\chi^2}\right) = \frac{12}{n}.$$

The skewness and excess kurtosis of the chi-square distribution would rapidly converge to zero when the degrees of freedom increases. Which is not surprising, since we know that the distribution function of a properly standardized sample variance S^2 would converge to a normal distribution (see Sect. 5.3.4.1).

3. **Property 3:** Let Z be standard normally distributed, $Z \sim \mathcal{N}(0, 1)$, let V_n^2 be chi-square distributed with n degrees of freedom, $V_n^2 \sim \chi_n^2$, and assume that Z and V_n^2 are independent. The distribution function of the ratio of this standard normal random variable and the square root of a chi-square $Z/(V_n/\sqrt{n})$ has a so-called Student t-distribution with n degrees of freedom. The Student t PDF is given by

$$f_t(x) = \frac{\Gamma((n+1)/2)}{\Gamma(n/2)\sqrt{n\pi}}\left(1 + \frac{x^2}{n}\right)^{-(n+1)/2} \qquad x \in \mathbb{R}.$$

Note that Student's t-density is symmetric around zero. The symmetry implies the following relation for the pth quantile: $x_p(f_t) = -x_{1-p}(f_t)$. This is similar to the quantiles of the normal distribution (i.e. $z_p = -z_{1-p}$). A graphical representation of a t-probability distribution with only two degrees of freedom is presented in Fig. 5.2 together with a standard normal probability distribution. We will see the use of the t-distribution when we calculate confidence intervals for the mean of normally distributed random variables.

The first four moments of the t-distribution with n degrees of freedom are given by

$$\mu(f_t) = 0, \quad \sigma^2(f_t) = \frac{n}{n-2}, \quad \gamma_1(f_t) = 0, \quad \gamma_2(f_t) = \frac{6}{n-4}$$

Thus the Student t-density with $n = 1$ or $n = 2$ does not have a finite variance. The tails would go to zero when $|x|$ converges to ∞, but the tails multiplied by x^2

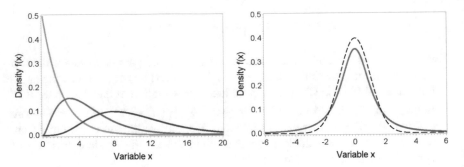

Fig. 5.2 Left: chi-square densities (light gray curve $df = 2$; gray curve $df = 5$; dark gray curve $df = 10$. Right: t-density with $df = 2$ (gray curve) and standard normal density (black dotted curve)

do not go to zero or not fast enough to make the area under the curve finite. The skewness is zero only when the degrees of freedom are larger than 3. To have a finite kurtosis the degrees of freedom should be larger than 4.

5.4.1 Confidence Intervals for Normal Populations

The three properties regarding i.i.d. normally distributed random variables, as discussed in the previous section, are useful to obtain confidence intervals for the mean μ and variance σ^2, without needing a large sample size n. To illustrate this we will first consider the mean μ. Property 1 provides that $(\bar{X} - \mu)/\left(\sigma/\sqrt{n}\right)$ is standard normal distributed. Since we do not know the standard deviation σ in practice, we need to estimate this. One option is to take the sample standard deviation S as we discussed in Sect. 5.3.5. We also know from property 2 that $V_{n-1}^2 = (n-1)S^2/\sigma^2 \sim \chi_{n-1}^2$. Then rewriting the random variable $(\bar{X} - \mu)/\left(S/\sqrt{n}\right)$ into

$$\frac{\bar{X} - \mu}{S/\sqrt{n}} = \frac{(\bar{X} - \mu)/\left(\sigma/\sqrt{n}\right)}{\sqrt{(n-1)S^2/\sigma^2}/\sqrt{n-1}} = \frac{Z}{V_{n-1}/\sqrt{n-1}},$$

with Z standard normally distributed. Now using property 3, we see that $(\bar{X} - \mu)/\left(S/\sqrt{n}\right)$ has a Student t-distribution with $n - 1$ degrees of freedom. This means that we can use the quantile values of the Student t-distribution to formulate confidence intervals on μ. If $x_p\,(f_t)$ is the pth quantile of the Student t-distribution with $n - 1$ degrees of freedom, the $1 - 2p$ confidence interval for μ (with $p < 0.5$) is now

$$\left(\bar{X} - x_{1-p}\,(f_t)\,\frac{S}{\sqrt{n}},\ \bar{X} + x_{1-p}\,(f_t)\,\frac{S}{\sqrt{n}}\right] \tag{5.6}$$

Note that the only restriction on the sample size n, is that it should be larger than 2, otherwise we cannot calculate a sample variance. Furthermore, this interval is very similar to asymptotic confidence intervals of the form $(T_n - z_{1-p}\tau_n, T_n + z_{1-p}\tau_n]$, where we used the quantiles of the standard normal distribution (i.e., $x_{1-p}(f_t)$ was replaced by z_{1-p}). When the sample size is increasing, the quantile of the Student t-distribution with $n - 1$ degrees of freedom converges to the quantile of the standard normal distribution. The quantiles of the t-distribution are already quite close to the same quantiles of the normal distribution when sample sizes are larger than one hundred.

Calculating the confidence interval in Eq. (5.6) for the number of hours per week of television watching for the school data does not lead to anything else than what we reported in Sect. 5.3.5. The 95% confidence interval is (14.14, 14.32]. The sample size is so large that the t-quantile is equal to the normal quantile. Indeed, $x_{0.975}(f_t)$ with 50,068 degrees of freedom is equal to 1.960011. Since the form of the confidence interval in Eq. (5.6) is the same as the form of asymptotic confidence intervals, the non-normality of the underlying data of television watching has become irrelevant when the sample size is large.

As we saw in Sect. 5.3.5, the asymptotic $1 - 2p$ confidence interval for the variance σ^2 is given by $(S^2 - z_{1-p}\hat{\tau}_n, S^2 - z_{1-p}\hat{\tau}_n]$, where the estimated standard error $\hat{\tau}_n$ is given by

$$\hat{\tau}_n = S^2\sqrt{[b_2 + 2]/n},$$

and b_2 is the sample excess kurtosis. An alternative confidence interval for the variance σ^2 can be created under the assumption that $X_1, X_2, \ldots, X_n$ are i.i.d. normally distributed, $X_i \sim N(\mu, \sigma^2)$. This will make use of property 2 from the previous subsection and we will show that this alternative confidence interval has a different form than the asymptotic confidence interval.

We know that $V_{n-1}^2 = (n - 1)S^2/\sigma^2 \sim \chi_{n-1}^2$, which implies that we can directly use the chi-square distribution. Let $x_p(f_{\chi^2})$ be the pth quantile of the chi-square distribution with $n - 1$ degrees of freedom. Then we obtain $\Pr(V_{n-1}^2 \leq x_p(f_{\chi^2})) = p$. Rewriting this probability results into $\Pr(\sigma^2 \geq (n - 1)S^2/x_p(f_{\chi^2})) = p$. Additionally, we have $\Pr(V_{n-1}^2 > x_{1-p}(f_{\chi^2})) = p$, which results in the probability $\Pr(\sigma^2 < (n - 1)S^2/x_{1-p}(f_{\chi^2})) = p$. Thus a $1 - 2p$ confidence interval for the variance σ^2 is now given by

$$[(n - 1)S^2/x_{1-p}(f_{\chi^2}), (n - 1)S^2/x_p(f_{\chi^2})) \tag{5.7}$$

Note that the form of this confidence interval is no longer of the form "estimate plus and minus a constant times the standard error", but there is still a resemblance. The asymptotic confidence interval can be written into the form of $(c_n S^2, C_n S^2]$, with $c_n = 1 - z_{1-p}\sqrt{[b_2 + 2]/n}$ and $C_n = 1 + z_{1-p}\sqrt{[b_2 + 2]/n}$. Thus the asymptotic confidence interval is of the same form as the interval in Eq. (5.7), but it uses different constants to multiply the sample variance.

The advantage of the confidence interval in Eq. (5.7) over the asymptotic confidence interval for the variance is that it does not require large sample sizes. It will be an appropriate confidence interval[9] for any sample size larger than 2, if X_1, $X_2, \ldots, X_n$ are i.i.d normally distributed. The asymptotic confidence interval does not require the assumption of normality, but requires large datasets.

Using the number of hours per week of television watching in the school data, we can also calculate the 95% confidence interval if we assume that the distribution for the hours of watching television is normal. The following R code provides the results:

```
> n <- dim(schooldata)
> n
[1] 50069   13
> v <- var(schooldata$TV)
> v
[1] 108.9057
qchi_up <- qchisq(0.975,n[1]-1)
> qchi_up
[1] 50690.11
> qchi_low <- qchisq(0.025,n[1]-1)
> qchi_low
[1] 49449.68
> lcl <- (n[1]-1)*v/qchi_up
> lcl
[1] 107.5692
> ucl <- (n[1]-1)*v/qchi_low
> ucl
[1] 110.2675
```

The results are somewhat different from the interval (106.64, 111.17] obtained in Sect. 5.3.5, while the sample size is quite large. The reason is that the underlying distribution of the number of hours per week of television watching is far from normal. More precisely, the issue is the excess kurtosis, which deviates from zero. If b_2 is close to zero, the constants $(n-1)/x_{1-p}(f_{\chi^2})$ and $(n-1)/x_p(f_{\chi^2})$ used in Eq. (5.7) would be close to the constants $c_n = 1 - z_{1-p}\sqrt{2/n}$ and $c_n = 1 - z_{1-p}\sqrt{2/n}$, respectively, where b_2 is taken zero. Thus in the situation of the data on television watching, we expect that the asymptotic confidence interval is preferred over the confidence interval based on normality, since the excess kurtosis is not close to zero. It was estimated at $b_2 = 3.653064$, see Sect. 5.3.5.1.

Confidence intervals for the population variance σ^2 immediately result in confidence intervals on the population standard deviation σ by just taking the square root.

---

[9] Appropriate confidence intervals means that the confidence interval contains the parameter of interest with the correct level of confidence. If we construct 95% confidence intervals, we would like the probability that the parameter is inside the confidence interval to be equal to 95%. If we would use asymptotic confidence intervals, and the normal approximation is not close yet due to relatively small sample sizes, the confidence level could deviate from 95%.

5.4.2 Lognormally Distributed Populations

The lognormal distribution function—see also Chap. 4—is strongly related to the normal distribution function, since $\log(X)$ is normal $\mathcal{N}(\mu, \sigma^2)$ when X is lognormally $\mathcal{LN}(\mu, \sigma^2)$ distributed. This makes the analysis of data that represents realizations of lognormally distributed random variables relatively easy, since we can just take the logarithmic transformation and use sample statistics on the transformed data.

Thus if $X_1, X_2, \ldots, X_n$ are i.i.d. lognormally distributed, the sample average and sample variance of the transformed random variables $\log(X_1)$, $\log(X_2), \ldots, \log(X_n)$, with $\log$ the natural logarithm, are given by

$$\bar{X}_{\log} = \frac{1}{n} \sum_{i=1}^{n} \log(X_i) \qquad S_{\log}^2 = \frac{1}{n-1} \sum_{i=1}^{n} \left(\log(X_i) - \bar{X}_{\log}\right)^2. \qquad (5.8)$$

These sample statistics are unbiased estimates of μ and σ^2, respectively. Even stronger, using the results from Sect. 5.4, the distribution function of $\bar{X}_{\log}$ is normally distributed $\mathcal{N}\left(\mu, \sigma^2/n\right)$ and $(n-1) S_{\log}^2/\sigma^2$ has a chi-square distribution function with $n-1$ degrees of freedom. This also implies that the geometric average $\exp\left(\bar{X}_{\log}\right) = \prod_{i=1}^{n} \sqrt[n]{X_i}$ is lognormally distributed.

5.5 Methods of Estimation

In the previous sections we introduced several sample statistics and discussed some of their characteristics. We also demonstrated that these sample statistics can be viewed as estimators of certain population properties. For instance, the sample average $\bar{X}$ can be used as an estimator of the population mean $\mu(f)$. However, it would be much nicer if we could construct procedures that would allow us to directly estimate the parameters of population distributions, as opposed to estimating only characteristics of the distributions. Below we discuss two such approaches, *method of moments estimation* (MME) and *maximum likelihood estimation* (MLE). The latter is considered the better one in most cases and is more generic. Note that both estimation methods provide us with parameter estimates that are themselves simply functions of our sample data: hence, estimators obtained using either MME or MLE are themselves sample statics (T_n) and we can study their distribution functions (we will do so when we study the standard error of MLE estimates in Sect. 5.5.2.3).

5.5.1   Method of Moments

The method of moments connects strongly to the theory on sample statistics T_n from Chap. 2. Assume that the population density f_θ depends on a set of parameters $\theta = (\theta_1, \theta_2, \ldots, \theta_m)^T$. For example, the normal density has two parameters $\theta_1 = \mu$ and $\theta_2 = \sigma^2$. Estimates of these parameters can then be obtained by computing m or more central moments. As discussed earlier, the central moments of a random variable X, which has a density f_θ, are defined by

$$\mu_r(f_\theta) = \mathbb{E}(X - \mu(f_\theta))^r = \int_{\mathbb{R}} (x - \mu(f_\theta))^r f_\theta(x)\, dx,$$

with $\mu(f_\theta) = \mathbb{E}(X)$ the mean value. For discrete random variables we would have

$$\mu_r(f_\theta) = \mathbb{E}(X - \mu(f_\theta))^r = \sum_{k=0}^{\theta} (k - \mu(f_\theta))^r f_\theta(k),$$

where $f_\theta(k)$ represents the probability that X is equal to k.

The moments clearly depend on the parameters θ and the moments can be estimated with the sample moments $M_r = \frac{1}{n} \sum_{i=1}^{n} (X_i - \bar{X})^r$, with $M_1 = \bar{X}$ and $\bar{X}$ the sample average. If we equate the sample moments M_r to the centralized population moments μ_r, we create a system of equations that can possible be solved for θ.

Thus when executing the method of moments, we are looking for parameters $\theta = (\theta_1, \theta_2, \ldots, \theta_m)^T$ that satisfy the following equations $\bar{X} = \mu(f_\theta)$ and $M_r = \mu_r(f_\theta)$ for $r = 2, 3, \ldots, m$. The solution $\tilde{\theta} = (\tilde{\theta}_1, \tilde{\theta}_2, \ldots, \tilde{\theta}_m)^T$ is called the *method of moments estimator*. Note that the second sample moment M_2 is equal to $(n - 1)S^2/n$ and is thus not an unbiased estimator of $\mu_r(f_\theta)$, since we already now that S^2 is unbiased.

5.5.1.1   Example: Lognormal Distribution

To illustrate the method of moments, consider the random variable X that is lognormally distributed, i.e. $X \sim \mathcal{LN}(\mu, \sigma^2)$; see Chap. 4 for more details. This means that the population density is given by $f_L(x) = \phi\left([\log(x) - \mu]/\sigma\right)/[x\sigma]$, when $x > 0$ and zero otherwise. This density contains only two parameters $\theta_1 = \mu$ and $\theta_2 = \sigma^2$, and we only need to solve the two equations

$$\begin{aligned} \bar{X} &= \mu(f_L) = \exp\left(\mu + 0.5\sigma^2\right) \\ M_2 &= \sigma^2(f_L) = \exp\left(2\mu + \sigma^2\right)\left(\exp\left(\sigma^2\right) - 1\right) \end{aligned} \tag{5.9}$$

Using the relationship that $\sigma^2(f_L)/\mu^2(f_L) = \exp(\sigma^2) - 1$, we obtain that σ^2 can be estimated by

$$\tilde{\sigma}^2 = \log\left(1 + \frac{M_2}{\bar{X}^2}\right) = \log\left(\bar{X}^2 + M_2\right) - 2\log\left(\bar{X}\right).$$

Note that the sample average is always positive, i.e. $\bar{X} > 0$. Using the estimator $\tilde{\sigma}^2$ in the first equation in Eq. (5.9), we obtain the moment estimator for μ, which is given by

$$\tilde{\mu} = \log\left(\bar{X}\right) - 0.5\left[\log\left(\bar{X}^2 + M_2\right) - 2\log\left(\bar{X}\right)\right] \tag{5.10}$$
$$= 2\log\left(\bar{X}\right) - 0.5\log\left(\bar{X}^2 + M_2\right). \tag{5.11}$$

We could have taken an alternative approach and considered the set of transformed random variables $\log(X_1), \log(X_2), \ldots, \log(X_n)$. These random variables are normally distributed with mean μ and variance σ^2. The moment estimators in this setting are now $\tilde{\mu} = \bar{X}_{\log}$ and $\tilde{\sigma}^2 = (n-1)S_{\log}^2/n$, where $\bar{X}_{\log}$ and $S_{\log}^2$ are defined in Eq. (5.8). This shows that the moment estimators on transformed data are different from the moment estimators on the original data. This non-uniqueness issue is considered a real disadvantage of the moment estimators.

5.5.1.2 MME Calculation in R for Lognormal

To further illustrate the approach we will calculate, using R, the estimates for the time spent in front of the television in the dataset `high-school.csv` that we have already encountered. In reality this variable TV cannot come from a lognormal distribution, since a lognormal distribution does not produce zeros. Hence, we will calculate the estimates using the data from the children that do watch television, since this may possibly be lognormally distributed. To do this we first create a new dataset `new_data` and subsequently calculate the summary statistics for TV in the new dataset:

```
new_data <- schooldata[schooldata$TV>0, ]
> summary(new_data$TV)
   Min. 1st Qu. Median  Mean 3rd Qu.   Max.
   1.00    7.00  12.00  14.46   20.00  70.00
```

We can now compute the sample moments that we need to estimate the population parameters:

```
> n <- nrow(new_data)
> n
[1] 49255
> Xbar <- mean(new_data$TV)
> Xbar
[1] 14.4643
> M2 <- (n-1)*var(new_data$TV)/n
> M2
[1] 107.302
```

Next, we can use the results in the previous section to estimate the population parameters:

```
> mu <- 2*log(Xbar)-0.5*log(Xbar^2+M2)
> mu
[1] 2.464677
> sigma2 <- log(Xbar^2+M2)-2*log(Xbar)
> sigma2
[1] 0.414013
```

Alternatively, we can calculate the moment estimates on the log transformed observations. The results are:

```
> logx <- log(new_data$TV)
> Xlog <- mean(logx)
> Xlog
[1] 2.409657
> Slog <- var(logx)
> Slog
[1] 0.6048436
```

These estimates are quite different from the moment estimates used on the original (non-transformed) data; in particular, the variance is different. This shows that the moment estimators may greatly vary when transformations are used. It should be noted that both estimates are in principle appropriate.

5.5.2   Maximum Likelihood Estimation

The maximum likelihood approach is probably best explained in the context of a set of random variable $X_1, X_2, \ldots, X_n$ having an i.i.d. Bernoulli $\mathscr{B}(p)$ distribution. In this setting we would like to estimate the parameter p. If we obtain a realization x_1, $x_2, \ldots, x_n$, we could ask how likely it is that we observe this set of results for given values of p. This so-called *likelihood* of the data is given by

$$L(p) = p^{x_1}(1-p)^{1-x_1} p^{x_2}(1-p)^{1-x_2} \cdots p^{x_n}(1-p)^{1-x_n} \qquad (5.12)$$

Indeed, the term $p^{x_i}(1-p)^{1-x_i}$ is the probability that the random variable X_i will attain the realization x_i, i.e. $P(X_i = x_i) = p^{x_i}(1-p)^{1-x_i}$, with $x_i \in \{0, 1\}$. Thus the product $L(p)$ in Eq. (5.12) represents the probability that $X_1, X_2, \ldots, X_n$ will attain the realization $x_1, x_2, \ldots, x_n$. Note that we fully exploit the i.i.d. assumption: we assume identical distributions for each random variable X_i (and we thus work with parameters p as opposed to p_i), and the probability of the joint observations is given by the product over the probability of the individual observations; the latter is possible by virtue of the independence assumption.

Since the likelihood is a function of the parameter p, we can search for a p that would maximize the likelihood. The parameter p that would maximize the probability in Eq. (5.12) is the parameter that is most likely to have produced the realizations. The *maximum likelihood* estimate $\hat{p}$ is the value that maximizes $L(p)$ in Eq. (5.12).

To obtain this maximum, it is more convenient to take the logarithm of the likelihood $\ell(p)$ given by

$$\ell(p) = \sum_{i=1}^{n} [x_i \log(p) + (1 - x_i) \log(1 - p)].$$

The logarithm of a function achieves its maximum value at the same points as the function itself, but is easier to work with analytically as we can replace the multiplications by sums that are easier to differentiate.

As we know from Calculus, the maximum of a function can be obtained by taking the derivative and then equating it to zero and solving the equation for the variable of interest. Taking the derivative of $\ell(p)$ with respect to p and equating it to zero results in

$$\ell'(p) = \sum_{i=1}^{n} \left[\frac{x_i}{p} - \frac{1 - x_i}{1 - p}\right] = 0 \iff (1 - p)\sum_{i=1}^{n} x_i - np + p\sum_{i=1}^{n} x_i = 0 \iff np = \sum_{i=1}^{n} x_i$$

Thus the maximum likelihood estimate is now given by $\hat{p} = \bar{x}$. Since any value $\bar{x} - \varepsilon$, $\varepsilon > 0$, for p in the derivative gives a positive value the solution is a maximum. The ML estimator is now $\hat{p} = \bar{X}$, which is the same as the MME, since the first moment of a Bernoulli random variable is p.

The ML approach can be generalized to any population density or probability mass function f_θ. If $X_1, X_2, \ldots, X_n$ are i.i.d. with density f_θ, the likelihood function is given by $L(\theta | X_1, X_2, \ldots, X_n) = \prod_{i=1}^{n} f_\theta(X_i)$ and the log likelihood function is given by

$$\ell_\theta \equiv \ell(\theta | X_1, X_2, \ldots, X_n) = \sum_{i=1}^{n} \log f_\theta(X_i). \tag{5.13}$$

The maximum likelihood estimator $\hat{\theta} = (\hat{\theta}_1, \hat{\theta}_2, \ldots, \hat{\theta}_n)^T$ is the set of parameters that maximizes the log likelihood function in Eq. (5.13). It is considered a (vector of) random variable(s), since it is a (set of) function(s) of the random variables X_1, $X_2, \ldots, X_n$. It can often be determined by solving the set of equations given by

$$\frac{\partial}{\partial \theta_k} \ell_\theta = \sum_{i=1}^{n} \left[\left(\frac{\partial}{\partial \theta_k} f_\theta(X_i)\right) / f_\theta(X_i)\right] = 0, \tag{5.14}$$

with $\partial \ell_\theta / \partial \theta_k$ indicating the derivative of ℓ_θ with respect to θ_k. This set of equations is often referred to as the *likelihood equations*. The solution $\hat{\theta}$ of Eq. (5.14) does not always result in a closed form expression, which means that we have to resort to numerical approaches if we want to determine the MLE on data.

5.5.2.1   Example: Lognormal Distribution

Recall that the population density is given by $f_L(x) = \phi\left(\left[\log(x) - \mu\right]/\sigma\right)/[x\sigma]$, when $x > 0$ and zero otherwise, and that the set of parameters that we want to estimate is given by $\theta_1 = \mu$ and $\theta_2 = \sigma^2$. Note also that the standard normal density is given by $\phi(x) = \exp(-x^2/2)/\sqrt{2\pi}$. The log likelihood in this setting is equal to

$$\sum_{i=1}^{n} \log f_L(X_i) = \sum_{i=1}^{n}\left[-\frac{(\log(X_i) - \mu)^2}{2\sigma^2} - \log\sqrt{2\pi} - \log(X_i) - \log(\sigma)\right]$$

To maximize this function it should be noted that the terms $\log\sqrt{2\pi}$ and $\log(X_i)$ can essentially be ignored, since they will be the same whatever we choose for μ and σ. Taking the derivatives with respect to μ and σ and equating them to zero we obtain the following equations

$$\sum_{i=1}^{n}\left[\frac{\log(X_i) - \mu}{\sigma^2}\right] = 0 \quad\text{and}\quad \sum_{i-1}^{n}\left[\frac{(\log(X_i) - \mu)^2}{\sigma^3} - \frac{1}{\sigma}\right] = 0$$

Solving these equations leads to the solutions $\hat{\mu} = \bar{X}_{\log} = \sum_{i=1}^{n}\log(X_i)$ and $\hat{\sigma}^2 = \frac{1}{n}\sum_{i=1}^{n}(\log(X_i) - \bar{X}_{\log})^2 = (n-1)S_{\log}^2/n$ provided earlier in Eq. (5.8). These solutions are the MME of the logarithmically transformed random variables.[10] This implies that the MLE for σ^2 is not unbiased, since the expected value of $\hat{\sigma}^2$ is equal to $E(\hat{\sigma}^2) = E((n-1)S_{\log}^2/n) = (n-1)\sigma^2/n$, although this bias vanishes when the sample size gets large.

Using the ML estimates for the density parameters to estimate the population parameters $\mu(f_L)$ or $\sigma^2(f_L)$ in the logarithm scale does not provide unbiased estimator for the mean of the population in the original scale, since the MLE estimate $\exp(\hat{\mu} + 0.5\hat{\sigma}^2)$ is unequal to the expected value of the sample mean $\bar{X}$. This is often considered a drawback of the MLE. It may estimate the parameters of the PDF or PMF unbiasedly, but they do not always estimate the population mean and variance unbiasedly. This would be different from the moment estimators applied to the original scale of the observations.

5.5.2.2   MLE Calculation in R for Lognormal

To determine the maximum likelihood estimates on a real dataset for a certain density f_θ, we can program the log likelihood ℓ_θ ourselves and then use the function `mle` from the `stats` package to maximize the log likelihood function. This can be done using the following R code:

---

[10] Note that the MLE and MME are not always the same. This depends on the particular density.

```
> library(stats4)
> minuslogl <- function(mu, sigma) {
+   densities <- dlnorm(new_data$TV, meanlog=mu, sdlog=sigma)
+   -sum(log(densities))
+ }
> mle(minuslogl, start=list(mu=10, sigma=5))

Call:
mle(minuslogl = minuslogl, start = list(mu = 10, sigma = 5))

Coefficients:
      mu       sigma
2.4096583 0.7777092
```

Note that the `mle` procedure in R provides the standard deviation $\hat{\sigma}$ instead of the variance $\hat{\sigma}^2$. The reason is that the `lnorm` is provided with a standard deviation. Squaring the value 0.7777092 results in 0.6048.

It should be noted that there also exists other packages that do not require you to formulate the function yourself. One such package is called `fitdistr(x, distr="name")`, where you put the name of the distribution in place of `"name"`. It requires the installation of the package MASS, after which you can execute the following R code:

```
> library(MASS)
> fitdistr(new_data$TV, densfun="lognormal")
     meanlog          sdlog
  2.409656712    0.777709043
 (0.003504225)  (0.002477861)
```

The advantage of this direct approach is that the estimates come with an estimated standard error (i.e., the estimated standard deviation of the estimators), similar to what we discussed for the sample average and sample variance.

5.5.2.3 Standard Error of MLE

The standard errors of the maximum likelihood estimators are calculated based on the variance of the large sample distribution (or asymptotics) of the maximum likelihood estimators. Under certain regularity conditions, it can be shown that $\sqrt{n}(\hat{\theta} - \theta)$ converges to a normal distribution $\mathcal{N}\left(0, I^{-1}(\theta)\right)$, with $\hat{\theta}$ the MLE and $I(\theta)$ the so-called Fisher information.[11] We will illustrate the proof, without being formal. For convenience purposes, we will assume that the density f_θ has only one parameter instead of multiple parameters $\theta_1, \theta_2, \ldots, \theta_m$.

The first derivative of the log likelihood, which is also referred to as the *score function* $S(\theta)$, is given by

---

[11] The theory of maximum likelihood estimation was developed by Sir Ronald Fisher.

$$S_n(\theta) \equiv \frac{d}{d\theta}\ell_\theta = \sum_{i=1}^{n} \frac{d}{d\theta} \log f_\theta(X_i) = \sum_{i=1}^{n} \frac{f_\theta'(X_i)}{f_\theta(X_i)}$$

The expectation of the score function is zero (under certain regularity conditions), since $\mathbb{E}[f_\theta'(X_i)/f_\theta(X_i)] = \int_\mathbb{R} f_\theta'(x)dx = \frac{d}{d\theta}\int_\mathbb{R} f_\theta(x)dx = 0$. This implies that $\mathbb{E}[S_n(\theta)] = 0$. The variance of the score function is now given by $\mathbb{E}[S_n^2(\theta)] = n\int_\mathbb{R}[(f_\theta'(x))^2/f_\theta(x)]dx$. We will now define the Fisher information $I(\theta)$ by

$$I(\theta) = \mathbb{E}\left[\frac{d}{d\theta}\log f_\theta(X)\right]^2 = \int_\mathbb{R}[(f_\theta'(x))^2/f_\theta(x)]dx. \qquad (5.15)$$

with X having density f_θ.

The score function is a sum of independent random variables, which means that if we standardize the score function appropriately, the central limit theorem tells us that the distribution function of the score function will converge to a normal distribution function. Thus $S_n(\theta)/\sqrt{nI(\theta)}$ converges to a standard normal random variable $Z \sim \mathcal{N}(0, 1)$. It can also be shown that the Fisher information is equal to the minus expectation of the derivative of the score function with respect to the parameter θ, i.e.

$$nI(\theta) = -\mathbb{E}\left(S_n'(\theta)\right)$$

with $S_n'(\theta) = \frac{d}{d\theta}S_n(\theta)$. This derivative is also a sum of independent random variables; thus properly standardized it will converge to a standard normal random variable. This also implies that $S_n'(\theta)/n$ will converge to the Fisher information $I(\theta)$. Now using a first-order Taylor expansion for the score function $S_n(\hat{\theta})$ that is evaluated in the ML estimator $\hat{\theta}$, we obtain that $S_n(\hat{\theta}) \approx S_n(\theta) + (\hat{\theta} - \theta)S_n'(\theta)$. But the score function in the MLE estimator is zero, i.e. $S_n(\hat{\theta}) = 0$, since the MLE estimator $\hat{\theta}$ maximizes the likelihood. This means that

$$\sqrt{n}\left(\hat{\theta} - \theta\right) \approx -\frac{\sqrt{n}S_n(\theta)}{S_n'(\theta)} = -\frac{1}{\sqrt{I(\theta)}}\frac{S_n(\theta)}{\sqrt{nI(\theta)}}\frac{nI(\theta)}{S_n'(\theta)}$$

Since $nI(\theta)/S_n'(\theta)$ coverges to 1 and $S_n(\theta)/\sqrt{nI(\theta)}$ converges to $Z \sim \mathcal{N}(0, 1)$, we obtain that $\sqrt{n}(\hat{\theta} - \theta)$ converges to $\mathcal{N}(0, 1/I(\theta))$. Thus, we see that the asymptotic variance of the maximum likelihood estimator is $I^{-1}(\theta)$. This implies that the approximate standard error of the maximum likelihood estimator is now $SE(\hat{\theta}) = 1/\sqrt{nI(\theta)}$. Since we have an estimator $\hat{\theta}$ of the parameter θ, the standard error is estimated with

$$\hat{SE}(\hat{\theta}) = \frac{1}{\sqrt{nI(\hat{\theta})}}.$$

This asymptotic standard error can be used in the calculation of confidence intervals on θ using the theory of Sect. 5.3.5. When $T_n = \hat{\theta}$ is the MLE for θ having an estimated standard error $\hat{\tau}_n = 1/\sqrt{nI(\hat{\theta})}$ (using the notation of Sects. 5.3.5 and 5.4.1), the $1 - 2p$ asymptotic confidence interval is

$$(\hat{\theta} - z_{1-p}/\sqrt{nI(\hat{\theta})}, \hat{\theta} + z_{1-p}/\sqrt{nI(\hat{\theta})}].$$

To illustrate this theory on asymptotic standard errors, let $X_1, X_2, \ldots, X_n$ be i.i.d. Poisson $\mathscr{P}(\lambda)$ distributed. Recall that the population mean $\mu(f) = \lambda$ and the population variance $\sigma^2(f) = \lambda$. The maximum likelihood estimator for $\theta = \lambda$, which is also the moment estimator, is given by the sample average $\hat{\lambda} = \bar{X}$. Indeed, the log likelihood is $\ell_\theta = \sum_{i=1}^{n}[X_i \log(\lambda) - \log(X_i!) - \lambda]$. Equating the derivative with respect to λ to zero results into $\sum[(X_i/\lambda) - 1] = 0$. Solving this for λ gives $\hat{\lambda} = \bar{X}$. The asymptotic distribution of the sample average has been discussed in Sect. 5.3.4. It was demonstrated that the distribution of $\sqrt{n}(\bar{X} - \mu(f))/\sigma(f)$ converges to a standard normal distribution. Applying this to the Poisson distributed random variables, we see that $\sqrt{n}(\bar{X} - \lambda)$ converges to $N(0, \lambda)$.

Using Eq. (5.15), the Fisher information for the Poisson distribution is given by $I(\lambda) = \mathbb{E}[(X/\lambda) - 1]^2$. This is equal to $\mathbb{E}[(X - \lambda)/\lambda]^2 = 1/\lambda$. Thus indeed, the asymptotic variance of the MLE is given by $I^{-1}(\lambda) = \lambda$ and the standard error of $\bar{X}$ is equal to $\sqrt{\lambda/n}$.

Problems

5.1 Consider the exponential CDF $F_E(x) = 1 - \exp(-\lambda x)$, for $x > 0$ and otherwise equal to zero. Now let X be distributed according to this exponential distribution.

1. Determine the relative standard deviation of X.
2. Determine the pth quantile.
3. Determine the skewness and kurtosis.
4. Given realizations 0.05, 0.20, 1.72, 0.61, 0.24, 0.79, 0.13, 0.59, 0.26, 0.54 from random variables $X_1, \ldots, X_{10}$ which we assume to be distributed i.i.d. exponentially, use R to compute the maximum likelihood estimate of λ.

5.2 Consider the data of the approximately 50,000 school children listed in `high-school.csv`.

1. Calculate for each numerical variable the average, variance, skewness, and kurtosis.
2. Calculate for each numerical variable a 95% confidence interval on the population mean and on the population variance.
3. Calculate for each numerical variable the 0.20th quantile.
4. Calculate the proportion of children that do not play sport and calculate a 95% confidence interval on the population proportion.

5.3 Consider the Bernoulli distribution $\Pr(X = 1) = p = 1 - \Pr(X = 0)$.

1. Now let $X_1, X_2, \ldots, X_{10}$ be i.i.d. $\mathscr{B}(p)$ distributed with $p = 0.2$. Then calculate the sample average $\bar{X} = (X_1 + X_2 + \cdots + X_{10})/10$ and the standardized sample average $(\bar{X} - 0.2)/\sqrt{0.016}$. Simulate this standardized sample average 10,000 times and create a histogram of the 10,000 sample averages. What is the average and variance? Is the distribution approximately standard normal?
2. Simulate such a sample 10,000 times and calculate for each simulation the asymptotic 95% confidence interval for p using the sample average. Count how often the true parameter p is contained in these 10,000 confidence intervals.
3. Now let $X_1, X_2, \ldots, X_{50}$ be i.i.d. $\mathscr{B}(p)$ distributed with $p = 0.2$. Then calculate the sample average $\bar{X} = (X_1 + X_2 + \cdots + X_{50})/50$ and the standardized sample average $(\bar{X} - 0.2)/\sqrt{0.0032}$. Simulate this standardized sample average 10,000 times and create a histogram of the 10,000 sample averages. What is the average and variance? Is the distribution approximately standard normal?

5.4 Consider again the exponential CDF $F_E(x) = 1 - \exp(-\lambda x)$ for $x > 0$ and otherwise equal to zero.

1. Now let X_1, X_2, and X_3 be i.i.d. F_E distributed with $\lambda = 0.5$. Then calculate a 95% asymptotic confidence interval on λ^{-1} using the sample average $\bar{X} = (X_1 + X_2 + X_3)/3$. Simulate these confidence intervals 10,000 times and count how often the true parameter $\lambda^{-1} = 2$ is contained in these 10,000 confidence intervals.
2. Now draw $X_1, X_2, \ldots, X_{20}$ from the exponential distribution with $\lambda = 0.5$ and calculate a 95% asymptotic confidence interval on λ^{-1} using the sample average $\bar{X} = (X_1 + X_2 + \cdots + X_{20})/20$. Simulate these confidence intervals 10,000 times and count how often the true parameter $\lambda^{-1} = 2$ is contained in these 10,000 confidence intervals.
3. Explain the difference in the results from parts 1 and 2.
4. Now draw $X_1, X_2, \ldots, X_{20}$ from the exponential distribution with $\lambda = 0.5$ and calculate a 95% asymptotic confidence interval on λ^{-2} using the sample variance $S^2 = \sum_{i=1}^{20}(X_i - \bar{X})^2/19$. Simulate these confidence intervals 10,000 times and count how often the true parameter $\lambda^{-2} = 4$ is contained in these 10,000 confidence intervals.
5. As parts 2 and 4 both provide confidence intervals on λ, which approach would you prefer ($\bar{X}$ or S^2)? Explain why.

5.5 Consider the exponential CDF $F_E(x) = 1 - \exp(-\lambda(x - \eta))$ for $x > \eta$ and otherwise equal to zero. Assume that the random variables $X_1, X_2, \ldots, X_n$ are i.i.d. exponentially $\exp(\eta, \lambda)$ distributed.

1. Determine the moment estimator for λ in case $\eta = 0$.
2. Determine the maximum likelihood estimator for λ in case $\eta = 0$.
3. Determine the moment estimators for λ and η.
4. Determine the maximum likelihood estimators for λ and η.

5.6 Consider the data on approximately 50,000 children at high schools in the Netherlands and focus on the variable SPORTS. Create a data set that contains only positive values (i.e., eliminate the zeros). Assume that the data on SPORTS are from a gamma distribution with density function

$$f_G(x) = \frac{\beta^\alpha}{\Gamma(\alpha)} x^{\alpha-1} \exp(-\beta x),$$

when $x > 0$ and otherwise equal to zero.

1. Determine mathematically the moment estimators if $X_1, X_2, \ldots, X_n$ are i.i.d. gamma $G(\alpha, \beta)$ distributed.
2. Use the results from part 1 to determine the moment estimates for α and β based on the SPORTS data.
3. Use R to compute the maximum likelihood estimates for α and β.
4. Sample 100 times 1,000 children from the data and calculate the moment estimates and the maximum likelihood estimates for each of the 100 draws. Per estimation method, calculate the average of the estimates for α and β and the standard deviation of these estimates. Can you make a choice on which method of estimation is preferred for α and β?

Reference

B. Patrick, *Probability and Measure* (A Wiley-Interscience Publication, Wiley, Hoboken, 1995)

Chapter 6
Multiple Random Variables

6.1 Introduction

Up to now we have mainly focussed on the analysis of a single variable. We have discussed probability density functions (PDFs), probability mass functions (PMFs), and distribution functions (CDFs) as descriptions of the population values for such a single variable and connected these functions to a single random variable. These probability functions were functions with a single argument $x \in \mathbb{R}$. For instance, the CDF $F_\theta(x)$ was defined for all $x \in \mathbb{R}$. In this chapter we will extend the concept of probability functions to multiple arguments, say (x, y), that would represent multiple random variables.

Implicitly, we have already discussed multiple random variables. First of all, we discussed the occurrence of multiple events in Chap. 3. We will see in this chapter that this relates to bivariate binary random variables, X and Y. Secondly, we discussed multiple random variables that originated from (simple random) sampling from a population in Chap. 5. Indeed, we considered the set of random variables $X_1, X_2, \ldots, X_n$ being i.i.d. with distribution function F_θ. With this assumption the *joint* distribution function of all these random variables simultaneously is fully determined by F_θ. This was most explicit when we discussed maximum likelihood estimation: here we saw that the joint likelihood of the variables—by virtue of the fact that they were independent—was the product of the likelihoods of each of the individual random variables. Thus the likelihood function was nothing else than the product of the PDFs.

In this chapter we study distribution functions of multiple random variables in a bit more detail, in particular bivariate distribution functions. Here we will mainly discuss that $(X_1, Y_1), (X_2, Y_2), \ldots, (X_n, Y_n)$ are i.i.d. and the random variables X_i and Y_i are typically not independent. We start by introducing some more theory: we discuss joint PMFs, PDFs, and CDFs of multiple random variables (also called *multivariate probability functions*). After discussing this theory, we introduce properties of multivariate distribution functions that relate to the dependency of X_i and Y_i, also referred to as *measures of association*. We discuss several estimators for estimation

© Springer Nature Switzerland AG 2022
M. Kaptein and E. van den Heuvel, *Statistics for Data Scientists*, Undergraduate Topics in Computer Science, https://doi.org/10.1007/978-3-030-10531-0_6

of these measures of association and discuss how to construct confidence intervals. Finally, we demonstrate how they can be calculated with R. Hence, in this chapter we will study dependency between variables. We will cover:

- Joint probability functions of discrete and continuous random variables
- Properties of multivariate distributions: measures of association
- Estimators for measures of association
- Confidence intervals for measures of association
- Some other sample statistics that quantify associations

Note that this chapter is a mix of theory and practice. Although we do not discuss the theory of multiple distributions extensively in this book,[1] we do provide a lot of information on measures of associations for bivariate random variables. We discuss associations as a population parameter but also in the form of sample statistics. Only in the final section do we illustrate how to calculate measures of association on real data.

Admittedly, this chapter is relatively long. However, it is structured such that parts of the chapter can easily be skipped without missing the main points. Section 6.2 introduces multivariate distribution functions and is essential to understand the material in this chapter. Section 6.3 details how multivariate distribution functions can be constructed; this section can easily be skipped by readers who are not interested in the underlying theory. Section 6.4 provides the main properties of multivariate distributions which we deem essential theory. Section 6.5 extends the essential theory, by providing various measures of association. In Sects. 6.6 and 6.7 we discuss estimators of measures of association: An introductory course might only cover Sects. 6.5.1 and 6.6.1 and skip the other sections. Section 6.8 details how we can use R to compute various measures of association.

6.2   Multivariate Distributions

In Chap. 4 we introduced random variables and their probability distributions. We discussed how a single random variable X with outcome values in $\mathbb{R}$ has a cumulative density function F_X if the probability that X is observed in the interval $(-\infty, x]$ is given by the distribution function F_X, i.e., $\Pr(X \leq x) = F_X(x)$. This was true for both discrete and continuous random variables.[2]

This notion can easily be extended to functions of more than one variable (see both Chaps. 3 and 4). We say that random variables X and Y have a *joint* distribution function F_{XY} if the probability that X is observed in the interval $(-\infty, x]$ *and* the

---

[1] For a more extensive theoretical discussion on multivariate distribution functions see Ross (2014); Nelsen (2007).

[2] Note that we have changed the notation a bit to indicate that the CDF F_X belongs to the random variable X. We do this because we will introduce multiple random variables hereafter, each having their own CDF that does not necessarily have to be identical for all random variables.

probability that Y is observed in the interval $(-\infty, y]$ is given by the function F_{XY}, i.e., $\Pr(X \leq x, Y \leq y) = F_{XY}(x, y)$. The joint distribution function of two random variables is also called a *bivariate distribution function*. We can extend this beyond two variables and consider $F_{X_1 X_2 \cdots X_K}$, with $F_{X_1 X_2 \cdots X_K}(x_1, x_2, \ldots, x_K) = \Pr(X_1 \leq x_1, X_2 \leq x_2, \ldots, X_K \leq x_K)$ the joint distribution function of K random variables $X_1, X_2, \ldots, X_K$. We often call this joint distribution function a *multivariate distribution function*.

The joint distribution function contains all the information on how the random variables are related to each other, i.e., how the random variables are *dependent* on each other. In practice, we may say that variables are "co-related", a term not used frequently anymore, but it was used by Sir Francis Galton, who was among the first to discuss co-relation in depth. If one variable increases and the other variable also increases (on average) or if one variable increases while the other variable decreases (on average) they are said to co-relate. For instance, taller people typically have a larger weight, a relationship we are well aware of, but it was first identified by Galton who studied co-relations among anthropometric data; see for instance Galton (1889).

6.2.1 Definition of Independence

Two random variables X and Y are called *independent* when the bivariate distribution function is equal to the product of the *marginal* distribution functions.[3] In mathematics, independence of X and Y holds when $F_{XY}(x, y) = F_X(x) F_Y(y)$ for all $(x, y) \in \mathbb{R} \times \mathbb{R}$, with F_X the CDF of X and F_Y the CDF of Y (also called the marginal distribution functions in this context of bivariate random variables). The random variables X and Y are called dependent when they are not independent. The random variables $X_1, X_2, \ldots, X_K$ are called *mutually* independent when the joint distribution function is the product of the marginal distribution functions, i.e., $F_{X_1 X_2 \cdots X_K}(x_1, x_2, \ldots, x_K) = F_{X_1}(x_1) F_{X_2}(x_2) \cdots F_{X_K}(x_K)$ for all $x_k \in \mathbb{R}$.[4] If in case we assume that all distribution functions are identical, i.e., $F_{X_1}(x) = F_{X_2}(x) = \cdots = F_{X_K}(x) = F(x)$ for all x, then we have the concept of $X_1, X_2, \ldots, X_K$ being i.i.d. with distribution function F (as we studied in Chap. 5).

It is important to realize that dependency can occur within a single population unit or between multiple population units. For single units, the random variables typically represent different variables, like height and weight for a single person (which was

---

[3] Note that this definition is very similar to the definition of independence of events as discussed in Chap. 3.

[4] Note that there exists examples of *pairwise independence* without having mutual independence. In other words, we may have that X and Y, X and Z, and Y and Z are (pairwise) independent, but X, Y, and Z are not mutually independent. Thus we may have $F_{XY}(x, y) = F_X(x) F_Y(y)$, $F_{XZ}(x, z) = F_X(x) F_Z(z)$, and $F_{YZ}(y, z) = F_Y(y) F_Z(z)$ for all x, y, and z, but we may not have $F_{XYZ}(x, y, z) = F_X(x) F_Y(y) F_Z(z)$ for all x, y, and z. Pairwise independence is thus weaker than mutual independence. When we talk about independence among multiple random variables we mean mutual independence.

studied by Francis Galton) or the two dimensions of the face in the face-data discussed in Chap. 1. For multiple units, as we studied in Chap. 4 (although we assumed independence), the random variables may typically represent the same variable, like the height of siblings. Here the dependence is introduced by an overarching unit or cluster (i.e., family), that indicates that specific results belong to each other. In this case, dependence occurs when the results (e.g., height) for members of the same family are closer to each other than results taken from members of different families. We discussed this in Chap. 2 for cluster random sampling. The theory in this chapter applies to both situations, and the context will make it clear whether we study single or multiple units.

In Chap. 4 we introduced distribution functions through PDFs or PMFs and connected distribution functions to random variables, essentially showing that all these concepts are strongly related. The same concepts hold for multivariate random variables. In the following two sections we will demonstrate this, first for discrete and then for continuous random variables. Then in the third subsection we will discuss how we can construct joint PDFs, PMFs, or CDFs.

6.2.2 Discrete Random Variables

If X and Y are both discrete random variables, than the *joint* probability mass function (PMF) is:

$$f_{XY}(x, y) = \Pr(X = x, Y = y)$$

This function of two variables x and y gives the probability of the occurrence of the events $X = x$ *and* $Y = y$ happening simultaneously. Here we will assume again that the values of x and y are both elements of the natural numbers $\mathbb{N}$. Some or many combinations of pairs (x, y) may not occur, which implies that the probability for these values is zero, i.e., $f_{XY}(x, y) = 0$. Since this bivariate PMF represents the probability of the occurrence of events, all the probabilities should add up to 1. Thus we have

$$\sum_{(x,y)\in\mathbb{N}\times\mathbb{N}} f_{XY}(x, y) = \sum_{x=0}^{\infty}\sum_{y=0}^{\infty} f_{XY}(x, y) = \sum_{y=0}^{\infty}\sum_{x=0}^{\infty} f_{XY}(x, y) = 1. \qquad (6.1)$$

The *marginal* PMF of Y is given by $f_Y(y) = \sum_{x=0}^{\infty} f_{XY}(x, y)$ and the *marginal* PMF of X is given by $f_X(x) = \sum_{y=0}^{\infty} f_{XY}(x, y)$. They indicate the PMF of Y aggregated over the possible choices of X and of X aggregated over the possible choices of Y, respectively. Thus, the joint PMF of two variables X and Y also allows us to obtain the (marginal) PMFs of the single variables X and Y. This also means that the theory of Chaps. 3, 4, and 5 applies to the single random variable X and Y. Note that we can extend the bivariate setting to more variables, for example to three variables: $f_{XYZ}(x, y, z) = \Pr(X = x, Y = y, Z = z)$. The marginal PMF of X is then $f_X(x) = \sum_{y=0}^{\infty} \sum_{z=0}^{\infty} f_{XYZ}(x, y, z)$. For Y and Z we can do something similar.

The joint CDF F_{XY} of the random variables X and Y is now provided in the same way as we did for single random variables:

$$F_{XY}(x, y) = \sum_{k=0}^{x} \sum_{l=0}^{y} f_{XY}(k, l),$$

for every x and y in $\mathbb{N}$. Now we accumulate both the x and y variables. It can be seen that this leads to the definition of the bivariate distribution function we discussed above: $F_{XY}(x, y) = \sum_{k=0}^{x} \sum_{l=0}^{y} f_{XY}(k, l) = \sum_{k=0}^{x} \sum_{l=0}^{y} \Pr(X = k, Y = l) = \Pr(X \leq x, Y \leq y)$ using the rules for adding up probabilities when the intersections of the events are empty.

We can also define the *conditional* PMF of X given $Y = y$:

$$f_{X|Y}(x|y) = \Pr(X = x|Y = y) = \frac{\Pr(X = x, Y = y)}{\Pr(Y = y)} = \frac{f_{XY}(x, y)}{f_Y(y)},$$

when $f_Y(y) > 0$ (and similarly for Y given $X = x$). Note that we have already covered this concept in Chap. 3 for individual *events* and here we extend this to distributions. Analogously, we define the conditional density $f_{X|Y}(x|y)$ to be equal to 0 when $f_Y(y)$ is equal to zero. The conditional PMF $f_{X|Y}$ gives the probability of $X = x$ given a specific choice of $Y = y$. For instance, if X represents a favourite subject at school and Y represents gender, the conditional probability is the probability that a particular gender, say female, likes a specific subject, say mathematics, as her favourite subject (among all females).

When the random variables X and Y are independent, the conditional PMF becomes equal to the marginal PMF. To see this, we will first illustrate that independence means that $f_{XY}(x, y) = f_X(x) f_Y(y)$. Indeed,

$$
\begin{aligned}
f_{XY}(x, y) &= \Pr(X = x, Y = y) \\
&= \Pr(X \leq x, Y \leq y) - \Pr(X \leq x, Y \leq y - 1) \\
&\quad - \Pr(X \leq x - 1, Y \leq y) + \Pr(X \leq x - 1, Y \leq y - 1) \\
&= F_{XY}(x, y) - F_{XY}(x, y - 1) - F_{XY}(x - 1, y) + F_{XY}(x - 1, y - 1) \\
&= F_X(x) F_Y(y) - F_X(x - 1) F_Y(y) - F_X(x) F_Y(y - 1) + F_X(x - 1) F_Y(y - 1) \\
&= (F_X(x) - F_X(x - 1))(F_Y(y) - F_Y(y - 1)) \\
&= \Pr(X = x) \Pr(Y = y) \\
&= f_X(x) f_Y(y)
\end{aligned}
$$

Now using this relation in the definition of conditional PMFs, we immediately obtain that $f_{X|Y}(x|y) = f_X(x)$ when X and Y are independent.

To give concrete examples of bivariate discrete distribution functions, we will consider the distribution function of two binary random variables and link this with our theory in Chap. 3 and we will introduce the multinomial distribution function as a multivariate extension of the binomial distribution function.

6.2.2.1   Two Binary Random Variables

Consider the following joint PMF for random variables X and Y where $x \in \{0, 1\}$ and $y \in \{0, 1\}$:

$$
f_{XY}(x, y) = \begin{cases} p_{00} & \text{if } x = 0 \text{ and } y = 0 \\ p_{01} & \text{if } x = 0 \text{ and } y = 1 \\ p_{10} & \text{if } x = 1 \text{ and } y = 0 \\ p_{11} & \text{if } x = 1 \text{ and } y = 1 \end{cases} \tag{6.2}
$$

with $p_{00} \geq 0$, $p_{01} \geq 0$, $p_{10} \geq 0$, and $p_{11} \geq 0$ unknown parameters such that $p_{00} + p_{01} + p_{10} + p_{11} = 1$. For example, the parameters can be given by

$$
f_{XY}(x, y) = \begin{cases} \frac{338}{763} & \text{if } x = 0 \text{ and } y = 0 \\ \frac{77}{763} & \text{if } x = 0 \text{ and } y = 1 \\ \frac{256}{763} & \text{if } x = 1 \text{ and } y = 0 \\ \frac{9}{763} & \text{if } x = 1 \text{ and } y = 1 \end{cases}
$$

These probabilities were provided in (contingency) Table 3.2 of Chap. 3, where the variable X indicates the gender of a participant ($X = 1$ means male and $X = 0$ means female) and the variable Y represents the occurrence of Dupuytren disease ($Y = 1$ means presence of Dupuytren disease and $Y = 0$ means absence of Dupuytren disease). It may be shown (see exercises) that the variables X and Y in this situation are not independent. The measures of risk that were discussed in Chap. 3 would all hold for binary random variables defined by Eq. (6.2).

6.2.2.2   The Multinomial Distribution

The multinomial distribution function occurs naturally if we sample without replacement from a population with K classes. For instance, the quality of a product may be divided into five categories: very low, low, okay, high, very high. When we consider a sample of size n, we may keep track of the number of products in each of the K categories, i.e., we observe random variables $X_1, X_2, \ldots, X_K$, with $K = 5$. The joint PDF for $X_1, X_2, \ldots, X_K$ is the multinomial PDF given by

$$f_{X_1 X_2 \ldots X_K}(x_1, x_2, \ldots, x_K) = \frac{n!}{x_1! x_2! \cdots x_K!} p_1^{x_1} p_2^{x_2} \cdots p_K^{x_K} \tag{6.3}$$

with $x_k \in \{0, 1, 2, \ldots, n\}$, $n = x_1 + x_2 + \cdots + x_K$, $p_k > 0$, and $p_1 + p_2 + \cdots + p_K = 1$. The probability p_k represents the percentage of units (or products in our example) in the population (the full production of products) that has quality level k.

Note that this PMF is a generalization of the binomial PMF. If we consider $K = 2$, the multinomial PMF reduces to the binomial PDF. However, for the binomial PMF we usually do not mention or incorporate x_2 in the PMF, as this is automatically defined through $x_2 = n - x_1$, and only focus on the number of events X_1.

6.2.3 Continuous Random Variables

The theory we have just discussed for discrete PMFs can also be extended to the PDFs of continuous random variables. Let's denote the joint PDF of two continuous random variables X and Y by $f_{XY}(x, y)$. Similar to PDFs $f_X(x)$ of single random variables X, this does not represent probabilities. The probability $\Pr(X = x, Y = y)$ is equal to zero, while $f_{XY}(x, y)$ may be positive, i.e., there exist x and y, such that $f_{XY}(x, y) > \Pr(X = x, Y = y) = 0$.

The joint CDF for X and Y is now defined by

$$F_{XY}(x, y) = \int_{-\infty}^{x} \int_{-\infty}^{y} f_{XY}(u, v) du dv. \tag{6.4}$$

This means that we integrate out the area under the function $f_{XY}(u, v)$ over the interval $(-\infty, x]$ for u and $(-\infty, y]$ for v. This can be considered even more generally, if we consider any set $A \subset \mathbb{R} \times \mathbb{R}$ and define the probability that $(X, Y) \in A$ by

$$\Pr\left((X, Y) \in A\right) = \iint_A f_{XY}(u, v) du dv \tag{6.5}$$

In the special case that $A = (-\infty, x] \times (-\infty, y]$ we obtain $\Pr(X \le x, Y \le y) = F_{XY}(x, y)$, which is similar to what we discussed in Chap. 4 on single random variables.

The marginal PDF for continuous random variables is now obtained by integration instead of summation that we used for discrete random variables:

$$f_X(x) = \int_{-\infty}^{\infty} f_{XY}(x, y) dy \quad \text{and} \quad f_Y(y) = \int_{-\infty}^{\infty} f_{XY}(x, y) dx. \tag{6.6}$$

The marginal CDFs can be obtained through the marginal PDFs, using the integral of the PDF over the interval $(-\infty, x]$ like we did in Chap. 4, but it can also be obtained through the joint CDF by letting y or x converge to infinity: $F_X(x) = \lim_{y \to \infty} F_{XY}(x, y)$ and $F_Y(y) = \lim_{x \to \infty} F_{XY}(x, y)$.

The conditional densities are defined in the same way as for discrete random variables. They are given by

$$f_{X|Y}(x|y) = \frac{f_{XY}(x, y)}{f_Y(y)} \quad \text{and} \quad f_{Y|X}(y|x) = \frac{f_{XY}(x, y)}{f_X(x)}. \tag{6.7}$$

Note that the conditional PDF $f_{X|Y}(x|y)$ is only defined in values for y that gives $f_Y(y) > 0$ and conditional PDF $f_{Y|X}(y|x)$ is only defined in x for which $f_X(x) > 0$. When $f_X(x) = 0$ or $f_Y(y) = 0$ the respective conditional PDFs are defined to be zero. If the random variables X and Y are independent, the joint density is the product of the marginal densities, i.e., $f_{XY}(x, y) = f_Y(y) f_Y(y)$ for all x and y in $\mathbb{R}$.

6.2.3.1   Bivariate Normal Density Function

A well-know bivariate continuous distribution function is the bivariate normal distribution function. The PDF is given by:

$$f(x, y) = \frac{1}{2\pi\sigma_X\sigma_Y\sqrt{1 - \rho^2}} \exp\left(-\frac{z_1^2 - 2\rho z_1 z_2 + z_2^2}{2(1 - \rho^2)}\right) \tag{6.8}$$

with $z_1 = (x - \mu_X)/\sigma_X$ the standardized normal variable for X and $z_2 = (y - \mu_Y)/\sigma_Y$ the standardized normal variable for Y. Note that when the parameter ρ is equal to zero we obtain that the bivariate normal PDF is the product of the normal PDF of X and the normal PDF of Y. Thus when $\rho = 0$, the normal random variables X and Y are independent. However, when $\rho \neq 0$, the normal random variables X and Y are dependent. The parameter ρ is called the *correlation coefficient* and is contained within the interval $[-1, 1]$.

Note that for distribution functions of more than two random variables, say $(X_1, X_2, \ldots, X_K)$ we often resort to using vector/matrix notation. For example, when considering the joint normal distribution function of K random variables we denote its mean by a vector $\boldsymbol{\mu} = (\mu_1, \mu_2, \ldots, \mu_K)^T$ of length K, and a $K \times K$ covariance matrix Σ. The diagonal of this matrix contains the variances of the random variables, with σ_k^2 the variance of variable X_k placed at the kth row and column in Σ. We will not discuss this in more detail, but it is good to be aware that sometimes a very shorthand notation is used to denote the joint distribution function of a large number of random variables.

6.3 Constructing Bivariate Probability Distributions

Dependency between random variables has been studied extensively in the literature and it is still a highly important topic within statistics and data science. Researchers have tried to develop specific and special classes of multivariate and bivariate distribution functions. Here we will describe just a few approaches to constructing families of CDFs, without having the intention of being complete. There are several books discussing the topic in much more detail Samuel et al. (2004); Johnson et al. (1997); Balakrishnan and Lai (2009). We will focus on bivariate random variables, although some of the concepts can easily be extended to higher dimensions. The concepts are somewhat mathematical, but it shows very well the complexities and opportunities of creating bivariate CDFs. Such families can then be used in practice when they describe the bivariate data appropriately. Thus they are not just theoretical, they can be made practical. We will connect or illustrate some of these CDFs later in this chapter when we discuss measures of association.

6.3.1 Using Sums of Random Variables

Dependent random variables are sometimes constructed by combining random variables. Let U, V, and W be three independent random variables and define $X = W + U$ and $Y = W + V$. Since both X and Y share the same random variable W, the random variables X and Y must be dependent. If we only observe the random variables X and Y in practice, the random variable W is sometimes referred to as *latent variable*. Note that this approach to constructing bivariate CDFs creates a very large set of bivariate CDFs, since we have not specified the underlying CDFs of U, V, and W. Thus this approach gives a lot of flexibility.

This way of constructing dependent random variables has been applied, for instance, to normal random variables, Poisson random variables, and binomial random variables. It is particular convenient for these distributions, since they have the property that when we add up independent random variables they stay within the same family of distributions. For the normal distribution function we have already seen this in property 1 of Sect. 5.4, but this property also holds true for Poisson random variables and for binomial random variables under specific conditions. Thus when U, V, and W are all normal, Poisson, or binomial, the marginal CDFs of X and Y are also normal, Poisson, or Binomial, respectively.

When $U \sim \mathcal{N}(\mu_U, \sigma_U^2)$, $V \sim \mathcal{N}(\mu_V, \sigma_V^2)$, and $W \sim \mathcal{N}(\mu_W, \sigma_W^2)$, it can be shown that X and Y are bivariate normally distributed with

$$\mu_X = \mu_W + \mu_U$$
$$\mu_Y = \mu_W + \mu_V$$
$$\sigma_X^2 = \sigma_W^2 + \sigma_U^2$$
$$\sigma_Y^2 = \sigma_W^2 + \sigma_V^2$$
$$\rho = \frac{\sigma_W^2}{\sqrt{(\sigma_W^2 + \sigma_U^2)(\sigma_W^2 + \sigma_V^2)}}.$$

When $U \sim \mathscr{P}(\lambda_U)$, $V \sim \mathscr{P}(\lambda_V)$, and $W \sim \mathscr{P}(\lambda_W)$, it can be shown that X is Poisson distributed with parameter $\lambda_W + \lambda_U$ and Y is Poisson distributed with parameter $\lambda_W + \lambda_V$. The bivariate PDF is then given by

$$f_{XY}(x, y) = \exp(-[\lambda_U + \lambda_V + \lambda_W]) \frac{\lambda_U^x \lambda_V^y}{x!y!} \sum_{k=0}^{\min(x,y)} \frac{x!y!}{(x-k)!(y-k)!k!} \left(\frac{\lambda_W}{\lambda_U \lambda_V}\right)^k.$$

A similar construction is available for binomial random variables, but is outside the scope of this book.

It may be obvious to see that this method of constructing a bivariate CDF can be extended easily to more than just two random variables. Then each random variable contains the same component W. This form has been used extensively in the classical theory of measurement reliability, in particular with normally distributed random variables. The random variable W represents a "true" value of something that is measured, and the other random variables U and V represent measurement errors. Thus in the bivariate case, the same unit is just measured twice.

6.3.2 Using the Farlie–Gumbel–Morgenstern Family of Distributions

When X and Y are two random variables with marginal CDFs F_X and F_Y, respectively, we can create a bivariate CDF in the following way:

$$F_{XY}(x, y) = F_X(x) F_Y(y)(1 + \alpha(1 - F_X(x))(1 - F_Y(x))).$$

The parameter α should be within $[-1, 1]$ and when it is equal to zero, the random variables X and Y are independent. The parameter α may be seen as the dependency parameter and the larger the absolute value $|\alpha|$ the stronger the dependency. This class of distribution functions is referred to as the one-parameter Farlie–Gumbel–Morgenstern (FGM) distribution functions (Schucany et al. 1978). The class has been extended to a generalized FGM class, but this is outside the scope of the book. It can be shown that the marginal distribution functions are given by F_X and F_Y, respectively. For instance, $\lim_{y \to \infty} F_{XY}(x, y) = F_X(x) \cdot 1 \cdot (1 + \alpha(1 -$

$F_X(x)) \cdot 0) = F_X(x)$. The bivariate PDF is given by

$$f_{XY}(x, y) = f_X(x)f_Y(y)(1 + \alpha(1 - 2F_X(x))(1 - 2F_Y(x))).$$

The advantage of this class is that it may be used with many different choices for F_X and F_Y. For instance, if we choose F_X and F_Y normal CDFs, we have created a bivariate CDF for X and Y that has marginal normal CDFs, but which is not equal to the bivariate normal PDF given by the bivariate PDF in Eq. (6.8) from the previous section. Thus, when X and Y are marginally normally distributed, there are different ways of creating dependencies between X and Y that are different from the bivariate PDF in Eq. (6.8).

The FGM and its generalizations have been widely studied in the literature. They have been applied to areas like hydrology, but have not found their way into practice. The reason is that it can only model weak dependencies between random variables X and Y; see Sect. 6.5.

6.3.3 Using Mixtures of Probability Distributions

Although a very generic formulation can be provided for this class of distribution functions (Marshall and Olkin 1988), we would like to introduce a smaller class that is defined through conditional PDFs. Let Z be a random variable and conditionally on this random variable we assume that X and Y are independent distributed. Thus the joint PDF of X and Y given Z, which is denoted by $f_{XY|Z}(x, y|z)$, is now given by

$$f_{XY|Z}(x, y|z) = f_{X|Z}(x|z)f_{Y|Z}(y|z) = \frac{f_{XZ}(x, z)f_{YZ}(y, z)}{f_Z^2(z)}.$$

Then the joint PDF for X and Y is obtained by

$$f_{XY}(x, y) = \int_{-\infty}^{\infty} f_{X|Z}(x|z)f_{Y|Z}(y|z)f_Z(z)dz.$$

These bivariate mixtures of probability distributions have been applied to many different applications. They have been used in survival analysis, where the outcomes X and Y represent time to failure. They have also found their way into medical and epidemiological sciences, where multiple measurements over time are collected on individuals. The outcomes X and Y represent two observations (either discrete or continuous) on the same individual. The random variable Z is used to drive the dependency between the repeated measures. We will illustrate this with two examples.

Assume that Z is normally distributed with mean μ_Z and variance σ_Z^2 and conditional PDFs $f_{X|Z}(x|z)$ and $f_{X|Z}(x|z)$ are given by $f_{X|Z}(x|z) = \phi((x - \mu_1 - z)/\sigma_1)/\sigma_1$ and $f_{Y|Z}(y|z) = \phi((y - \mu_2 - z)/\sigma_2)/\sigma_2$, with ϕ the standard normal

PDF, then $f_{XY}(x, y)$ is equal to Eq. (6.8) with

$$\mu_X = \mu_Z + \mu_1$$
$$\mu_Y = \mu_Z + \mu_2$$
$$\sigma_X^2 = \sigma_Z^2 + \sigma_1^2$$
$$\sigma_Y^2 = \sigma_Z^2 + \sigma_2^2$$
$$\rho = \frac{\sigma_Z^2}{\sqrt{(\sigma_Z^2 + \sigma_1^2)(\sigma_Z^2 + \sigma_2^2)}}.$$

In this example, the mixture distribution is just another way of indicating that X and Y are equal to $X = W + U$ and $Y = W + V$, respectively, with U, V, and W independent normally distributed random variables ($U \sim \mathcal{N}(\mu_1, \sigma_1^2)$, $V \sim \mathcal{N}(\mu_2, \sigma_2^2)$, and $W \sim \mathcal{N}(\mu_Z, \sigma_Z^2)$). This shows that some of the approaches are not unique.

Another interesting example, where we do not deal with sums of random variables, is when we take X and Y as binary variables and Z standard normally distributed. We assume that $f_{X|Z}(1|z) = 1 - f_{X|Z}(0|z) = \Phi(\alpha_X + \beta_X z)$ and $f_{Y|Z}(1|z) = 1 - f_{Y|Z}(0|z) = \Phi(\alpha_Y + \beta_Y z)$, with α_X, β_X, α_Y, β_Y unknown parameters and Φ the standard normal CDF. Then the marginal PMFs $f_X(x)$ and $f_Y(y)$ are given by

$$f_X(1) = 1 - f_X(0) = \Phi\left(\alpha_X/\sqrt{1 + \beta_X^2}\right)$$
$$f_Y(1) = 1 - f_Y(0) = \Phi\left(\alpha_Y/\sqrt{1 + \beta_Y^2}\right),$$

while the joint PDF for X and Y is given by Eq. (6.2) with

$$p_{00} = \int_{-\infty}^{\infty} (1 - \Phi(\alpha_X + \beta_X z))(1 - \Phi(\alpha_Y + \beta_Y z))\phi(z)dz$$
$$p_{01} = \int_{-\infty}^{\infty} (1 - \Phi(\alpha_X + \beta_X z))\Phi(\alpha_Y + \beta_Y z)\phi(z)dz$$
$$p_{10} = \int_{-\infty}^{\infty} \Phi(\alpha_X + \beta_X z)(1 - \Phi(\alpha_Y + \beta_Y z))\phi(z)dz$$
$$p_{11} = \int_{-\infty}^{\infty} \Phi(\alpha_X + \beta_X z)\Phi(\alpha_Y + \beta_Y z)\phi(z)dz$$

Binary models like these are often used in the analysis of questionnaire data, with X and Y representing questions or items to which you can answer only yes (or correct) and no (or incorrect). This can easily be extended to many more binary questions. The variable Z represents the ability of a person who takes the questionnaire. If the ability is high for a person, correctly answering the questions is no problem. The parameters α_X and α_Y are considered the difficulty parameters for the two items.

The larger the parameters the more difficult it is to correctly answer the questions. The parameters β_X and β_Y are called the discrimination parameters, which tells us how well we can distinguish between different abilities of groups of people.

6.3.4 Using the Fréchet Family of Distributions

Maurice Fréchet demonstrated that any bivariate CDF $F_{XY}(x, y)$ can be bounded from below and from above using the marginal CDFs $F_X(x)$ and $F_Y(y)$ (Plackett 1965). The boundaries are given by

$$\max\{F_X(x) + F_Y(y) - 1, 0\} \le F_{XY}(x, y) \le \min\{F_X(x), F_Y(y)\}$$

The boundary functions $F_{BL}(x, y) = \max\{F_X(x) + F_Y(y) - 1, 0\}$ and $F_{BU}(x, y) = \min\{F_X(x), F_Y(y)\}$ are both CDFs themselves. The two boundary CDFs can then be taken to form a one-parameter class of CDFs of the form

$$F_{LU}(x, y) = \lambda F_{BL}(x, y) + (1 - \lambda)F_{BU}(x, y), \quad \lambda \in [0, 1].$$

It should be noted that the marginal CDFs for X and Y are F_X and F_Y, respectively. Furthermore, the parameter λ indicates the strength of the dependence, even though this class does not contain the independent setting. There is no λ value that leads to $F_{LU}(x, y) = F_X(x)F_Y(y)$. Therefore, it has been extended to also include the independence case, but we do not discuss this here. The class of CDFs above (and its extensions) is referred to as the Fréchet class of distribution functions.

The Fréchet class of CDFs has been used to analyze bivariate dependent Poisson distributed data from a production environment (Fatahi et al. 2012) and (its extension) to financial applications (Yang et al. 2009).

6.4 Properties of Multivariate Distributions

Now that we have generalized single random variables to bivariate and multivariate random variables, we can examine some properties of the multiple variables and their joint distribution functions. We will examine the expectation and variance—which is called *covariance* in this case—of multivariate random variables, similar to the univariate case. We will also introduce a standardized measure of the covariance—the *correlation*—which quantifies the linear dependence between two variables. Furthermore, we will discuss a few alternative measures of association.

6.4.1 Expectations

For single random variables we have discussed (central) moments, as these moments relate the PDF or PMFs to the population characteristics. Now that we have dependent multiple random variables we can study expectations of functions of these variables to quantify population values that represent the way that random variables are related. Here we discuss some important examples for discrete random variables; for continuous random variables the definitions are similar, although the summation signs are replaced by integrations. We will focus again on bivariate random variables.

Although we study a joint PMF $f_{XY}(x, y)$, we might still be interested in the expected value of just one of the random variables involved, for example Y. In this case we can easily obtain the marginal distribution function of Y using $f_Y(y) = \sum_{x=0}^{\infty} f_{XY}(x, y)$ as stated before, and subsequently compute the expectation $\mathbb{E}(Y)$, as we have seen in Chap. 4. Thus the results we discussed in Chap. 4 remain applicable for both X and Y separately.

However, we may now also investigate moments of one variable, say Y again, given a specific result of the other variable, say $X = x$. This means that we are interested in the *conditional* expectation of $\psi(Y)$ given $X = x$, i.e., $\mathbb{E}(\psi(Y)|X = x)$. Here ψ is a function that can be chosen to our liking. Given a joint PMF this can be computed as follows:

$$\mathbb{E}(\psi(Y)|X = x) = \sum_{y=0}^{\infty} \psi(y) f_{Y|X}(y|x)$$

which quantifies the mean value of $\psi(Y)$ given a specific choice of $X = x$. Note that this expectation is thus a function of x. When we choose $\psi(y) = y$, we obtain the expected value or mean of Y conditionally on $X = x$ and it is given by $\mu_Y(x) = \mathbb{E}(Y|X = x)$. Since this is a function of x, we may study the random variable $\mu_Y(X) = \mathbb{E}(Y|X = X)$ and in particular the mean or expected value of this random variable. This expected value is given by

$$\mathbb{E}[\mu_Y(X)] = \sum_{x=0}^{\infty} \mu_Y(x) f_X(x)$$

$$= \sum_{x=0}^{\infty} \mathbb{E}(Y|X = x) f_X(x)$$

$$= \sum_{x=0}^{\infty} \sum_{y=0}^{\infty} y f_{Y|X}(y|x) f_X(x)$$

$$= \sum_{x=0}^{\infty} \sum_{y=0}^{\infty} y f_{XY}(x, y)$$

$$= \sum_{y=0}^{\infty} y f_Y(y) = \mu_Y.$$

Note that we have used that $\sum_{x=0}^{\infty} y f_{XY}(x, y) = y f_Y(y)$. The result that $\mathbb{E}[\mu_Y(X)] = \mu_Y$ may not be surprising, as $\mu_Y(x)$ represents the mean value for Y when $X = x$ and if we then average out all the specific mean values for Y over all x, weighted with the marginal probability $f_X(x)$, we should obtain the mean value of Y.

We may also like to know how much Y would vary if we know that $X = x$. This variability may provide us information on how well we can predict y if we have observed x. If this variability is very small, we know that the value y should be close to the conditional mean $\mu_Y(x)$. Thus we would like to study the conditional variance of Y given that $X = x$. This conditional variance, denoted by $\mathsf{VAR}(Y|X = x)$ is given by

$$\mathsf{VAR}(Y|X = x) = \mathbb{E}((Y - \mu_Y(x))^2 | X = x) = \sum_{y=0}^{\infty} (y - \mu_Y(x))^2 f_{Y|X}(y|x).$$

Note that we calculate the variance around the conditional mean $\mu_Y(x)$, and not around μ_Y, since $\mu_Y(x)$ is the expected value for Y when we condition on or know that $X = x$. This conditional variance is also a function of x, and we may denote it by $\sigma_Y^2(x)$. If we average out all these variances over all x, like we did with the conditional mean, we obtain

$$\mathbb{E}\sigma_Y^2(X) = \sum_{x=0}^{\infty} \sigma_Y^2(x) f_X(x)$$

$$= \sum_{x=0}^{\infty} \sum_{y=0}^{\infty} (y - \mu_Y(x))^2 f_{Y|X}(y|x) f_X(x)$$

$$= \sum_{x=0}^{\infty} \sum_{y=0}^{\infty} [(y - \mu_Y)^2 + 2(y - \mu_Y)(\mu_Y - \mu_Y(x)) + (\mu_Y - \mu_Y(x))^2] f_{XY}(x, y)$$

$$= \sum_{y=0}^{\infty} (y - \mu_Y)^2 f_Y(y) - \sum_{x=0}^{\infty} (\mu_Y(x) - \mu_Y)^2 f_X(x)$$

$$= \sigma_Y^2 - \mathsf{VAR}(\mu_Y(X)).$$

Thus, when we average out all the conditional variances of Y given $X = x$, we do not obtain the variance of Y, but we obtain less than the variance of Y. This is not surprising, since we have eliminated the variability in Y that is induced by X. More precisely, the variability in the conditional mean is what we have eliminated. If we rearrange the equality above, we obtain a well-known relation in statistics on variances, which is equal to

$$\mathsf{VAR}(Y) = \mathbb{E}[\mathsf{VAR}(Y|X = X)] + \mathsf{VAR}(\mathbb{E}(Y|X = X)).$$

Note that we have seen something similar in Chap. 2 when we looked at stratified sampling (see Eq. (2.6)). The total variability in a single variable Y is the sum of the variability in Y within strata and the variability between strata in the means of Y.

If we wish to study how strongly X and Y are related to each other, we would need to investigate the expectations of functions of X and Y, say $g(x, y)$. For this general function, the expectation is given by:

$$\mathbb{E}[g(X, Y)] = \sum_{x=0}^{\infty} \sum_{y=0}^{\infty} g(x, y) f_{XY}(x, y)$$

Note that we already have investigated an example of this type when we discussed the conditional variance. Here $g(x, y)$ was selected equal to $g(x, y) = (y - \mu_Y(x))^2$. Another example, which we will use later in this chapter, is $g(x, y) = xy$. It is easy to show that if X and Y are independent then $\mathbb{E}(XY) = \mathbb{E}(X)\mathbb{E}(Y)$, which we have already discussed in Chap. 4.

6.4.2 Covariances

We have talked extensively about the variance of a random variable, and about ways of estimating variances. We now introduce this concept for two random variables. We define the covariance of X and Y by

$$\mathsf{COV}(X, Y) = \sigma_{XY} = \mathbb{E}[(X - \mathbb{E}(X))(Y - \mathbb{E}(Y))].$$

By using the calculation rules for expectation in Sect. 4.10 in Chap. 4 we have used the following algebraic relations:

$$\mathbb{E}\big[(X - \mathbb{E}(X))(Y - \mathbb{E}(Y))\big] = \mathbb{E}\big[XY - X(\mathbb{E}(Y)) - (\mathbb{E}(X))Y + (\mathbb{E}(X))(\mathbb{E}(Y))\big]$$
$$= \mathbb{E}(XY) - (\mathbb{E}(X))(\mathbb{E}(Y)) - (\mathbb{E}(X))(\mathbb{E}(Y)) + (\mathbb{E}(X))(\mathbb{E}(Y))$$
$$= \mathbb{E}(XY) - (\mathbb{E}(X))(\mathbb{E}(Y)).$$

The covariance has the following properties:

1. $\text{COV}(X, X) = \text{VAR}(X) = \sigma_X^2$.
2. If X and Y are independent, then $\text{COV}(X, Y) = 0$.[5]
3. $\text{COV}(X, Y) = \text{COV}(Y, X)$.
4. $\text{COV}(aX, Y) = a\text{COV}(X, Y)$.
5. $\text{COV}(X + c, Y) = \text{COV}(X, Y)$.
6. $\text{COV}(X + Y, Z) = \text{COV}(X, Z) + \text{COV}(Y, Z)$.

Now that we have the concept of covariance, we can extend the calculation rules on the variances of adding and subtracting two random variables that we discussed in Sect. 4.10 under the assumption of independence. Irrespective of the underlying CDFs for X and Y, the variances of $X + Y$ and $X - Y$ are now given by

1. $\text{VAR}(X + Y) = \text{VAR}(X) + 2\text{COV}(X, Y) + \text{VAR}(Y)$.
2. $\text{VAR}(X - Y) = \text{VAR}(X) - 2\text{COV}(X, Y) + \text{VAR}(Y)$.

If we have independence between X and Y, we obtain that $\text{COV}(X, Y) = 0$ and the two rules are reduced to the rule we mentioned in Sect. 4.10.

6.4.2.1 Covariance of Bivariate Normal Distributions

To illustrate the calculation of covariances, we will assume that X and Y arc bivariate normally distributed, with PDF given by Eq. (6.8). This means that $\mathbb{E}(X) = \mu_X$, $\mathbb{E}(Y) = \mu_Y$, $\text{VAR}(X) = \sigma_X^2$, and $\text{VAR}(Y) = \sigma_Y^2$. To determine the covariance between X and Y, it may be more convenient to introduce the standardized random variables $Z_1 = (X - \mu_X)/\sigma_X$ and $Z_2 = (Y - \mu_Y)/\sigma_Y$ and to note that the PDF for these standardized random variables is simply $\exp\{-(z_1^2 - 2\rho z_1 z_2 + z_2^2)/(2(1 - \rho^2))\}/(\sqrt{2\pi(1 - \rho^2)})$. The covariance of X and Y can now be determined by

---

[5] Note that the converse is not necessarily true: two random variables can have a covariance of 0, and still be dependent. A simple example is to take X as a standard normal random variable and $Y = X^2$. It is clear that X and Y are dependent, since Y is a function of X, and that $\text{COV}(X, Y) = \mathbb{E}(XY) - \mathbb{E}X\mathbb{E}Y = \mathbb{E}X^3 = 0$.

$$
\begin{aligned}
\text{COV}(X, Y) &= \mathbb{E}(X - \mu_X)(Y - \mu_Y) \\
&= \sigma_X \sigma_Y \mathbb{E}[Z_1 Z_2] \\
&= \sigma_X \sigma_Y \int_{-\infty}^{\infty} \int_{-\infty}^{\infty} z_1 z_2 \frac{1}{2\pi(1-\rho^2)} \exp\left\{ -\frac{z_1^2 - 2\rho z_1 z_2 + z_2^2}{2(1-\rho^2)} \right\} dz_1 dz_2 \\
&= \sigma_X \sigma_Y \int_{-\infty}^{\infty} z_2 \frac{1}{\sqrt{2\pi}} \exp\left\{ -\frac{z_2^2}{2} \right\} \int_{-\infty}^{\infty} z_1 \frac{1}{\sqrt{2\pi(1-\rho^2)}} \exp\left\{ -\frac{(z_1 - \rho z_2)^2}{2(1-\rho^2)} \right\} dz_1 dz_2 \\
&= \sigma_X \sigma_Y \int_{-\infty}^{\infty} \rho z_2^2 \frac{1}{\sqrt{2\pi}} \exp\left\{ -\frac{z_2^2}{2} \right\} dz_2 \\
&= \rho \sigma_X \sigma_Y.
\end{aligned}
$$

For the bivariate normal distribution function we have a very nice property that does not hold for many other bivariate CDFs. The distribution function of any linear combination of X and Y, say $aX + bY$, has a normal distribution function with mean $\mu = a\mu_X + b\mu_Y$ and variance $\sigma^2 = a^2 \sigma_X^2 + 2ab\rho\sigma_X\sigma_Y + b^2 \sigma_Y^2$.

6.4.2.2 Covariance of Bivariate Lognormal Distributions

The bivariate lognormal CDF is implicitly defined through the bivariate normal distribution. If we assume that (X, Y) is bivariate normal, with parameters μ_X, μ_Y, σ_X, σ_Y, and ρ, then $(\exp(X), \exp(Y))$ has a bivariate lognormal distribution function. To determine the covariance of $(\exp(X), \exp(Y))$, recall that $X + Y$ has a normal distribution function with mean $\mu_X + \mu_Y$ and variance $\sigma_X^2 + 2\rho\sigma_X\sigma_Y + \sigma_Y^2$. The covariance of $(\exp(X), \exp(Y))$ is now equal to

$$
\begin{aligned}
\text{COV}(\exp(X), \exp(Y)) &= \mathbb{E}[\exp(X) \exp(Y)] - \mathbb{E}[\exp(X)]\mathbb{E}[\exp(Y)] \\
&= \mathbb{E}[\exp(X + Y)] - \exp(\mu_X + 0.5\sigma_X^2) \exp(\mu_Y + 0.5\sigma_Y^2) \\
&= \exp(\mu_X + \mu_Y + 0.5(\sigma_X^2 + 2\rho\sigma_X\sigma_Y + \sigma_Y^2)) \\
&\quad - \exp(\mu_X + 0.5\sigma_X^2) \exp(\mu_Y + 0.5\sigma_Y^2) \\
&= \exp(\mu_X + 0.5\sigma_X^2 + \mu_Y + 0.5\sigma_Y^2)(\exp(\rho\sigma_X\sigma_Y) - 1) \\
&= \mathbb{E}[\exp(X)]\mathbb{E}[\exp(Y)](\exp(\rho\sigma_X\sigma_Y) - 1).
\end{aligned}
$$

This covariance is quite different from the covariance $\rho\sigma_X\sigma_Y = \text{COV}(X, Y)$ in the logarithmic scale. Thus the covariance is affected by transformations.

6.4.2.3 Covariance for Sums of Random Variables

In a more general setting, we may consider covariances of bivariate distribution functions that are created from sums of random variables, as we discussed in Sect. 6.3. If we assume that $X = W + U$ and $Y = W + V$, with U, V, and W mutually independent having means μ_U, μ_V and μ_W, and variances σ_U^2, σ_V^2, and σ_W^2, respectively, the covariance of X and Y can be calculated as well. Using the calculation rules of Sect. 4.10, the means of X and Y are given by $\mu_X = \mu_W + \mu_U$ and $\mu_Y = \mu_W + \mu_V$, respectively. Then the covariance between X and Y is

$$
\begin{aligned}
\mathsf{COV}(X, Y) &= \mathbb{E}[(X - \mu_X)(Y - \mu_Y)] \\
&= \mathbb{E}[(W - \mu_W + U - \mu_U)(W - \mu_W + V - \mu_V)] \\
&= \mathsf{VAR}(W) + \mathbb{E}[(W - \mu_W)(V - \mu_V)] + \mathbb{E}[(U - \mu_U)(W - \mu_W)] \\
&\quad + \mathbb{E}[(U - \mu_U)(V - \mu_V)] \\
&= \sigma_W^2.
\end{aligned}
$$

We have not made any assumptions about the distribution function of U, V, and W: we have only assumed that they had finite variances. Thus, irrespective of the underlying distributions, the covariance of X and Y is the variance of W. For instance, for a Poisson distributed random variable W, i.e., $W \sim \mathscr{P}(\lambda_W)$, the covariance of X and Y is $\mathsf{VAR}(W) = \lambda_W$, irrespective of the CDFs of U and V.

6.4.2.4 Covariance of Farlie–Gumbel–Morgenstern Distributions

If we assume that the bivariate CDF of X and Y is from the Farlie–Gumbel–Morgenstern family, the covariance can be written as

$$
\begin{aligned}
\mathsf{COV}(X, Y) &= \mathbb{E}(XY) - \mathbb{E}(X)\mathbb{E}(Y) \\
&= \int_{-\infty}^{\infty} \int_{-\infty}^{\infty} xy f_X(x) f_Y(y) \left(1 + \alpha(1 - 2F_X(x))(1 - 2F_Y(y))\right) dx\,dy - \mu_X \mu_Y \\
&= \alpha \mathbb{E}[X(1 - 2F_X(X))]\mathbb{E}[Y(1 - 2F_Y(Y))].
\end{aligned}
$$

Now assume that F_X and F_Y are exponential CDFs with parameters λ_X and λ_Y, respectively. Then the expectations in the covariance can be calculated explicitly. We would obtain $\mathbb{E}[X(1 - 2F_X(X))] = -[2\lambda_X]^{-1}$ and $\mathbb{E}[Y(1 - 2F_Y(Y))] = -[2\lambda_Y]^{-1}$, respectively. This means that the Farlie–Gumbel–Morgenstern CDF with marginal exponential CDFs leads to a covariance of $-\alpha/[4\lambda_X \lambda_Y]$.

6.4.2.5 Covariance of Bivariate Mixture Distributions

If we assume that X and Y are conditionally independent given the random variable $Z = z$, for all values of z, we can also establish the covariance. First of all, we will show that $\mathbb{E}(XY)$ is equal to $\mathbb{E}[\mu_{XY}(Z)]$, with $\mu_{XY}(z) = \mathbb{E}(XY|Z = z)$. Indeed, we have

$$
\begin{aligned}
\mathbb{E}[\mu_{XY}(Z)] &= \int_{-\infty}^{\infty} \mu_{XY}(z) f_Z(z) dz \\
&= \int_{-\infty}^{\infty} \int_{-\infty}^{\infty} \int_{-\infty}^{\infty} xy f_{XY|Z}(x, y|z) f_Z(z) dx dy dz \\
&= \int_{-\infty}^{\infty} \int_{-\infty}^{\infty} \int_{-\infty}^{\infty} xy f_{XYZ}(x, y, z) dx dy dz \\
&= \int_{-\infty}^{\infty} \int_{-\infty}^{\infty} xy f_{XY}(x, y) dx dy \\
&= \mathbb{E}(XY).
\end{aligned}
$$

Because X and Y are independent conditionally on $Z = z$, we also obtain that $\mathbb{E}(XY|Z = z) = \mathbb{E}(X|Z = z)\mathbb{E}(Y|Z = z) = \mu_X(z)\mu_Y(z)$. These two results now lead to a covariance for X and Y that is given by

$$
\begin{aligned}
\mathsf{COV}(X, Y) &= \mathbb{E}(XY) - \mathbb{E}(X)\mathbb{E}(Y) \\
&= \mathbb{E}[\mu_{XY}(Z)] - \mu_X \mu_Y \\
&= \mathbb{E}[\mu_X(Z)\mu_Y(Z)] - \mu_X \mu_Y \\
&= \mathbb{E}[(\mu_X(Z) - \mu_X)(\mu_Y(Z) - \mu_Y)] \\
&= \mathsf{COV}(\mu_X(Z), \mu_Y(Z)).
\end{aligned}
$$

Thus the covariance of X and Y is equal to the covariance of random variables $\mu_X(Z)$ and $\mu_Y(Z)$. If we assume that X and Y are identically distributed (conditionally on Z), we would have that $\mu_X(z) = \mu_Y(z) = \mu(z)$, which would lead to $\mathsf{COV}(X, Y) = \mathsf{VAR}(\mu(Z))$.

Now assume that Z is a random variable with outcomes in $(0, \infty)$, e.g., a lognormal or exponential random variable, with mean μ_Z and variance σ_Z^2. Furthermore, assume that the conditional distribution function of X given $Z = z$ is Poisson with parameter $z\lambda_X$, $X|Z = z \sim \mathscr{P}(z\lambda_X)$, and the conditional distribution function of Y given $Z = z$ is Poisson with parameter $z\lambda_Y$, $Y|Z = z \sim \mathscr{P}(z\lambda_Y)$; then the conditional expectations $\mu_X(z)$ and $\mu_Y(z)$ are given by $\mu_X(z) = z\lambda_X$ and $\mu_Y(z) = z\lambda_Y$, respectively. This implies that $\mu_X = \mathbb{E}(\mu_X(Z)) = \lambda_X \mathbb{E}(Z) = \lambda_X \mu_Z$ and $\mu_Y = \lambda_Y \mu_Z$. Then the covariance of X and Y is given by

$$
\mathsf{COV}(\mu_X(Z), \mu_Y(Z)) = \mathbb{E}[\mu_X(Z)\mu_Y(Z)] - \mu_X \mu_Y = \lambda_X \lambda_Y \left(\mathbb{E}(Z^2) - \mu_Z^2 \right) = \lambda_X \lambda_Y \sigma_Z^2.
$$

Note that the Poisson assumption is not really relevant in the calculation of this covariance. The relevant part is that $\mu_X(z) = z\lambda_X$ and $\mu_Y(z) = z\lambda_Y$. Thus any alternative CDF for which it make sense to assume $\mu_X(z) = z\lambda_X$ and $\mu_Y(z) = z\lambda_Y$, for instance the exponential distribution, would give the same covariance of $\text{COV}(X, Y) = \lambda_X \lambda_Y \sigma_Z^2$.

6.5 Measures of Association

In the previous subsection we focused on covariances of several bivariate distributions. Some of them were studied in a very generic form, without specifying F_X and F_Y. However, covariances are not easily comparable across different families of bivariate CDFs. One of the reasons is that covariances are affected by variance parameters. For instance, the covariance of a bivariate normal CDF was given by $\rho\sigma_X\sigma_Y$ and depends strongly on the variability of X and Y. Thus a covariance of 10 and 100 could be explained by any of the three parameters ρ, σ_X, and σ_Y. Thus the incomparability is caused by a lack of standardization, implying that we would like a measure that has no unit or that is dimensionless. For the covariance of $\rho\sigma_X\sigma_Y$ it seems reasonable to standardize with the standard deviations of X and Y. However, another reason for the incomparability is that dependency between X and Y can take different forms and therefore standardization may become setting-specific, which would not help comparability. In the setting of bivariate Poisson mixture distributions we saw that the covariance was given by $\lambda_X \lambda_Y \sigma_Z^2$, with λ_X and λ_Y mean parameters and σ_Z^2 a variance parameter. Here we would like to eliminate the mean parameters λ_X and λ_Y, since they provide no information on the dependency. They will be there even if there is independence. Simply using the standard deviations of X and Y, like we suggested for the normal case, does not eliminate λ_X and λ_Y. The underlying mixture distribution is quite different from the bivariate normal distribution and therefore the covariances are not directly comparable between these CDFs. We just have to accept that covariance is but one measure that captures dependency.

In this section we will discuss different measures of association that will capture in some way the dependency between X and Y. Only one of them uses the covariance. In Chap. 3 we already discussed three measures of association for binary random variables: risk difference, relative risk, and odds ratio. None of them were defined through covariances, but rather through their conditional probabilities. We will see in this section that we can also define measures of association through their joint CDF.

6.5.1 Pearson's Correlation Coefficient

As we just mentioned, one way of standardizing the covariance is to divide the covariance by the standard deviation of both X and Y. This standardization makes sense, since we would study the covariance of the random variables $Z_X = (X - \mu_X)/\sigma_X$ and $Z_Y = (Y - \mu_Y)/\sigma_Y$. Both standardized random variables have a variance of one. The covariance of the standardized random variables is referred to as Pearson's *correlation coefficient* and it is often denoted by ρ_P. It is given by:

$$\rho_P = \text{CORR}(X, Y) = \text{COV}(Z_X, Z_Y) = \mathbb{E}\left(\frac{X - \mu_X}{\sigma_X}\right)\left(\frac{Y - \mu_Y}{\sigma_Y}\right) = \frac{\text{COV}(X, Y)}{\sqrt{\text{VAR}(X)\text{VAR}(Y)}}.$$

Pearson's correlation coefficient has some nice properties:

1. $\text{CORR}(X, Y) \in [-1, 1]$
2. If $\text{CORR}(X, Y) = 1$, then $Y = aX + b$, where $a > 0$
3. If $\text{CORR}(X, Y) = -1$, then $Y = aX + b$, where $a < 0$
4. $\text{CORR}(aX + b, cY + d) = \text{CORR}(X, Y)$ for $a, c > 0$ or $a, c < 0$.

Pearson's correlation coefficient is a way of quantifying how two variables "correlate". If Pearson's correlation is positive, the two variables X and Y move in the same direction. If X is increasing then Y should be increasing as well. This is typically seen in the anthropometric data of Sir Francis Galton. A taller person is typically heavier than a shorter person. If Pearson's correlation is negative, the random variables move in opposite directions. A typical example is that an increase in age reduces cognitive abilities. When Pearson's correlation coefficient is zero the random variables are called *uncorrelated*.[6] The following rule of thumb is often used to qualify the strength of Pearson's correlation coefficient (Hinkle et al. 2003):

$$0.90 < |\rho_P| \leq 1.00 \text{ Very strong correlation}$$
$$0.70 < |\rho_P| \leq 0.90 \text{ Strong correlation}$$
$$0.50 < |\rho_P| \leq 0.70 \text{ Moderate correlation}$$
$$0.30 < |\rho_P| \leq 0.50 \text{ Low correlation}$$
$$0 \leq |\rho_P| \leq 0.30 \text{ Negligible correlation.}$$

6.5.1.1 Correlation for Bivariate Normal Distributions

It should be noted that Pearson's correlation coefficient nicely works for bivariate normal CDFs, since the $\text{CORR}(X, Y)$ for bivariate normally distributed random variables X and Y is equal to the PDF parameter ρ, i.e., $\rho_P = \text{CORR}(X, Y) = \rho$. Thus Pearson's correlation coefficient is only a function of the parameter ρ that indicates the strength of the dependence between X and Y. This is not the case for all the other covariances (see below) that we discussed in the previous section. Pearson's correlation coefficient may not be a function of the dependence parameter alone.

---

[6] Recall that uncorrelated does not mean independent.

Fig. 6.1 Pearson's
correlation coefficient for
bivariate lognormal
distributions

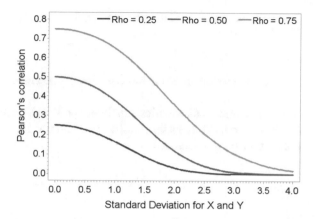

6.5.1.2 Correlation for Bivariate Lognormal Distributions

We already saw that the covariance of two lognormally distributed random variables
is given by $\mathbb{E}[\exp(X)]\mathbb{E}[\exp(Y)](\exp(\rho\sigma_X\sigma_Y) - 1)$, with (X, Y) bivariate normal.
Since the variances of $\exp(X)$ and $\exp(Y)$ are given by $(\mathbb{E}[\exp(X)])^2(\exp(\sigma_X^2) -$
$1)$ and $(\mathbb{E}[\exp(Y)])^2(\exp(\sigma_Y^2) - 1)$, respectively, Pearson's correlation coefficient
becomes

$$\rho_P = \frac{\exp(\rho\sigma_X\sigma_Y) - 1}{\sqrt{(\exp(\sigma_X^2) - 1)(\exp(\sigma_Y^2) - 1)}}.$$

When the variances of X and Y are the same, say $\sigma_X^2 = \sigma_Y^2 = \sigma^2$, Pearson's correlation coefficient reduces to $\rho_P = [\exp(\rho\sigma^2) - 1]/[\exp(\sigma^2) - 1]$. This correlation is
a decreasing function in σ^2. When σ^2 is close to zero the correlation is ρ and when
σ^2 becomes very large the correlations converges to zero. On the other hand, the
correlation is relatively constant for the variance $\sigma^2 \leq 1$ (see Fig. 6.1).

6.5.1.3 Correlation for Sums of Random Variables

Here we assumed that $X = W + U$ and $Y = W + V$, with U, V, and W mutually
independently distributed, with means μ_U, μ_V and μ_W, and variances σ_U^2, σ_V^2, and σ_W^2,
respectively. The variances of X and Y are now $\sigma_X^2 = \sigma_W^2 + \sigma_U^2$ and $\sigma_Y^2 = \sigma_W^2 + \sigma_V^2$,
respectively, using the calculation rules of Sect. 4.10. Pearson's correlation coefficient
now becomes equal to

$$\rho_P = \frac{\sigma_W^2}{\sqrt{(\sigma_W^2 + \sigma_U^2)(\sigma_W^2 + \sigma_V^2)}}. \tag{6.9}$$

This correlation coefficient is often referred to as the *intraclass correlation coefficient* (ICC).

6.5.1.4 Correlation of Farlie–Gumbel–Morgenstern Distributions

Since the marginal distribution functions for X and Y are equal to F_X and F_Y for the Farlie–Gumbel–Morgenstern family of distribution functions, respectively, Pearson's correlation simply becomes

$$\rho_P = \frac{\alpha \mathbb{E}[X(1 - 2F_X(X))]\mathbb{E}[Y(1 - 2F_Y(Y))]}{\sigma_X \sigma_Y}.$$

It has been shown that this correlation coefficient can never be larger than $1/3$ (Schucany et al. 1978). Thus Pearson's correlation coefficient for the FGM family is bounded by $-1/3 \leq \rho_P \leq 1/3$. This implies that the FGM family of CDFs may be useful for applications where the random variables are weakly correlated.

In the case that F_X and F_Y are exponential CDFs with parameters λ_X and λ_Y, respectively, the covariance was provided by $-\alpha/[4\lambda_X \lambda_Y]$. Since the standard deviations σ_X and σ_Y are given by λ_X^{-1} and λ_Y^{-1}, respectively, Pearson's correlation coefficient becomes equal to $\rho_P = -\alpha/4$. Thus, in the case of exponential CDFs in the FGM family, Pearson's correlation is only a function of the dependence parameter.

6.5.1.5 Correlation of Bivariate Mixture Distributions

The marginal CDFs for X and Y are determined through the conditional PDFs:

$$f_X(x) = \int_{-\infty}^{\infty} f_{X|Z}(x|z) f_Z(z) dz \quad \text{and} \quad f_Y(y) = \int_{-\infty}^{\infty} f_{Y|Z}(y|z) f_Z(z) dz.$$

If Z is a discrete random variable, integration is replaced by summation. Thus the variances of X and Y are now given by (see Sect. 6.4.1)

$$\text{VAR}(X) = \mathbb{E}[\text{VAR}(X|Z = Z)] + \text{VAR}(\mu_X(Z))$$
$$\text{VAR}(Y) = \mathbb{E}[\text{VAR}(Y|Z = Z)] + \text{VAR}(\mu_Y(Z)).$$

Pearson's correlation coefficient is then equal to

$$\rho_P = \frac{\text{COV}(\mu_X(Z), \mu_Y(Z))}{\sqrt{(\mathbb{E}[\text{VAR}(X|Z = Z)] + \text{VAR}(\mu_X(Z)))(\mathbb{E}[\text{VAR}(Y|Z = Z)] + \text{VAR}(\mu_Y(Z)))}}.$$

To illustrate this with an example, assume again that Z is a random variable with outcomes in $(0, \infty)$, having mean μ_Z and variance σ_Z^2, and assume that the conditional distribution function of X given $Z = z$ is Poisson with parameter $z\lambda_X$, $X|Z = z \sim$

$\mathscr{P}(z\lambda_X)$ and the conditional distribution function of Y given $Z = z$ is Poisson with parameter $z\lambda_Y$, $Y|Z = z \sim \mathscr{P}(z\lambda_Y)$. The covariance was obtained by $\text{COV}(X, Y) = \lambda_X \lambda_Y \sigma_Z^2$. The conditional variance of X given Z is given by $\text{VAR}(X|Z = z) = z\lambda_X$, which is the conditional mean of the Poisson distribution. Thus $\mathbb{E}(\text{VAR}(X|Z = Z)) = \mu_Z \lambda_X$. The variance of $\mu_X(Z)$ is

$$\text{VAR}(\mu_X(Z)) = \text{VAR}(Z\lambda_X) = \lambda_X^2 \text{VAR}(Z) = \lambda_X^2 \sigma_Z^2.$$

Putting the pieces together, we obtain that $\mathbb{E}[\text{VAR}(X|Z=Z)]+\text{VAR}(\mu_X(Z)) = \mu_Z \lambda_X + \sigma_Z^2 \lambda_X^2$. For Y we obtain the same relation $\mathbb{E}[\text{VAR}(Y|Z = Z)] + \text{VAR}(\mu_Y(Z)) = \mu_Z \lambda_Y + \sigma_Z^2 \lambda_Y^2$. Hence, Pearson's correlation coefficient now becomes equal to

$$\rho_P = \frac{\sigma_Z^2}{\sqrt{(\sigma_Z^2 + \mu_Z \lambda_X^{-1})(\sigma_Z^2 + \mu_Z \lambda_Y^{-1})}}.$$

6.5.2 Kendall's Tau Correlation

Pearson's correlation coefficient is just one measure of association that quantifies the dependency between X and Y. We already knew this, because we have already defined specific measures of association for two binary random variables X and Y in Chap. 3. In this section we will discuss Kendall's tau, which is a measure of *concordance*. A concordance measure is similar to correlation. To understand this alternative approach to dependency between X and Y, we consider two independent draws from F_{XY}, say (X_1, Y_1) and (X_2, Y_2). Independence means here that X_1 and X_2 are independent and Y_1 and Y_2 are independent. The two pairs of random variables (X_1, Y_1) and (X_2, Y_2) are called *concordant* when $(X_2 - X_1)(Y_2 - Y_1) > 0$ and *discordant* when $(X_2 - X_1)(Y_2 - Y_1) < 0$. If a pair is concordant or discordant, it means that the change in the two random variables $D_X = X_2 - X_1$ (from X_1 to X_2) and $D_Y = Y_2 - Y_1$ (from Y_1 to Y_2) are dependent. Concordance refers to a "positive" dependency, which means that the direction of change in X is the same as the direction of change in Y. Discordance means a "negative" dependence, where the direction of change in X is opposite to the direction of change in Y.

Kendall's tau measures the strength between the dependency of D_X and D_Y and can be defined in terms of probabilities:

$$\begin{aligned}
\tau_K &= \text{Pr}\left((X_2 - X_1)(Y_2 - Y_1) > 0\right) - \text{Pr}\left((X_2 - X_1)(Y_2 - Y_1) < 0\right) \\
&= 2\,\text{Pr}\left((X_2 - X_1)(Y_2 - Y_1) > 0\right) - 1 \\
&= 4\,\text{Pr}(X_1 < X_2, Y_1 < Y_2) - 1.
\end{aligned}$$

Note that Maurice Kendall defined his correlation coefficient on data (thus as an estimator and not as a parameter), but the population parameter it represents is obvious: see Kendall (1938) and Sect. 6.6. Kendall's tau is also referred to as a correlation coefficient and it varies between -1 and 1. It should be noted that the minimum -1 is attained when F_{XY} is equal to the Fréchet lower bound, i.e., $F_{XY}(x, y) = \max\{F_X(x) + F_Y(y) - 1, 0\}$. The value 1 is attained when F_{XY} is equal to the Fréchet upper bound, i.e., $F_{XY}(x, y) = \min\{F_X(x), F_Y(y)\}$ (see Fuchs et al. 2018).

When X and Y are independent, Kendall's tau is equal to zero. Indeed, if X and Y are independent, then all four random variables are independent, since we assumed that the two pairs (X_1, Y_1) and (X_2, Y_2) were already independent. This makes D_X and D_Y independent and this implies that the probability that $D_X D_Y > 0$ is equal to 0.5, which makes Kendall's tau equal to zero. Furthermore, Pearson's correlation coefficient of D_X and D_Y is given by $\mathsf{CORR}(D_X, D_Y) = \mathsf{CORR}(X, Y)$. Thus Pearson's correlation coefficient also measures the concordance or discordance in a certain way.

We have seen that Pearson's correlation coefficient fits very well with the bivariate normal distribution, but Kendall's tau fits very well with the continuous FGM distributions and the Fréchet family of distribution functions. Irrespective of the marginal CDFs F_X and F_Y, Kendall's tau for a bivariate FGM distribution is equal to $\tau_K = 2\alpha/9$ (Fuchs et al. 2018) and for the Fréchet family Kendall's tau is $\tau_K = (2\lambda - 1)/3$ (Nelsen 2007). Thus Kendall's tau correlation coefficient is only a function of the dependency parameter in the FGM and Fréchet families of distribution functions when we deal with continuous random variables X and Y. This general characteristic for Kendall's tau does not happen for Pearson's correlation coefficient, but it does happen for special choices of F_X and F_Y (see Sect. 6.5.1). For bivariate normal distribution functions Kendall's tau is given by $\tau_K = 2\arcsin(\rho)/\pi$, with ρ the parameter of the bivariate normal distribution. Since ρ is equal to Pearson's correlation coefficient ρ_P for the normal distribution, the relation $\tau_K = 2\arcsin(\rho)/\pi$ links Kendall's tau to Pearson's correlation coefficient for normal data. Finally, for the bivariate mixture distributions Kendall's tau must be studied case by case to determine which parameters of the joint distribution function are included in τ_K.

6.5.3 Spearman's Rho Correlation

Kendall's tau was defined through probabilities using two i.i.d. pairs (X_1, Y_1) and (X_2, Y_2) having CDF F_{XY}. It measured the concordance between X and Y. There is an alternative approach that relates to concordance and to Pearson's correlation coefficient. This approach is Spearman's rho. To define Spearman's rho in terms of the joint CDF, we need to use three i.i.d. pairs (X_1, Y_1), (X_2, Y_2), and (X_3, Y_3) of random variables having joint distribution function F_{XY}, but we do not need to use all six random variables. Spearman's rho correlation is defined by the following probability:

$$\rho_S = 2\left[\Pr((X_1 - X_2)(Y_3 - Y_1) > 0) - \Pr((X_1 - X_2)(Y_3 - Y_1) < 0)\right].$$

This definition of concordance did not come from Spearman, since Spearman introduced only the estimator for this dependency parameter (see Sect. 6.6.3). The current formulation was established later to understand what population parameter Spearman's estimator was trying to capture. This is not obvious from Spearman's estimator, like it was for Kendall's tau. We will see that the estimator is strongly related to Pearson's correlation, while its definition in terms of F_{XY} is quite similar to that of Kendall's tau, but now using three pairs. Note that Spearman's rho quantifies concordance of change in one dimension from the first to the second observation and change in the second dimension from the first to the third observation. Thus when X and Y are independent, Spearman's rho correlation coefficient becomes equal to zero. Furthermore, it ranges from -1 to 1, like Kendall's tau and Pearson's correlation coefficient. However, Spearman's rho may be quite different from Kendall's tau. The following results can be shown mathematically (Nelsen 2007):

$$-1 \leq 3\tau_K - 2\rho_S \leq 1,$$

and there exists CDFs for which these boundaries are attainable. Thus Spearman's rho and Kendall's tau can differ substantially in a population. They can also both differ strongly from Pearson's correlation. Thus in this sense they all represent a different way of capturing the dependency between X and Y.

Similar to Kendall's tau, Spearman's rho correlation also fits very well with the FGM and Fréchet families of distribution functions. For the FGM family of CDFs, Spearman's rho is equal to $\rho_S = \alpha/3$ and for the Fréchet family of CDFs, Spearman's rho is equal to $\rho_S = [2\lambda - 1]/3$ (Nelsen 2007). Thus Spearman's rho and Kendall's tau both estimate the same function of the dependency parameter λ in the Fréchet family of distributions, but different functions for the FGM family. For the bivariate normal distribution function, Spearman's rho is equal to $\rho_S = 6\arcsin(\rho/2)/\pi$, with ρ the correlation parameter of the bivariate normal distribution (Moran 1948). Finally, for the bivariate mixture distribution functions Spearman's rho must be studied case by case to determine which parameters of the joint distribution function are included in ρ_S.

6.5.4   Cohen's Kappa Statistic

Pearson's correlation coefficient, Kenadall's tau, and Spearman's rho are all measures of correlation and concordance, but in 1960, Jacob Cohen published another measure of association, referred to as a measure of *agreement*. It measures how much two observers or raters agree on the evaluation of the same set of n items or units into K exhaustive and mutually exclusive classes (Cohen 1960). Since there is also a chance of accidentally classifying units in the same category by the raters, he created a measure that quantifies how well the raters agree on classification that is corrected for accidental or chance agreement. This chance agreement has strong similarities with grade corrections in multiple choice exams.

The bivariate random variables X and Y, which represents two readings on the same unit, both take their values in $\{1, 2, 3, \ldots, K\}$. The joint PDF is fully defined by the set of PDF parameters $p_{11}, p_{12}, \ldots, p_{1K}, p_{21}, p_{22}, \ldots, p_{2K}, \ldots, p_{K1}, p_{K2}, \ldots, p_{KK}$ given by

$$\Pr(X = x, Y = y) = p_{xy} \geq 0, \quad \text{and} \quad \sum_{x=0}^{K} \sum_{y=0}^{K} p_{xy} = 1.$$

Thus the probabilities make up a $K \times K$ contingency table, similar to the 2×2 contingency table in Table 3.1 that we used for diagnostic tests. The probability p_{xy} represents the probability that a unit is classified by the first rater in class x and by the second rater in class y. Thus, the probability p_{kk} indicates the probability that both raters classify a unit in the same class k. This implies that when $p_{xy} = 0$ for every $x \neq y$, there will be no difference in classification between the two raters.

Thus one relevant measure for agreement is the probability $p_O = \sum_{k=1}^{K} p_{kk}$. It represents the probability that both raters classify units in the same classes. When it is equal to 1, there is perfect agreement. On the other hand, when the ratings are independent, i.e., $\Pr(X = x, Y = y) = f_X(x) f_Y(y)$, the expected probability that both raters classify a unit in the same class k is equal to $f_X(k) f_Y(k)$. Thus, based on independent ratings, we expect a probability of correctly classifying units in the same classes to be equal to $p_E = \sum_{k=1}^{K} f_X(k) f_Y(k)$. Thus it may not be fair to contribute all correct classifications to the raters, because some of them may have happened just by chance. Cohen therefore proposed the kappa statistic, given by

$$\kappa_C = \frac{p_O - p_E}{1 - p_E}.$$

When there is perfect agreement ($p_O = 1$), the kappa statistic reaches its maximum at the value of one ($\kappa_C = 1$), but when the ratings are independent, the kappa statistic is equal to $\kappa_C = 0$. Although the kappa statistic can become smaller than zero, in particular when there is discordance, in most practical settings, the kappa will be between 0 and 1 ($\kappa_C \in [0, 1]$). The following criteria are sometimes used to qualify the agreement:

$$0.80 < \kappa_C \leq 1.00 \text{ High agreement}$$
$$0.60 < \kappa_C \leq 0.80 \text{ Substantial agreement}$$
$$0.40 < \kappa_C \leq 0.60 \text{ Moderate agreement}$$
$$0.20 < \kappa_C \leq 0.40 \text{ Fair agreement}$$
$$0 < \kappa_C \leq 0.20 \text{ Poor agreement.}$$

The kappa statistic has been applied to different applications, including medical and engineering settings. Although it was typically developed for nominal data, it has been applied to ordinal random variables as well, including binary random variables.

For ordinal random variables there exists a weighted version of Cohen's kappa. When X and Y are further apart, i.e., $|X - Y| > 1$, this seems to be a more serious misclassification then when X and Y just differ one class, i.e., $|X - Y| = 1$. This weighted version would address this issue of larger differences in misclassification. However, it is outside the scope of our book.

6.6 Estimators of Measures of Association

In Chap. 5 we discussed moment and maximum likelihood estimation for the parameters of univariate CDFs. Both approaches exist also for the estimation of the parameters of bivariate and multivariate CDFs, including estimation of the dependency parameters. However, MLE of the dependency parameter can be quite cumbersome when we deal with some of the more general families of joint CDFs (e.g., the FGM CDFs, the Fréchet CDFs, and the mixtures of CDFs). Thus, in this section we will focus on how we can estimate the dependency between X and Y using mostly moment estimators, knowing that this may not always be optimal, it does provide some insight in the association of X and Y.

6.6.1 Pearson's Correlation Coefficient

Pearson's correlation coefficient was defined through the covariance of X and Y and the covariance was represented by an expectation of a function of X and Y. This means that Pearson's correlation coefficient represents a population characteristic. Similar to our approach of estimating population means and variances, where we defined a sample variant of the population characteristic, we can also do this for population covariances and correlations.

Now assume that we have observed the pairs of observations (X_1, Y_1), (X_2, Y_2), $\ldots$, (X_n, Y_n) on n units, being i.i.d. $(X_i, Y_i) \sim F_{XY}$. This means that we assume that $(X_1, Y_1), (X_2, Y_2), \ldots, (X_n, Y_n)$ are n independent copies of (X, Y), the same way as we assumed for a single random variable. Since the pairs are identically distributed, the covariance $\text{COV}(X_i, Y_i)$ is equal to $\text{COV}(X, Y)$ for all units i.

The sample covariance, S_{XY}, is then defined by

$$S_{XY} = \frac{1}{n-1} \sum_{i=1}^{n} (X_i - \bar{X})(Y_i - \bar{Y}),$$

with $\bar{X}$ and $\bar{Y}$ the sample average of variable X and Y, respectively. The sample covariance is an unbiased estimator for the population covariance $\text{COV}(X, Y)$. This follows from the following observation:

$$\mathbb{E}(S_{XY}) = \frac{1}{n-1} \sum_{i=1}^{n} \mathbb{E}[(X_i - \bar{X})(Y_i - \bar{Y})]$$

$$= \frac{1}{n-1} \sum_{i=1}^{n} \mathbb{E}[(X_i - \mu_X + \mu_X - \bar{X})(Y_i - \mu_Y + \mu_Y - \bar{Y})]$$

$$= \frac{1}{n-1} \sum_{i=1}^{n} \mathbb{E}[(X_i - \mu_X)(Y_i - \mu_Y) - (X_i - \mu_X)(\bar{Y} - \mu_Y)$$
$$- (\bar{X} - \mu_X)(Y_i - \mu_Y) + (\bar{X} - \mu_X)(\bar{Y} - \mu_Y)]$$

$$= \frac{n}{n-1} \text{COV}(X, Y) - 2\frac{1}{n(n-1)} \sum_{i=1}^{n} \mathbb{E}[(X_i - \mu_X)(Y_i - \mu_Y)]$$

$$+ \frac{1}{n(n-1)} \sum_{i=1}^{n} \mathbb{E}[(X_i - \mu_X)(Y_i - \mu_Y)]$$

$$= \text{COV}(X, Y).$$

Note that we have used the independence between pairs of variables so that $\mathbb{E}[(X_i - \mu_X)(Y_j - \mu_Y)] = \mathbb{E}(X_j - \mu_X)\mathbb{E}(Y_i - \mu_Y) = 0$, when $i \neq j$.

The sample version of Pearson's correlation, r_P, also called the *product-moment correlation coefficient* or sample correlation coefficient, can now be computed by substituting the sample covariance and the sample variances S_X^2 and S_Y^2 in the definition of ρ_P. Thus an estimate of Pearson's correlation coefficient is

$$r_P = \frac{S_{XY}}{S_X S_Y}. \tag{6.10}$$

By rewriting r_P, we can calculate the product-moment correlation coefficient in a number of equivalent ways. Here are a few:

$$r_P = \frac{S_{XY}}{S_X S_Y}$$

$$= \frac{\sum_{i=1}^{n}(X_i - \bar{X})(Y_i - \bar{Y})}{(n-1)S_X S_Y}$$

$$= \frac{\sum_{i=1}^{n}(X_i - \bar{X})(Y_i - \bar{Y})}{\sqrt{\sum_{i=1}^{n}(X_i - \bar{X})^2 \sum_{i=1}^{n}(Y_i - \bar{Y})^2}}$$

$$= \frac{\sum_{i=1}^{n} X_i Y_i - n\bar{X}\bar{Y}}{(n-1)S_X S_Y}$$

$$= \frac{n \sum_{i=1}^{n} X_i Y_i - \sum_{i=1}^{n} X_i \sum_{i=1}^{n} Y_i}{\sqrt{n \sum_{i=1}^{n} X_i^2 - (\sum_{i=1}^{n} X_i)^2} \sqrt{n \sum_{i=1}^{n} Y_i^2 - (\sum_{i=1}^{n} Y_i)^2}}.$$

The product-moment correlation r_P is typically not unbiased, which means that $\mathbb{E}(r_P) \neq \rho_P$. The reason is that both the numerator and the denominator are random variables and the expectation of a ratio is not equal to the ratio of the expectations. Furthermore, the expectation of the numerator in r_P is unbiased for the numerator in the definition of Pearson's correlation ρ_P. Calculating the bias is difficult or even impossible. However, there are a few things known about the estimator r_P. First of all, if $\rho_P = 0$ and the pairs (X_i, Y_i) are i.i.d. bivariate normally distributed, the distribution function of r_P is related to the t-distribution. Under these conditions, it can be shown that the CDF of $r_P \sqrt{n-2}/\sqrt{1-r_P}$ has the CDF of a t-distribution with $n-2$ degrees of freedom. For any value of ρ_P, but still assuming that the pairs (X_i, Y_i) are bivariate normally distributed, Sir Ronald Fisher determined the exact PDF of r_P (Fisher 1915), and demonstrated that $z_{r_P} = 0.5[\log(1 + r_P) - \log(1 - r_P)]$ is approximately normally distributed with mean $0.5[\log(1 + \rho_P) - \log(1 - \rho_P)]$ and variance $1/(n-3)$ (Fisher et al. 1921). This transformation is referred to as the *Fisher z-transformation*. Finally, the asymptotic distribution of r_P is normal when the pairs (X_i, Y_i) are i.i.d. F_{XY} with finite fourth central moments. More specifically, $\sqrt{n}(r_P - \rho_P) \sim \mathcal{N}(0, n(1 - \rho_P^2)^2/(n-3))$, see Bonett and Wright (2000).

Knowledge of the distribution function of the product-moment correlation coefficient can be used to create confidence intervals. Under the assumption of normality, it is common to use the Fisher z-transformation and calculate the $100\%(1 - \alpha)$ confidence interval by

$$\left(\frac{1}{2} \log \left(\frac{1 + r_P}{1 - r_P} \right) - \frac{z_{1-\alpha/2}}{\sqrt{n-3}}, \frac{1}{2} \log \left(\frac{1 + r_P}{1 - r_P} \right) + \frac{z_{1-\alpha/2}}{\sqrt{n-3}} \right],$$

with z_{1-p} the pth upper quantile of the standard normal distribution function. These limits can then be transformed back to the original scale using the inverse transformation $[\exp\{2x\} - 1]/[\exp\{2x\} + 1]$ of the Fisher z-transformation. Thus the confidence interval in the original scale is

$$\left(\frac{\exp\{2[z_{r_P} - z_{1-\alpha/2}/\sqrt{n-3}]\} - 1}{\exp\{2[z_{r_P} - z_{1-\alpha/2}/\sqrt{n-3}]\} + 1}, \frac{\exp\{2[z_{r_P} + z_{1-\alpha/2}/\sqrt{n-3}]\} - 1}{\exp\{2[z_{r_P} + z_{1-\alpha/2}/\sqrt{n-3}]\} + 1} \right]$$

If the observed data is not normal and the sample size is relatively large, the asymptotic confidence interval from Sect. 5.3.5 may be applied directly on the product-moment estimator, i.e.,

$$\left(r_P - \frac{z_{1-\alpha/2}(1 - r_P^2)}{\sqrt{n - 3}}, r_P + \frac{z_{1-\alpha/2}(1 - r_P^2)}{\sqrt{n - 3}} \right].$$

As an alternative to this large sample approach, Fisher z-transformation is also frequently applied when the underlying data is not normally distributed.

6.6.1.1   Estimation of Dependency Parameters

The product-moment correlation can be used to determine information on dependency parameters. For instance, it directly estimates the parameter ρ in the bivariate normal CDF. Additionally, if the CDFs are constructed through sums of random variables $X = W + U$ and $Y = W + V$, the dependency is determined by the variance of W. If this variance is zero, there is no correlation. The product-moment correlation is a direct estimator of what was called the intraclass correlation coefficient (see Eq. 6.9). Together with the sample variances S_X^2 and S_Y^2, which are estimators for $\sigma_X^2 = \sigma_W^2 + \sigma_U^2$ and $\sigma_Y^2 = \sigma_W^2 + \sigma_V^2$, respectively, we may obtain an estimator for the variance σ_W^2 of W, using $r_P S_X S_Y$. In another situation, the bivariate CDF F_{XY} may be part of the FGM family with exponential marginal distribution functions, the dependency parameter α can be estimated by $r_P \bar{X} \bar{Y}$. Indeed, r_P estimates $\alpha \lambda_X^{-1} \lambda_Y^{-1}$ and the sample averages $\bar{X}$ and $\bar{Y}$ estimate the parameters λ_X^{-1} and λ_Y^{-1}, respectively.

6.6.2   Kendall's Tau Correlation Coefficient

An estimator of Kendall's tau correlation can be defined by

$$r_K = \frac{1}{n(n - 1)} \sum_{i=1}^{n} \sum_{j=1}^{n} \mathrm{sgn}(X_j - X_i)\mathrm{sgn}(Y_j - Y_i)$$

$$= \frac{2}{n(n - 1)} \sum_{i=1}^{n-1} \sum_{j=i+1}^{n} \mathrm{sgn}(X_j - X_i)\mathrm{sgn}(Y_j - Y_i),$$

with the sign function $\mathrm{sgn}(x)$ defined by

$$\mathrm{sgn}(x) = \begin{cases} 1 \text{ if } x > 0 \\ 0 \text{ if } x = 0 \\ -1 \text{ if } x < 0. \end{cases}$$

Thus the estimator depends only on the signs of $X_j - X_i$ and $Y_j - Y_i$, which implies that the estimator is independent of monotone transformations (increasing or decreasing functions) of the data. If we apply the estimator on $(\psi_1(X_1), \psi_2(Y_1))$, $(\psi_1(X_2), \psi_2(Y_2)), \ldots, (\psi_1(X_n), \psi_2(Y_n))$, when both ψ_1 and ψ_2 are increasing or both decreasing functions, we obtain the exact same estimator. This is easily seen, since the sign just compares the sizes without considering the differences in sizes.

Kendall showed that the distribution function of this estimator is approximately normal when sample size converges to infinity. Studying its moments is difficult, but the exact variance of r_K has been determined at $[2(2n + 5)]/[3n(n - 1)]$ (Kendall 1938), when the pairs are uncorrelated (i.e., $\tau_K = 0$). Furthermore, the variance of r_K is bounded from above with $\mathsf{VAR}(r_K) \leq 2(1 - \tau_K^2)/n$ (Long and Cliff 1997).

Several confidence intervals have been provided of which some are more easily calculated than others. Here we provide the $100\%(1 - \alpha)$ confidence interval based on the Fisher z-transformation. Although it may not be recommended for small sample sizes, for larger sample sizes it does provide good coverage (i.e., the confidence level of the interval is close to the intended level of $100\%(1 - \alpha)$) and it is relatively easy to calculate. The variance of r_K in the transformed Fisher z scale has been determined at $0.437/(n - 4)$ (Fieller et al. 1957). The $100\%(1 - \alpha)$ confidence interval is then determined by

$$\left(z_{r_K} - z_{1-\alpha/2}\sqrt{0.437/(n - 4)}, \ z_{r_K} + z_{1-\alpha/2}\sqrt{0.437/(n - 4)} \right],$$

with $z_{r_K} = 0.5[\log(1 + r_K) - \log(1 - r_K)]$ and z_p the pth quantile of the standard normal distribution. In the original scale, the confidence interval becomes equal to

$$\left(\frac{\exp(2[z_{r_K} - z_{1-\alpha/2}\sqrt{0.437/(n - 4)}]) - 1}{\exp(2[z_{r_K} - z_{1-\alpha/2}\sqrt{0.437/(n - 4)}]) + 1}, \ \frac{\exp(2[z_{r_K} + z_{1-\alpha/2}\sqrt{0.437/(n - 4)}]) - 1}{\exp(2[z_{r_K} + z_{1-\alpha/2}\sqrt{0.437/(n - 4)}]) + 1} \right]$$

Long and Cliff (1997) have suggested using a quantile from the t-distribution instead of the normal quantile $z_{1-\alpha/2}$, but they also showed that their Fisher z-transformed confidence interval was slightly conservative (gave confidence levels slightly higher than $100\%(1 - \alpha)$) for smaller sample sizes. As for larger sample sizes there will be hardly any difference between normal and t-quantiles, we believe that our choice would be more appropriate.

6.6.2.1 Estimation of Dependency Parameters

The estimator of Kendall's tau can be easily used to estimate the dependency parameter of the FGM and Fréchet family of distribution functions. The dependency parameter α of the FGM family can be estimated by $\hat{\alpha} = 9r_K/2$ and λ of the Fréchet family can be estimated by $\hat{\lambda} = [3r_K + 1]/2$. The confidence intervals on these parameters can be obtained by transforming the confidence limits in the same way. It should be noted that when these estimates are outside the range of the dependency parameters (i.e., $\hat{\alpha} \notin [-1, 1]$ and $\hat{\lambda} \notin [0, 1]$), the data may not support these families of distributions.

6.6.3   Spearman's Rho Correlation Coefficient

The estimator for Spearman's rho ρ_S is defined by the product-moment correlation coefficient on the ranks of the pairs of observations. It is referred to as *Spearman's rank correlation*. If we assume that we observe the pairs of random variables (X_1, Y_1), $(X_2, Y_2), \ldots, (X_n, Y_n)$, we can define the ranks for the x and y coordinates separately. The rank R_k^X of X_k is the position of X_k in the ordered values $X_{(1)}, X_{(2)}, \ldots, X_{(n)}$ from small to large. For instance, if we have observed five observations $x_1 = 5$, $x_2 = 1, x_3 = 7, x_4 = 2$, and $x_5 = 4$, then the observed ranks will be equal to $r_1^X = 4$, $r_2^X = 1, r_3^X = 5, r_4^X = 2$, and $r_5^X = 3$.[7] Note that if the same value occurs multiple times they all receive the same rank. For instance, if there were a sixth observation, $x_6 = 2$, then x_4 and x_6 both have the same value 2. This is called a *tie*. The ranks would now be given by $r_1^X = 5, r_2^X = 1, r_3^X = 6, r_4^X = 2.5, r_5^X = 4$, and $r_6^X = 2.5$. Since the ranks of x_4 and x_6 should be 2 and 3 or 3 and 2, they both get the average rank of 2.5. If we additionally define the rank R_k^Y for Y_k among the random variables $Y_{(1)}, Y_{(2)}, \ldots, Y_{(n)}$, we have translated the pair (X_k, Y_k) to a pair of ranks (R_k^X, R_k^Y). Then Spearman's rank correlation coefficient is defined by

$$
r_S = \frac{\sum_{i=1}^{n}(R_i^X - \bar{R}^X)(R_i^Y - \bar{R}^Y)}{\sqrt{\sum_{i=1}^{n}(R_i^X - \bar{R}^X)^2 \sum_{i=1}^{n}(R_i^Y - \bar{R}^Y)^2}},
$$

with $\bar{R}^X$ and $\bar{R}^Y$ the average ranks for the X and Y variables, i.e., $\bar{R}^X = \sum_{k=1}^{n} R_k^X/n$ and $\bar{R}^Y = \sum_{k=1}^{n} R_k^Y/n$. As the total number of ranks in a sample of n observations is equal to $n(n+1)/2$, the average ranks $\bar{R}^X$ and $\bar{R}^Y$ are equal to $\bar{R}^X = \bar{R}^Y = (n+1)/2$. Note that Spearman's rank correlation coefficient can also be written in a different form:

$$
r_S = 1 - 6 \sum_{i=1}^{n} \frac{(R_i^Y - R_i^X)^2}{n^3 - n}.
$$

Since Spearman's rank correlation depends on the ranks of the variables x and y, the estimator is independent of monotonic transformations (increasing or decreasing functions) of the data. Thus Spearman rho's estimator r_S on (X_1, Y_1), $(X_2, Y_2), \ldots, (X_n, Y_n)$ is exactly the same as Spearman rho's estimator r_S on the transformed pairs $(\psi_1(X_1), \psi_2(Y_1)), (\psi_1(X_2), \psi_2(Y_2)), \ldots, (\psi_1(X_n), \psi_2(Y_n))$, when ψ_1 and ψ_2 are both increasing or both decreasing functions. This is easily seen, since ranks only determine the position in the set of ordered values and do not take into account the levels, other than for comparison.

The CDF and some of its characteristics (like the moments) of r_S have been studied, mostly under the assumption of bivariate normal CDFs, but the distribution

---

[7] Thus, x_2 has the smallest value, and thus receives rank 1, while x_3 has the largest value, and thus receives rank 5.

function of r_S is not easy to determine in general. The reason is that the underlying joint distribution function of the pairs of random variables plays an important role in the distribution function of the ranks. To illustrate the complexity, consider for instance, $\Pr(R_n^X = n)$, which is given by

$$
\begin{aligned}
\Pr(R_n^X = n) &= \Pr(X_n = \max\{X_1, X_2, \ldots, X_n\}) \\
&= \Pr(X_n \geq X_1, X_n \geq X_2, \ldots, X_n \geq X_{n-1}) \\
&= \int_{-\infty}^{\infty} \Pr(X_n \geq X_1, X_n \geq X_2, \ldots, X_n \geq X_{n-1} | X_n = x) f_{X_n}(x) dx \\
&= \int_{-\infty}^{\infty} \Pr(X_1 \leq x, X_2 \leq x, \ldots, X_{n-1} \leq x | X_n = x) f_{X_n}(x) dx \\
&= \int_{-\infty}^{\infty} F_X^{n-1}(x) f_{X_n}(x) dx,
\end{aligned}
$$

with $f_{X_n} = f_X$, since $X_1, X_2, \ldots, X_n$ are i.i.d. with CDF F_X.

Although we may not prefer the use of Spearman's correlation coefficient over Pearson's product-moment estimator (see Sect. 6.6.4) under the assumption of normally and independently distributed pairs $(X_1, Y_1), (X_2, Y_2), \ldots, (X_n, Y_n)$, the mean and variance of r_S have been established under this setting (see Fieller and Pearson 1961). The mean is given by

$$
\rho_S = \mathbb{E}(r_S) = \frac{6}{(n+1)\pi} \left(\arcsin(\rho) + (n-2) \arcsin(\rho/2) \right) \approx \frac{6}{\pi} \arcsin(\rho/2).
$$

Thus the parameter ρ of the bivariate normal distribution can now be estimated by $\hat{\rho} = 2 \sin(\pi r_S / 6)$. The variance of r_S is a long series of powers of the correlation coefficient ρ, but we will provide the series up to power ρ^6. The variance is then approximately

$$
\text{VAR}(r_S) \approx \frac{1}{n} \left(1 - 1.1563465\rho^2 + 0.304743\rho^4 + 0.155286\rho^6 \right).
$$

Taking the square root of this variance and substituting $\hat{\rho}$ for ρ would provide an estimator of the standard error on the estimator r_S under the assumption of normality.

Different approaches of constructing $100\%(1-\alpha)$ confidence intervals on ρ_S through r_S has been suggested for $(X_1, Y_1), (X_2, Y_2), \ldots, (X_n, Y_n)$ being i.i.d. with CDF F_{XY}. First of all, there exists an asymptotic confidence interval applied directly on r_S, secondly there exists a confidence interval based on the t-distribution, and finally there exist confidence intervals based on the Fisher z-transformation. Since the asymptotic confidence interval on r_S directly seems to be liberal (the coverage of the interval is often lower than $100\%(1-\alpha)$) and the confidence interval based on the t-distribution is conservative (the coverage is typically larger than $100\%(1-\alpha)$), we only focus on the Fisher z-transformation, since this approach seems to have produced appropriate coverage: close to $100\%(1 - \alpha)$ for several settings (Caruso and Cliff 1997).

The Fisher z-transformation of r_S uses normal confidence intervals with a variance $S_{r_S}^2$ of r_S in this transformed scale. Different suggestions for this variance have been proposed: $S_{r_S}^2 = 1.06/(n - 3)$ (Fieller and Pearson 1961), $S_{r_S}^2 = (1 + r_S^2/2)/(n - 3)$ (Bonett and Wright 2000), and $S_{r_S}^2 = (n - 2)^{-1} + |z_{r_S}|/[6n + 4\sqrt{n}]$ (Caruso and Cliff 1997), with z_{r_S} the Fisher z-transformation of r_S, i.e., $z_{r_S} = 0.5[\log(1 + r_S) - \log(1 - r_S)]$. Based on the Fisher z-transformation, a $100\%(1 - \alpha)$ confidence interval is provided by

$$\left(z_{r_S} - z_{1-\alpha/2} S_{r_S}, z_{r_S} + z_{1-\alpha/2} S_{r_S} \right].$$

Taking the inverse Fisher z-transformation on these limits, a $100\%(1 - \alpha)$ confidence interval for ρ_S is constructed. This gives confidence limits equal to

$$\left(\frac{\exp(2[z_{r_S} - z_{1-\alpha/2} S_{r_S}]) - 1}{\exp(2[z_{r_S} - z_{1-\alpha/2} S_{r_S}]) + 1}, \frac{\exp(2[z_{r_S} + z_{1-\alpha/2} S_{r_S}]) - 1}{\exp(2[z_{r_S} + z_{1-\alpha/2} S_{r_S}]) + 1} \right].$$

For correlation coefficients less than or equal to 0.5, simulation studies with normally distributed random variables show that the Fisher z-transformed confidence intervals with the three different standard errors behave very similar (Bonett and Wright 2000). The coverage probabilities are around 95%, irrespective of the use of the three standard errors. When the correlation coefficient is larger than 0.5, Fieller and Pearson's standard error underestimates the variability of the estimator of Spearman's rho. Thus the coverage probability of the confidence interval with Fieller and Pearson's standard error is smaller than 95%. For correlation coefficients larger than 0.7, the standard error of Caruso and Cliff also starts to underestimate the variability of the estimator of Spearman's rho. For the standard error of the approach of Bonett and Wright underestimation occurs for correlation coefficients larger than 0.9. Thus, for normally distributed data, Bonett and Wright's approach is most appropriate, but less is known when the bivariate distribution is deviating from normality.

6.6.3.1 Estimation of Dependency Parameters

The estimator of Spearman's rho can be easily used to estimate the dependency parameter of the FGM and Fréchet family of distribution functions. The dependency parameter α of the FGM family can be estimated by $\hat{\alpha} = 3r_S$ and λ of the Fréchet family can be estimated by $\hat{\lambda} = [3r_S + 1]/2$. Confidence intervals can be constructed through confidence intervals on ρ_S, using the same functions. It should be noted that when these estimates are outside the range of the dependency parameters (i.e., $\hat{\alpha} \notin [-1, 1]$ and $\hat{\lambda} \notin [0, 1)$), the data may not support these families of distributions.

6.6.4 *Should We Use Pearson's Rho, Spearman's Rho or Kendall's Tau Correlation?*

Now that we have the availability of three measures of correlation (or concordance), the question arises: which one should we use on the data? This has been a question for a long time and different opinions exist in the literature, all providing arguments for choosing just one of them. In a way this is surprising, since they all describe a different aspect of the population and the difference can be quite large depending on the family of CDFs. Thus, if we know what measure we would like to quantify, the choice is easy. Nevertheless, here we will discuss a few facts and we hope to provide more insight into which one to use if there is no clear preference of choosing one of the population parameters. It should be mentioned upfront that there always exist exceptions, but general trends can be given.

The product-moment correlation provides a measure that is particularly suitable for *linear* associations between variables. To illustrate this we would like to show the famous example of Anscombe's quartet in Fig. 6.2. For each of the four figures the product-moment estimator is equal to ≈ 0.8. The figures clearly show that the "co-relation" or dependency between X and Y is quite different across the four figures, even though the product-moment correlation is 0.8 for all figures.

The product-moment estimator would be most suitable for the top left figure in Fig. 6.2. The reason is that these data seem to be close to bivariate normally

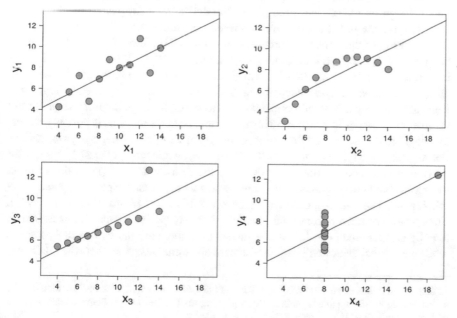

Fig. 6.2 Anscombe's quartet: an illustration of four different cases that all have the same product-moment correlation

distributed data and Pearson's correlation coefficient is the dependency parameter. Thus when the data are bivariate normally distributed, or when transformations can be applied that would make the data approximately bivariate normal, we suggest applying the product-moment estimator on the (transformed) data. It has been shown that Pearson's estimator r_P is more efficient (i.e., has a smaller asymptotic variance) than Spearman's estimator (Borkowf 2002).

If we do not have normally distributed data and a proper transformation of the data that would lead to approximately bivariate normal data is difficult to find, it is not obvious which of the three options to choose.[8] Often Pearson's product-moment estimator is immediately disqualified, since it is affected by transformations of the data. If none of the transformations leads to the top left figure in Fig. 6.2 it is almost impossible to know which transformation should then be used. Borkowf provides a theoretical example with bivariate uniform distribution functions in which Pearson's product-moment estimator is preferred over Spearman's rank correlation (Borkowf 2002), but for practical data this would be difficult to assess. Furthermore, Pearson's product-moment estimator is also sensitive to outliers (illustrated in the bottom left figure in Fig. 6.2). Spearman's rho and Kendall's tau estimators are less sensitive and may prevent a long investigation into the outlier to determine whether it should be removed or changed. Thus for non-normal bivariate data (that is difficult to transform into normally dsitributed data) it is probably best not to use Pearson's product-moment estimator.

When we have to choose between Spearman's rank or Kendall's tau estimator, is there a clear preference? The answer is no, but there are a few observations that may help guide the choice. First of all, Kendall's tau estimator has a clear interpretation. It measures the difference between concordant and discordant pairs. The interpretation of Spearman is more difficult, although it is often presented as a robust version of Pearson's product-moment estimator. Secondly, Kendall's tau estimator is computationally more intensive, since it requires the comparisons of pairs with all other pairs, which is not needed with Spearman's rank. Thus on really large data, Kendall's tau may lead to numerical issues. However, for large datasets Kendall's tau estimator is more efficient than Spearman's rank correlation. This is possibly easiest to recognize if we compare the variances of r_K and r_S in the Fisher z-transformation. For Kendall's tau this variance was $0.437/(n - 4)$, while it was at least $1/(n - 3)$ for Spearman's rank correlation.[9] Finally, when the data contain ties (i.e., values among the x variable and/or values among the y variable are equal), Spearman's rank correlation has a smaller standard error than Kendall's tau estimator (Puth et al. 2015). Thus when we have non-normal bivariate data, we prefer Kendall's tau estimator over Spearman when the data is continuous (and there are hardly any ties present), while we prefer Spearman's rank when there are some ties. Note that when there are

---

[8] For the two figures at the bottom in Fig. 6.2 it is not directly obvious which transformation may be suitable to create a figure that is similar to the top left figure. For the top right figure a transformation of Y to $\sqrt{Y}$ will most likely create a figure close to the top left.

[9] We should be careful here to look solely at the standard error, since Kendall's tau and Spearman's rho do not measure the same population parameter and Kendall's tau is often smaller than Spearman's rho.

a lot of ties, several adaptations of Kendall's tau exist to deal with the ties, and these are preferred over Kendall's tau and Spearman's rank.

Despite our guidance, it should be realized that quantifying the dependency of X and Y by just one measure seems to provide only limited information on its dependency and you should not trust just one number blindly. We recommend always visualizing your data and using this to either decide what correlation measure is most suitable directly on the data or to otherwise determine possible transformations to come to more linear dependencies in the transformed scale. It should be noted that in some cases it may even be better to try to capture the full joint distribution function F_{XY} (although this is outside the scope of this book). Indeed, the dependency is fully captured by F_{XY}, although it may be more cumbersome to demonstrate that the estimated joint CDF does describe the observed data properly.

6.6.5 Cohen's Kappa Statistic

If we collect pairs (X_1, Y_1), (X_2, Y_2), ..., (X_n, Y_n) of i.i.d. bivariate random variables as copies of (X, Y) that can take their values in $\{1, 2, \ldots, K\}$, the data can be summarized in a $K \times K$ contingency table (similar to the 2×2 contingency table in Table 3.1). In cell $X = x$ and $Y = y$ the number of pairs, say N_{xy}, from (X_1, Y_1), (X_2, Y_2), ..., (X_n, Y_n) with the combination (x, y) is reported. Thus the data is summarized as

$$N_{xy} = \sum_{i=1}^{n} 1_{\{x\}}(X_i) 1_{\{y\}}(Y_i),$$

with $1_A(x)$ the indicator variable that is equal to 1 if $x \in A$ and 0 otherwise. It should be noted that the sum of all numbers in the contingency table should add up to n: $\sum_{x=1}^{K} \sum_{y=1}^{K} N_{xy} = n$. We may also add up the numbers by row or by column, leading to

$$N_{x\cdot} = \sum_{y=1}^{K} N_{xy} \quad \text{and} \quad N_{\cdot y} = \sum_{x=1}^{K} N_{xy}.$$

The distribution function of N_{xy} is binomial with parameters n and $p_{xy} = \Pr(X = x, Y = y)$. Moreover, the multivariate PMF for $N_{11}, N_{12}, \ldots, N_{1K}, N_{21}, N_{22}, \ldots, N_{2K}$, ..., $N_{K1}, N_{K2}, \ldots, N_{KK}$ has a multinomial distribution function (see Chap. 4) given by

$$\Pr(N_{11} = n_{11}, N_{12} = n_{12}, \ldots, N_{KK} = n_{KK}) = n! \prod_{x=1}^{K} \prod_{y=1}^{K} \left(\frac{p_{xy}^{n_{xy}}}{n_{xy}!} \right).$$

Cohen's Kappa statistic is now estimated by

$$\hat{\kappa}_C = \frac{\hat{p}_O - \hat{p}_E}{1 - \hat{p}_E}, \quad \text{with} \quad \hat{p}_O = \frac{1}{n}\sum_{k=1}^{K} N_{kk} \quad \text{and} \quad \hat{p}_E = \frac{1}{n^2}\sum_{k=1}^{K} N_{k\cdot} N_{\cdot k}.$$

Based on the multinomial distribution, the standard error for the estimator of Cohen's kappa statistics can be approximated. The standard error is defined through

$$\text{VAR}(\hat{\kappa}_C) \approx \frac{p_O(1 - p_O)}{n(1 - p_E)^2}.$$

Based on this standard error an asymptotic $100\%(1 - \alpha)$ confidence interval can be created. This interval is given by

$$\left(\hat{\kappa}_C - z_{1-\alpha/2}\frac{\sqrt{\hat{p}_O(1 - \hat{p}_O)}}{(1 - \hat{p}_E)\sqrt{n}}, \hat{\kappa}_C + z_{1-\alpha/2}\frac{\sqrt{\hat{p}_O(1 - \hat{p}_O)}}{(1 - \hat{p}_E)\sqrt{n}} \right],$$

with z_p the pth quantile of the standard normal distribution function.

The simplicity of Cohen's kappa statistic has made it very popular in all kinds of sciences to quantify agreement. However, Cohen's kappa statistic should not be used when particular outcomes of X and Y are rare. Thus when the probability $\Pr(X = k)$ or $\Pr(Y = k)$ is small compared to the other levels $\{1, 2, .., k - 1, k + 1, \ldots, K\}$, the statistic may underestimate the agreement. To illustrate this we assume that $K = 2$ and provide two 2×2 contingency tables (see Table 6.1).

Cohen's kappa statistic is approximately equal to 0.625 and 0.350 for the first and second table, respectively. Thus the second table provides a fair agreement, while the first table gives a moderate agreement. However, the probability of agreement (p_O) is larger in the second table (0.91) than in the first table (0.85). Thus, there is less misclassification in the second table than in the first table, but agreement in the second table is seriously lower than in the first table. This is caused by the imbalance in the probability of observing outcomes 1 and 2. Thus, Cohen's kappa works best when each outcome has approximately the same probability of occurrence, or is at least not strongly imbalanced.

Table 6.1 Examples of Cohen's kappa statistic

	$Y = 1$	$Y = 2$	Total		$Y = 1$	$Y = 2$	Total
$X = 1$	20	5	25	$X = 1$	3	3	6
$X = 2$	10	65	75	$X = 2$	6	88	94
Total	30	70	100	Total	9	91	100

6.6.6 Risk Difference, Relative Risk, and Odds Ratio

Although we have already discussed risk difference, relative risk, and odds ratio in Chap. 3, we would like to recall them here so that we can also report the calculation of confidence intervals. We assume that the bivariate binary data $(X_i, Y_i) \in \{0, 1\} \times \{0, 1\}$ can be captured by the summary statistics

$$N_{xy} = \sum_{i=1}^{n} 1_{\{x\}}(X_i) 1_{\{y\}}(Y_i), \quad \text{for} \quad (x, y) \in \{0, 1\} \times \{0, 1\}.$$

Note that we have already used this type of notation for Cohen's kappa statistic, but with binary data it is more convenient to use values 0 and 1 for X and Y, instead of using the values 1 and 2.

If we assume that Y represents some kind of outcome and X represents some kind of exposure, with $X = 0$ the reference group, the estimators of the risk difference ($\hat{RD}$), relative risk ($\hat{RR}$), and odds ratio ($\hat{OR}$) are defined by (see also Chap. 3):

$$\hat{RD} = \frac{N_{11}}{N_{11} + N_{10}} - \frac{N_{01}}{N_{01} + N_{00}} = \frac{N_{11}N_{00} - N_{10}N_{01}}{(N_{11} + N_{10})(N_{01} + N_{00})}$$

$$\hat{RR} = \frac{N_{11}(N_{01} + N_{00})}{N_{01}(N_{11} + N_{10})}$$

$$\hat{OR} = \frac{N_{11}N_{00}}{N_{10}N_{01}}$$

The statistical literature contains many different approaches to constructing confidence intervals. It is not our intention to provide them all, but to provide one approach for each measure of association that is relatively easy to calculate and also has a good performance in many settings (i.e., the coverage of the confidence interval containing the population parameter is close to the intended confidence level $100\%(1 - \alpha)$). For the remainder it may be useful to introduce the notation $\hat{p}_0 = N_{01}/N_0$. and $\hat{p}_1 = N_{11}/N_1$. as the estimators of the probability of the event $Y = 1$ in group $X = 0$ and $X = 1$, respectively, with $N_{x.} = N_{x1} + N_{x0}$.

For the risk difference we provide the method of Wilson (Newcombe 1998). This method first calculates $100\%(1 - \alpha)$ confidence limits on the proportions of event $Y = 1$ for the two groups $X = 0$ and $X = 1$ separately and then uses these confidence limits in the confidence interval on the risk difference. The two separate $100\%(1 - \alpha)$ confidence intervals are given by

$$L_x = \frac{2N_{x\cdot}\hat{p}_x + z_{1-\alpha/2}^2 - z_{1-\alpha/2}\sqrt{4N_{x\cdot}\hat{p}_x(1 - \hat{p}_x) + z_{1-\alpha/2}^2}}{2(N_{x\cdot} + z_{1-\alpha/2}^2)},$$

$$U_x = \frac{2N_{x\cdot}\hat{p}_x + z_{1-\alpha/2}^2 + z_{1-\alpha/2}\sqrt{4N_{x\cdot}\hat{p}_x(1 - \hat{p}_x) + z_{1-\alpha/2}^2}}{2(N_{x\cdot} + z_{1-\alpha/2}^2)},$$

for $x \in \{0, 1\}$ and z_p the pth quantile of the standard normal distribution function. The $100\%(1 - \alpha)$ confidence interval on the risk difference is then given by

$$\left(\hat{RD} - \sqrt{(\hat{p}_1 - L_1)^2 + (U_0 - \hat{p}_0)^2}, \hat{RD} + \sqrt{(U_1 - \hat{p}_1)^2 + (\hat{p}_0 - L_0)^2}\right].$$

For the relative risk, the confidence interval is calculated in the logarithmic scale. The estimated standard error of the logarithmic transformed relative risk is given by

$$\hat{SE}_{RR} = \sqrt{\frac{1 - \hat{p}_0}{N_{0\cdot}\hat{p}_0} + \frac{1 - \hat{p}_1}{N_{1\cdot}\hat{p}_1}}.$$

Thus the asymptotic $100\%(1 - \alpha)$ confidence interval on the logarithmic transformation of the relative risk is given by

$$(L_{RR}, U_{RR}] = \left(\log(\hat{RR}) - z_{1-\alpha/2}\hat{SE}_{RR}, \log(\hat{RR}) + z_{1-\alpha/2}\hat{SE}_{RR}\right].$$

The $100\%(1 - \alpha)$ confidence interval on the relative risk is then calculated by transforming these confidence limits back to the original scale using $(\exp(L_{RR}), \exp(U_{RR})]$.

The confidence interval on the odds ratio is also calculated in the logarithmic scale using the standard error

$$\hat{SE}_{OR} = \sqrt{N_{00}^{-1} + N_{01}^{-1} + N_{10}^{-1} + N_{11}^{-1}}.$$

The $100\%(1 - \alpha)$ confidence interval on the odds ratio in the logarithmic scale is given by

$$(L_{OR}, U_{OR}] = \left(\log(\hat{OR}) - z_{1-\alpha/2}\hat{SE}_{OR}, \log(\hat{OR}) + z_{1-\alpha/2}\hat{SE}_{OR}\right].$$

Thus the $100\%(1 - \alpha)$ confidence interval on the odds ratio is then calculated by transforming these confidence limits back to the original scale using $(\exp(L_{OR}), \exp(U_{OR})]$.

6.7 Other Sample Statistics for Association

Chapter 3 presented three measures of association (risk difference, relative risk, and odds ratio) that were typically developed for 2×2 contingency tables. In Sect. 6.5 we presented three correlation measures (Pearson's, Spearman's, and Kendall's correlation) that are most suitable for interval and ratio data. We also introduced Cohen's kappa statistic for measuring agreement in contingency tables. These measures were all formulated in terms of population parameters and they capture some part (or under certain settings all) of the dependency. However, there are many more measures of association and this section will provide a few of them that have also been used in practice. For these measures the underlying population parameters are less known or have not been emphasized that much in the literature as the ones we already discussed.

We have organized this section by the different types of categorical data: nominal, ordinal, and binary. The measures for nominal and ordinal data also apply to binary settings, but we discuss a few measures that were invented only for binary data. For continuous (interval and ratio) data, we may resort to either one of the three different correlation measures discussed in the previous section (see also the discussion on its use in Sect. 6.6.4).

6.7.1 Nominal Association Statistics

This section discusses statistics that are particularly useful for quantifying the dependency between two nominal random variables X and Y. They measure the departure from dependency and they are all functions of Pearson's *chi-square statistic*. Although they can be applied to nominal variables, they have also been used for ordinal and binary random variables X and Y. In this setting we assume that X can take its value in $\{1, 2, \ldots, K\}$ and Y can take its value in $\{1, 2, \ldots, M\}$ and we assume that $(X_1, Y_1), (X_2, Y_2), \ldots, (X_n, Y_n)$ are i.i.d. with CDF F_{XY}.[10] The data can then be summarized by

$$N_{xy} = \sum_{i=1}^{n} 1_{\{x\}}(X_i) 1_{\{y\}}(Y_i), \quad \text{with} \quad (x, y) \in \{1, 2, \ldots, K\} \times \{1, 2, \ldots, M\}$$

The numbers N_{xy} represent the frequencies of the cells in a $K \times M$ contingency table, with K rows and M columns. Note that the number of levels (K and M) for X and Y may now be different. For instance, if we wish to study the dependency between gender (boys and girls) and their favourite subject at school (mathematics, English, history, etc.) in the high school data on first- and second-year students, we

---

[10] The CDF for i.i.d. nominal bivariate random variables is typically the multinominal distribution.

would have a 2×19 (or a 19×2) contingency table, since the study collected data on 19 different subjects.

Similar to the material on Cohen's kappa statistic, we may also determine the number of pairs (X_i, Y_i) for which the X variable is equal to x, irrespective of the value of Y. This will be denoted by $N_{x.}$. Something similar can be done for the variable Y, leading to $N_{.y}$. They represent the row and column totals in the contingency table:

$$N_{x.} = \sum_{y=1}^{M} N_{xy} \quad \text{and} \quad N_{.y} = \sum_{x=1}^{K} N_{xy}.$$

If X and Y are independent, the joint PMF is the product of the two marginal PMFs: $\Pr(X = x, Y = y) = \Pr(X = x) \Pr(Y = y)$ for all $(x, y) \in \{1, 2, \ldots, K\} \times \{1, 2, \ldots, M\}$. We have seen with Cohen's kappa statistic that the joint PMF $\Pr(X = x, Y = y)$ can be estimated by N_{xy}/n and that the marginal PMFs $\Pr(X = x)$ and $\Pr(Y = y)$ can be estimated by $N_{x.}/n$ and $N_{.y}/n$, respectively. Pearson's chi-square statistic quantifies the difference between the *observed numbers* N_{xy} and their *expected numbers* $N_{x.}N_{.y}/n$ based on independence. Thus a part of Pearson's chi-square statistic is $[N_{xy} - N_{x.}N_{.y}/n]^2$. The square is used to make all differences positive. However, Pearson normalized the squared differences with the expected numbers. The reason is that a small squared difference between N_{xy} and $N_{x.}N_{.y}/n$ when $N_{x.}N_{.y}/n$ is small is not the same as a small squared difference between N_{xy} and $N_{x.}N_{.y}/n$ when $N_{x.}N_{.y}/n$ is large. Thus, to provide a measure for the $K \times M$ contingency table, Pearson added all these normalized squared differences over all cells of the contingency table, leading to Pearson's chi-square statistic

$$\chi_P^2 = \sum_{x=1}^{K} \sum_{y=1}^{M} \frac{[N_{xy} - N_{x.}N_{.y}/n]^2}{N_{x.}N_{.y}/n} = \frac{1}{n} \sum_{x=1}^{K} \sum_{y=1}^{M} \frac{[nN_{xy} - N_{x.}N_{.y}]^2}{N_{x.}N_{.y}}. \tag{6.11}$$

It may be obvious that this method of constructing a statistic is always non-negative ($\chi_P^2 \geq 0$) and that it is equal to zero only when the observed frequencies in the cells of the contingency table are equal to the expected frequencies. Thus the larger the value for Pearson's chi-square statistic, the stronger the association between the two random variables X and Y. Since it calculates the differences between observed frequencies and frequencies under the assumption of independence, Pearson's chi-square represents a measure of *departure from independence*.

The CDF of Pearson's chi-square statistic has been studied and it has been demonstrated that it can be approximated with a chi-square distribution with $(K - 1)(M - 1)$ degrees of freedom when X and Y are independent.[11] That is why it is referred to as a chi-square statistic. Thus the PDF of Pearson's chi-square statistic has the form of the densities visualized in the left figure in Fig. 5.2.

---

[11] If independence is not satisfied, the CDF of Pearson's chi-square statistics is equal to a *non-central* chi-square distribution function. This distribution function has not been discussed in our book and it is outside our scope.

Table 6.2 Observed and expected frequencies for independence between gender and opinion on soccer

	Agree	Disagree	Total
Male	75 (66.7)	25 (33.3)	100
Female	25 (33.3)	25 (16.7)	50
Total	100	50	

To illustrate Pearson's chi-square statistic, we use a 2×2 contingency table where the variables X and Y are gender and their opinion on the statement "I like soccer", respectively. Men and women had to indicate whether they agree or disagree with the statement. Independence between X and Y would indicate that men and women both like soccer in the same ratios. Table 6.2 above gives the observed numbers of 100 males and 50 females. In brackets we have calculated the expected values using $N_{x.}N_{.y}/n$.

Based on the observed and expected results the chi-square statistic becomes: $\chi_P^2 = 1.042 + 2.083 + 2.083 + 4.167 = 9.375$. This value is quite large, if we consider that the statistic would follow a chi-square distribution with $df = 1$ degrees of freedom $(df = (2-1)(2-1))$ when sex and their opinions are independent. The probability that we observe a value larger than 3.84 is equal to only 5% if X and Y are independent.[12]

The advantage of Pearson's statistic is that it is applicable to any $K \times M$ contingency table, whether the random variables X and Y are binary, ordinal, or nominal. Note that Kendall's tau and Spearman's rho cannot be applied to nominal data, since they require that X and Y are ordinal. Thus, whenever one of X and Y is nominal, we may resort to Pearson's chi-square statistic. For this reason, Pearson's chi-square statistic is considered to measure *nominal association*. However, a disadvantage of Pearson's chi-square statistic is that it cannot be viewed as a correlation coefficient, like Kendall's tau and Spearman's rho, since it is not limited by the value 1. If we multiply all frequencies in the contingency table with a constant, Pearson's chi-square increases with the same constant. Thus Pearson's chi-square statistic is not properly normalized to be viewed as a proper association statistic.

Pearson normalized the statistic with the sample size n when he studied 2×2 contingency tables. This measure is also called Pearson's *squared phi-coefficient*:

$$\phi^2 = \frac{\chi_P^2}{n}.$$

For the setting of 2×2 contingency tables, Pearson's phi-coefficient ϕ is equal to the absolute value of Pearson's product-moment estimator applied on the binary pairs (X_i, Y_i). To see this we will rewrite Pearson's chi-square statistic defined

---

[12] The 95% and 99% quantiles of a chi-square distribution with 1 degrees of freedom are equal to 3.84146 and 6.63490, respectively.

in Eq. (6.11) and Pearson's product-moment estimator defined in Eq. (6.10) for the 2×2 contingency table. In the case of two categories $K = M = 2$, the numerator $[n N_{xy} - N_x. N_{.y}]^2$ in Eq. (6.11) is equal to $[N_{11} N_{22} - N_{12} N_{21}]^2$ for each combination of $(x, y) \in \{1, 2\} \times \{1, 2\}$. Thus the sums in Pearson's chi-square statistic then add up the denominators. This is equal to $n^2/[N_1. N_2. N_{.1} N_{.2}]$. Thus for binary X and Y, Pearson's chi-square statistic becomes:

$$\chi_P^2 = n \frac{[N_{11} N_{22} - N_{12} N_{21}]^2}{N_1. N_2. N_{.1} N_{.2}}.$$

This makes the phi-coefficient equal to $\phi = |N_{11} N_{22} - N_{12} N_{21}| / \sqrt{N_1. N_2. N_{.1} N_{.2}}$. Pearson's product-moment estimator in Eq. (6.10) can be written in terms of the contingency table frequencies N_{xy}. For binary random variables X and Y, the numerator $\sum_{i=1}^n (X_i - \bar{X})(Y_i - \bar{Y})$ in Eq. (6.10) is equal to $[N_{11} N_{22} - N_{12} N_{21}]/n$, see the equivalent forms of covariances. The two terms in the denominator are $\sum_{i=1}^n (X_i - \bar{X})^2 = N_2. N_1./n$ and $\sum_{i=1}^n (Y_i - \bar{Y})^2 = N_{.2} N_{.1}/n$. This makes Pearson's product-moment estimator ρ_P equal to $\rho_P = [N_{11} N_{22} - N_{12} N_{21}]/\sqrt{N_1. N_2. N_{.1} N_{.2}}$. The absolute value of ρ_P is now equal to ϕ.

For $K \times M$ contingency tables, Harald Cramér noticed that a normalization of Pearson's chi-square statistic with n was not ideal. He demonstrated that the maximum value of Pearson's chi-square statistic could become maximally equal to $n \min\{K - 1, M - 1\}$. Thus Cramér introduced the association measure V for general $K \times M$ contingency tables that would have its values in $[0, 1]$. It is defined by

$$V = \sqrt{\frac{\chi_P^2}{n \min\{K - 1, M - 1\}}}.$$

It is obvious that Cramér's V is equal to Pearson's ϕ coefficient when either the X or the Y variable has just two levels. Thus Cramér's V is the more general measure of nominal association, since it is properly normalized for all contingency tables and it reduces to Pearson's ϕ for 2×2 contingency tables.

Although the normalization is theoretically correct, researchers have criticized Cramér's V, since the maximum is only attained in artificial contingency tables. The maximum attainable value for V can be substantially smaller if we keep the row and column totals fixed to the totals that we have observed. Table 6.3 is an example of a table for which the marginal distribution limits the value of the (positive) association:

Pearson's ϕ is equal to $\phi = 0.11$ and there is no other 2×2 contingency table with the exact same row and column totals that results in a higher value of ϕ. Thus under the marginal constraints we cannot make the association any stronger (in this direction). However, the fact that it is not close to one does not necessarily imply that the relationship is very weak. This is because the marginal distribution "limits" the possible values of ϕ (or other measures of relationship strength). Nevertheless, for nominal random variables it does provide similar interpretations as the sample correlations and often ϕ or V are reported together with χ^2 and a p-value of the

Table 6.3 Example of 2×2 contingency table with maximal ϕ coefficient given the marginals

	Agree	Disagree	Total
Male	5	0	5
Female	40	5	45
Total	45	5	50

null-hypothesis test that there is no dependence between variables; null-hypothesis testing is a topic we will cover in the next chapter.

Although Pearson's chi-square, Pearson's phi-coefficient, and Cramér's V are meant for contingency tables, they can also be applied to continuous variables X and Y. In this case, X and Y must be transformed to categorical variables first. This can easily be done by forming non-overlapping intervals, like we also do for visualization of continuous data in histograms. For instance, we can create K intervals for the variable X: $(-\infty, \alpha_1], (\alpha_1, \alpha_2], \ldots, (\alpha_{K-2}, \alpha_{K-1}]$, and (α_{K-1}, ∞), with $\alpha_1, \alpha_2, \ldots, \alpha_K$ threshold values that are selected by ourselves, and M intervals for the variable Y: $(-\infty, \beta_1], (\beta_1, \beta_2], \ldots, (\beta_{M-2}, \beta_{M-1}]$, and (β_{M-1}, ∞), with $\beta_1, \beta_2, \ldots, \beta_M$ threshold values that are selected by ourselves. The continuous data $(X_1, Y_1), (X_2, Y_2), \ldots, (X_n, Y_n)$ can then be summarized by N_{xy}, which would then represent the number of pairs (X_i, Y_i) that falls in the set $(\alpha_{x-1}, \alpha_x] \times (\beta_{y-1}, \beta_y]$, with $\alpha_0 = \beta_0 = -\infty$ and $\alpha_K = \beta_M = \infty$. It should be noted though that the choice of the number of levels and the choice of thresholds can have a strong influence on the calculation of nominal associations. Nevertheless, they may be useful for continuous data as well, in particular when continuous data are already observed as a categorical variable (e.g. income).

6.7.2 Ordinal Association Statistics

Here we restrict ourselves to ordinal data for both X and Y and assume that the data can be summarized or represented by a $K \times M$ contingency table. Thus X and Y can take their value in $\{1, 2, \ldots, K\}$ and $\{1, 2, \ldots, M\}$, respectively. As an example, you may think of X and Y being severity ratings, like 1="very low", 2="low", 3="neutral", 4="high", and 5="very high", in, for instance, quality assessments or disease classifications. Potentially, Pearson's rho, Spearman's rho, and Kendall's tau estimators may seem to be suitable for this type of data, but they are not ideal. Pearson's rho will treat the values 1, 2, ..., K for X and the values 1, 2, ..., M for Y as numerical, while these numbers are somewhat arbitrary. Changing the values will lead to a different Pearson's rho estimate. Spearman's rho and Kendall's tau compare the ordinal values with each other, but in many comparisons the values cannot be ordered. If we observe the pairs $(1, 3), (3, 2), (3, 5), (1, 4), (2, 5)$, then pairs 1 and 4 and pairs 2 and 3 cannot be ordered in their x coordinate and pairs 3 and 5 cannot be ordered in their y coordinate. As we mentioned earlier, we observe ties in both the x and y coordinates. These ties affect the estimator as well as the calculation of its standard error. Here we will provide a few alternatives, closely related to Kendall's tau, but that do addresses the ties appropriately.

The issue with Kendall's tau is the use of the denominator $0.5n(n-1)$ that quantifies the number of pairs being compared. When there are many ties, the number of pairs that can really order the pairs is much lower. In the example of the five pairs $(1, 3)$, $(3, 2)$, $(3, 5)$, $(1, 4)$, $(2, 5)$, 10 pairs can potentially be compared, but for two comparisons we could not order the x coordinate and for one pair we could not order the y coordinate. Thus Kendall's tau can be simply adjusted by using the correct number of pairs that are being compared. This was developed by Goodman and Kruskal (1979) and is referred to as Goodman and Kruskal's gamma:

$$\gamma = \frac{\Pr((X_2 - X_1)(Y_2 - Y_1) > 0) - \Pr((X_2 - X_1)(Y_2 - Y_1) < 0)}{\Pr((X_2 - X_1)(Y_2 - Y_1) > 0) + \Pr((X_2 - X_1)(Y_2 - Y_1) < 0)}.$$

It can be estimated by

$$G = \frac{N_C - N_D}{N_C + N_D},$$

with N_C and N_D defined by

$$N_C = \sum_{i=1}^{n} \sum_{j=1}^{n} 1_{(0,\infty)}((X_j - X_i)(Y_j - Y_i))$$

$$N_D = \sum_{i=1}^{n} \sum_{j=1}^{n} 1_{(-\infty,0)}((X_j - X_i)(Y_j - Y_i)).$$

Here N_C and N_D represent the number of concordant and discordant pairs. It should be noted that the sample size n is typically larger than the sum of concordant and discordant pairs, i.e., $n > N_C + N_D$, due to the many ties in data of the $K \times M$ contingency table. Clearly, if there are no ties at all, Goodman and Kruskal's gamma reduces to Kendall's tau.

The distribution function of the estimator G has been studied under the assumption that $(X_1, Y_1), (X_2, Y_2), \ldots, (X_n, Y_n)$ are i.i.d. with distribution function F_{XY}. The distribution function of $\sqrt{n}(G - \gamma)$ is asymptotically normal with zero mean and a variance that is smaller than but close to (Goodman and Kruskal 1979):

$$\frac{2(1 - \gamma^2)}{\Pr((X_2 - X_1)(Y_2 - Y_1) > 0) + \Pr((X_2 - X_1)(Y_2 - Y_1) < 0)}.$$

This variance can be estimated with $2(1 - G^2)/(N_D + N_C)$. Thus, a $100\%(1 - \alpha)$ confidence interval on γ can be constructed using the theory of our asymptotic confidence intervals in Sect. 5.3.5. The confidence interval is then

$$\left(G - z_{1-\alpha/2}\sqrt{\frac{2(1 - G^2)}{N_D + N_C}}, G + z_{1-\alpha/2}\sqrt{\frac{2(1 - G^2)}{N_D + N_C}}, \right],$$

with z_p the pth quantile of the standard normal distribution function.

6.7.3 Binary Association Statistics

In 1901, Paul Jaccard published an index to classify ecological species (Jaccard 1901). The index tried to measure or quantify the *similarity* in a large number of attributes from two objects. The objects in Jaccard's work were mostly large areas of land (in Switzerland) and the attributes were the (presence and absence of) plant species that grew on the land. This is just one application, but it may be clear that many other applications could be formulated in which we would like to quantify similarity between objects, e.g., genetic similarity between two individuals, similarity of text from two manuscripts, or identification of individuals with biometrics (fingerprints, iris images, etc.).

After the introduction of the Jaccard measure of similarity, many more were developed (Choi et al. 2010). Over a period of 100 years, more than 70 indices have been proposed and discussed. Similarity measures are highly relevant in the field of data science, as they are not just used on attribute data of two objects: they are often applied to attribute data of many objects to group or identify clusters of objects. The goal is to group more similar objects together, such that objects in one cluster are much more similar than objects from different clusters.[13] So, there is a rich literature on similarity measures and we cannot possibly mention them all.

We will discuss two sets of similarity measures for 2×2 contingency tables that contain several well-known measures used in practice (see Gower and Legendre 1986). Here we may assume that we observed n binary attributes on two objects, leading to the n pairs $(X_1, Y_1), (X_2, Y_2), \ldots, (X_n, Y_n)$. In this type of application, the pairs may not be independent and/or identically distributed with just one CDF F_{XY}. Thus the analysis of the data may be considerably more complicated, but we will only focus on the summary data

$$N_{xy} = \sum_{i=1}^{n} 1_{\{x\}}(X_i) 1_{\{y\}}(Y_i), \quad \text{for} \quad (x, y) \in \{0, 1\} \times \{0, 1\}.$$

Note that for binary data we prefer the values $\{0, 1\}$ for X and Y, instead of $\{1, 2\}$. Here N_{11} is the number of attributes that both objects share and N_{00} is the number of attributes that both objects lack. N_{01} and N_{10} represent the numbers of objects that are present in one object but absent in the other object.

---

[13] Note that clustering of data is outside the scope of our book, but having a better understanding of the similarity measures may also help understand clustering techniques better.

Two sets of similarity indices were given by Gower and Legendre (1986)

$$S_\theta = \frac{N_{00} + N_{11}}{N_{00} + N_{11} + \theta[N_{01} + N_{10}]}$$

$$T_\theta = \frac{N_{11}}{N_{11} + \theta[N_{01} + N_{10}]},$$

with $\theta > 0$ a constant. The similarity measure S_θ focuses on the similarity of both the absence as well as the presence of attributes, since it uses both N_{00} and N_{11}, while the similarity measure T_θ focuses only on the presence of attributes. Thus when we wish to emphasize the similarity on attributes that are present, the set of T_θ is more appropriate than S_θ. Furthermore, both similarity measures are bounded from below with the value zero and from above with the value 1 ($S_\theta, T_\theta \in [0, 1]$). The closer the similarity measures get to one the more similar the two objects would be on the presence and absence of attributes. Therefore, the measures $1 - S_\theta$ and $1 - T_\theta$ are referred to as *dissimilarity measures*.

If we replace θ with specific values we obtain well-known similarity measures that have been published in the literature. The following list shows just a few options:

$$\text{Sokal \& Sneath(2)} : S_{0.5} = \frac{2[N_{00} + N_{11}]}{2[N_{00} + N_{11}] + N_{01} + N_{10}}$$

$$\text{Sokal \& Michener} : S_1 = \frac{N_{00} + N_{11}}{N_{00} + N_{11} + N_{01} + N_{10}}$$

$$\text{Roger \& Tanimoto} : S_2 = \frac{N_{00} + N_{11}}{N_{00} + N_{11} + 2[N_{01} + N_{10}]}$$

$$\text{Czekanowski} : T_{0.5} = \frac{2N_{11}}{2N_{11} + N_{01} + N_{10}}$$

$$\text{Jaccard} : T_1 = \frac{N_{11}}{N_{11} + N_{01} + N_{10}}$$

$$\text{Sokal \& Sneath(1)} : T_2 = \frac{N_{11}}{N_{11} + 2[N_{01} + N_{10}]}$$

Note that we already discussed the Sokal & Michener measure in the calculation of Cohen's kappa statistic, since it represents the probability of agreement (without correcting for chance agreement). It is also referred to as the Rand index (Albatineh 2010). The index by Czekanowski is sometimes referred to as the Sorenson & Dice index (Warrens 2008).

The similarity measures S_θ and T_θ are essentially functions of $[N_{00} + N_{11}]/[N_{01} + N_{10}]$ and $N_{11}/[N_{00} + N_{11}]$, respectively. These two measures have been proposed as similarity measures themselves:

$$\text{Sokal \& Sneath}(3) : S_{SS} = \frac{N_{00} + N_{11}}{N_{01} + N_{10}}$$

$$\text{Kulcynski} : T_K = \frac{N_{11}}{N_{01} + N_{10}}$$

Thus $S_\theta = \theta/[\theta + S_{SS}]$ and $T_\theta = \theta/[\theta + T_K]$. It should be noted that both S_{SS} and T_K are not properly normalized. They can be larger than 1 and are essentially unbounded, which makes them more difficult to interpret and to compare.

In a way, the similarity measures quantify the dependency between X and Y. The higher the value the stronger the dependency and more similar the two objects are. However, they are somewhat different from some other association measures for 2×2 contingency tables, like Yule's measure of association, one of the oldest measures of association (see below).

The dissimilarity measures can be considered a distance between the objects. Indeed, Gower and Legendre (1986) showed that the dissimilarity measures $1 - S_\theta$ and $1 - T_\theta$ are distance measures when $\theta \geq 1$ and $\sqrt{1 - S_\theta}$ and $\sqrt{1 - T_\theta}$ are distance measures when $\theta \geq 1/3$. Note that a distance measure d is defined by four characteristics: $d(a, b) \geq 0, d(a, a) = 0, d(a, b) = d(b, a)$, and $d(a, b) + d(b, c) \geq d(a, c)$, with a, b, and c representing arbitrary objects on which the distance is applied. The second condition requires us to look at dissimilarity measures instead of similarity measures.

Although all four conditions on the distance are important, it is the last condition that makes the dissimilarity measures as distances most attractive, in particular if classification of multiple objects is important. Indeed, we do not like the situation that two objects are unrelated while they are both related to the same object. In terms of social media, we do not want to use measures that could identify two people having no connection at all, but who do share a friend. In that setting they are related through their friend.

It is clear that ϕ is an alternative measure with values in $[0, 1]$. As it is not of the form $\sqrt{1 - S_\theta}$ or $\sqrt{1 - T_\theta}$, it is unknown if $1 - \phi$ or $\sqrt{1 - \phi}$ is a distance measure, like the proposed similarity measures for values of θ being large enough. Gower & Legendre demonstrated for Pearson's correlation coefficient ρ_P on the 2×2 contingency table that $1 - \rho_P$ is not a distance measure, but $\sqrt{1 - \rho_P}$ is a distance measure.

An alternative class of similarity measures, which is referred to as the $\mathscr{L}$ family of similarity indices, is defined by

$$S = \lambda + \mu(N_{00} + N_{11}), \tag{6.12}$$

where the parameters λ and μ can only be functions of the row and column totals, i.e., functions of $N_{0\cdot}, N_{1\cdot}, N_{\cdot 0}$, and $N_{\cdot 1}$. The Jaccard similarity measure is not contained in this family, but the Czekanowski index is contained in this family as well as the Sokal & Michener index (Albatineh 2010; Warrens 2008). This class also contains Cohen's kappa statistic. The parameters are then equal to

Sokal & Michener: $\lambda = 0$ and $\mu = \dfrac{1}{n}$

Czekanowski: $\lambda = 1 - \dfrac{1}{N_{1\cdot} + N_{\cdot 1}}$ and $\mu = \dfrac{1}{N_{1\cdot} + N_{\cdot 1}}$

Kappa: $\lambda = -\dfrac{N_{1\cdot}N_{\cdot 1} + N_{0\cdot}N_{\cdot 0}}{N_{1\cdot}N_{\cdot 0} + N_{\cdot 1}N_{0\cdot}}$ and $\mu = \dfrac{1}{N_{1\cdot}N_{\cdot 0} + N_{\cdot 1}N_{0\cdot}}.$

The mean and variance of the indices in this family of indices have been studied and they can be used to quantify confidence intervals, but this is outside the scope of our book.

Similar to the discussion of agreement, there is a discussion about whether the similarity measures should be corrected for similarity due to chance. For the $\mathscr{L}$ family, corrections have been proposed. A corrected index is of the form

$$CS = \frac{S - \mathbb{E}(S)}{1 - \mathbb{E}(S)},$$

with S the similarity measure in Eq. (6.12). Note that it is not always clear what the expectation of S is, since the setting in which the 2×2 contingency table is observed may change from setting to setting. Recall that $(X_1, Y_1), (X_2, Y_2), \ldots, (X_n, Y_n)$ are not (necessarily) i.i.d., which implies that different assumptions lead to different results for $\mathbb{E}(S)$. Nevertheless, it has been shown that several of the indices in this class all become equivalent after correction (Warrens 2008). This includes the Czekanowski index, the Sokal & Michener index, and Cohen's kappa statistics. Thus after correction these three indices are all equivalent.

The last measure that we would like to mention is Yule's Q statistic. It was termed the *coefficient of association*, which made sense at the time, as it is one of the oldest measures of association. Now that there are so many measures, we rather call it Yule's Q statistic. It is defined as

$$Q = \frac{N_{00}N_{11} - N_{01}N_{10}}{N_{00}N_{11} + N_{01}N_{10}}.$$

The measure ranges from -1 to $+1$ and when Q is zero there is no association, similar to Pearson's, Spearman's, and Kendall's correlation coefficients.

Yule's Q is different from the two families of similarity measures, since it is not of the form S_θ, T_θ, and S in Eq. (6.12). Neither is $1 - Q$ or $\sqrt{1 - Q}$ a distance measure (Gower and Legendre 1986), which makes it also different from Pearson's product-moment estimator on binary data. Yule's Q is a special case of Goodman & Kruskal's γ statistic applied to a 2×2 contingency table, but there is also a direct connection to the odds ratio, since Yule's Q can be rewritten as

$$Q = \frac{\hat{OR} - 1}{\hat{OR} + 1}. \tag{6.13}$$

Thus Yule's Q is a monotone function of the odds ratio, which makes Yule's Q an attractive measure. It transforms the odds ratio to a measure that is in line with correlation coefficients. But more importantly, if we randomly eliminate attributes from one object, say remove half of all the attributes from object 1, then both N_{11} and N_{10} would reduce by a factor 2, but Yule's Q would not reduce. Thus Yule's Q statistic is robust against the number of features that are present in one object, a characteristic that does not hold for the similarity measures. Confidence intervals on Yule's Q can easily be determined by using the confidence limits of the odds ratio and then substitute them in Eq. (6.13).

Having discussed many different measures of association or measures of similarity, it is not easy to choose among all the possible indices, in particular since there are so many. Here we provide a few simple directions, although we realize that each setting may require its own index to accomplish its specific goals. If the similarity on attributes is most important, we recommend a measure for the class T_θ, preferably with a parameter $\theta \geq 1/3$ to use a measure that can be viewed as a distance measure between the two objects. Clearly, when both the absence and presence of attributes are important, either S_θ or S in Eq. (6.12) can be used. If it should also represent a distance measure, S_θ with $\theta \geq 1/3$ seems most appropriate. When a chance corrected measure is more appropriate, the similarity measure S can be used. Depending on the way that the data is collected, the chance corrected measure reduces to one of seven measures (Gower and Legendre 1986). In some settings this could be Cohen's kappa statistic. Finally, when it is important that the measure is robust against the number of features that could be present, Yule's Q may seem an appropriate choice.

6.8 Exploring Multiple Variables Using R

Obviously, the measures of association we discussed above can easily be computed with R, either by using a package that has programmed some of the associations or otherwise by programming the associations ourself. We will illustrate the measures of association by using our different datasets. First we will study the correlation between the two (continuous) face dimensions in the face data. Secondly, we will study the association between two binary variables that we will create from two variables in the high-school data. We will study the association between watching television or not with computer use or not. Additionally, we will study some similarity measures on the potato data using their genetic profile. Finally, we will study the association between voting choice and education level from the voting data.

6.8.1 Associations Between Continuous Variables

To illustrate the associations for continuous variables we will use the face data. Recall that the first dimension (dim1) in the face data is the distance between the eyes and the second dimension (dim2) is the brow-to-chin ratio. Since both dimensions are continuous, the three correlation measures Pearson's rho, Spearman's rho, and Kendall's tau all apply nicely. They can all be calculated with the same R function cor(). After reading in the data, and calling it facedata, the following code was used to calculate all three correlation measures:

```
> n <- dim(facedata)
> n[1]
[1] 3628
> rho_P <- cor(facedata$dim1, facedata$dim2,
            method = c("pearson"))
> rho_P
[1] 0.7260428
> rho_S <- cor(facedata$dim1, facedata$dim2,
            method = c("spearman"))
> rho_S
[1] 0.7435
> tau_K <- cor(facedata$dim1, facedata$dim2,
            method = c("kendall"))
> tau_K
[1] 0.529564
```

As we can see, Pearson and Spearman's rho are relatively close to each other, but Kendall's tau is somewhat smaller. It is not uncommon that Kendall's tau is (substantially) lower than Spearman's rho (see Sect. 6.5.3). The results also show that the correlation coefficients are positive and quite large. Thus a larger brow-to-chin ratio co-occurs with a larger distance between the eyes. In Fig. 6.3 we have given the scatter plot for the two dimensions, where we can see how the two variables are co-related. The figure shows a strong relation, but it does not seem to be linear. This would possibly disqualify the use of Pearson's product-moment estimator, unless we

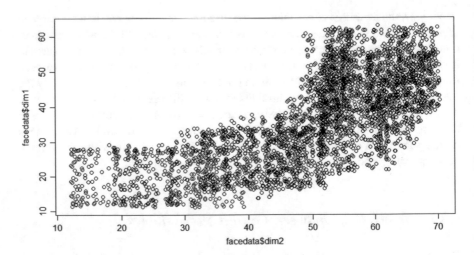

Fig. 6.3 Scatter plot between the two face dimensions in the face data

may find a suitable transformation of the two dimensions that would show a linear trend after transformation.

Confidence intervals can be calculated through the Fisher z-transformation. We have programmed the 95% confidence intervals ourselves, which are provided in the following code for all three correlation coefficients (even though Pearson's rho may not seem to be the most reasonable choice to quantify the association between the two face dimensions):

```
> za <- qnorm(0.975)
>
> z_P <- 0.5*(log(1+rho_P)-log(1-rho_P))
> LCL_FP <- z_P - za/sqrt(n[1]-3)
> UCL_FP <- z_P + za/sqrt(n[1]-3)
> LCL_P <- (exp(2*LCL_FP)-1)/(exp(2*LCL_FP)+1)
> UCL_P <- (exp(2*UCL_FP)-1)/(exp(2*UCL_FP)+1)
> LCL_P
[1] 0.7102827
> UCL_P
[1] 0.7410753
>
> z_S <- 0.5*(log(1+rho_S)-log(1-rho_S))
> LCL_FS <- z_S - za*sqrt(1.06/(n[1]-3))
> UCL_FS <- z_S + za*sqrt(1.06/(n[1]-3))
> LCL_S <- (exp(2*LCL_FS)-1)/(exp(2*LCL_FS)+1)
> UCL_S <- (exp(2*UCL_FS)-1)/(exp(2*UCL_FS)+1)
> LCL_S
[1] 0.7281344
> UCL_S
[1] 0.7581186
>
> z_K <- 0.5*(log(1+tau_K)-log(1-tau_K))
> LCL_FK <- z_K - za*sqrt(0.437/(n[1]-3))
> UCL_FK <- z_K + za*sqrt(0.437/(n[1]-3))
> LCL_K <- (exp(2*LCL_FK)-1)/(exp(2*LCL_FK)+1)
> UCL_K <- (exp(2*UCL_FK)-1)/(exp(2*UCL_FK)+1)
> LCL_K
[1] 0.5139033
> UCL_K
[1] 0.5448719
```

Due to the relatively large sample size, the confidence intervals are quite small. The three confidence intervals show the ranges of values (from the lower bound in the interval to the upper bound in the interval) for which the population parameters ρ_P, ρ_S, and τ_K may fall with 95% confidence. This also implies that it is not very likely that the two dimensions are uncorrelated, since the value 0 is far below the lower bounds. Thus it is unlikely that the two dimensions are independent. We will discuss this topic of hypothesis testing more formally and in more detail in Chap. 7.

6.8.2   Association Between Binary Variables

We discussed in previous sections several measures that could be used to quantify the dependency between binary variables. We discussed Cohen's kappa to understand agreement, association measures from Chap. 3 to quantify changes in risk, Pearson's chi-square statistic as a dependency measure, Pearson's ϕ coefficient as a correlation coefficient, Yule's Q measure as a measure of association, and several other measures of similarity. In principle, these measures can be calculated for any 2×2 contingency table, but it depends on the application which would make more or less sense. Here we will illustrate all of these measures.

6.8.2.1   Pearson's Chi-Square, Pearson's ϕ Coefficient, and Yule's Q

If we consider the binary variables on television watching and computer use from the data on high-school students, the most appropriate measures are Pearson's chi-square statistic, Pearson's ϕ coefficient, and Yule's Q statistic. The reason for choosing these measures is that we want to quantify the strength of the dependency between two different variables. There is no direct preference for an investigation of one binary variable conditionally on the level of the other binary variable. The two binary variables are viewed as two *outcomes* on the high-school children.

The following R code creates the 2×2 contingency table:

```
> x <- ifelse(schooldata$TV > 0, 1, 0)
> y <- ifelse(schooldata$COMPUTER > 0, 1, 0)
> xy<-table(x,y)
> xy
   y
x      0     1
  0   303   511
  1  5889 43366
```

Watching television (yes/no) is collected in the x variable and using the computer (yes/no) is collected in the y variable, but we could have interchanged this. The choice is essentially arbitrary. When $x = 1$ the student watches television and when $y = 1$ the student uses the computer. There are 303 students who do not watch television and do not use the computer. Pearson's chi-square statistic can easily be programmed with R statements, but we can also use an R function:

```
> chisq.test(xy, correct=FALSE)

    Pearson's Chi-squared test

data: xy
X-squared = 471.74, df = 1, p-value < 2.2e-16
```

Here we used the chi-square test in R, which calculates Pearson's chi-square statistic at 471.74. The option "`correct=FALSE`" is needed to avoid a correction on the calculation of Pearson's chi-square. This is called Yates correction, but we will not discuss this here. The R function also "tests" whether watching television and using the computer are dependent, but the topic of testing is postponed to Chap. 7. Pearson's chi-square is rather large, due to two elements. It will be large when the two variables are strongly dependent and when there is a weak dependency but now with a large sample size. To correct for the sample size of $n = 50,069$, Pearson's ϕ coefficient can be calculated. The ϕ coefficient is determined at 0.097 ($= \sqrt{(471.74/50,069)}$). It has a positive sign, since $303 \times 43,366$ is larger than $511 \times 5,889$, otherwise ϕ would be negative.

The estimate of the ϕ coefficient is not considered very large when we compare the result with the criteria or rules of thumb listed in the section on Pearson's correlation coefficient. The ϕ coefficient would indicate that the dependence between watching television and computer use can be neglected. However, Yule's Q statistic is estimated at 0.627. It can be calculated from the contingency table using the following R codes:

```
> N00<-303
> N10<-5889
> N01<-511
> N11<-43366
> Q<-(N00*N11-N10*N01)/(N00*N11+N10*N01)
> Q
[1] 0.6273149
```

Thus Yule's Q statistic indicates a moderate correlation between television watching and computer use when we apply the rules of thumb in Sect. 6.6.1. The difference in interpretation between Pearson's ϕ coefficient and Yule's Q is quite large in this example. The reason is a large imbalance in cell counts in the contingency table. More than 85% of the high-school students watch television and use a computer. This imbalance makes Pearson's ϕ coefficient somewhat less reliable. The ϕ coefficient can only range from -1 to $+1$ when there is no imbalance, i.e., N_{00} is of the same size as N_{11}. When there exist imbalances the range from -1 to 1 becomes (much) narrower, making it harder to apply the rules of thumb in Sect. 6.6.1. Thus, considering the three statistics for the current example, we conclude that television watching and computer use cannot be neglected. There exists a low to moderate dependency.

6.8.2.2   Risk Difference, Relative Risk, and Odds Ratio

To demonstrate the measures of association or measures of risk from Chap. 3 with their 95% confidence intervals, we will investigate whether the proportion of computer use between boys and girls is different. In this setting we have two well-defined groups of students (boys and girls) and a well-defined binary outcome (making use of a computer or not). The data on television watching and computer use could also have been used if we were interested in the difference in proportion of computer use for students who do watch television and students who do not watch television (or the other way around). In that case we would have a clear outcome variable being computer use and a clear subgroup variable being students who do or do not watch television. However, if we do not have a clear subgroup-outcome formulation, because we do not know if we should group students into watching television or should group students into using computers or not, it is more appropriate to quantify the dependency between television watching and computer use by the methods we discussed in Sect. 6.8.2.1.

The contingency table for gender and computer use can be constructed with the following R codes:

```
> s <-schooldata$GENDER
> y <- ifelse(schooldata$COMPUTER > 0, 1, 0)
> sy<-table(s,y)
> sy
      y
s        0      1
  Boy   1512  22958
  Girl  4680  20919
```

Thus the proportions of boys that use the computer is 93.82% ($= 22,958/(22,958 + 1,512)$) and the proportion of girls that use the computer is 81.72% ($= 20,919/ (20,919 + 4,680)$). Calculating the difference in proportion with their 95% confidence interval using R leads to the following codes:

```
> z<-qnorm(0.975,mean=0,sd=1,lower.tail=TRUE)
>
> N_boy=22958+1512
> p_boy=22958/N_boy
> p_boy
[1] 0.9382101
> L_boy <-(2*N_boy*p_boy+z^2-z*sqrt(4*N_boy*p_boy*(1-p_boy)+z^2))
    /(2*(N_boy+z^2))
> L_boy
[1] 0.935124
> U_boy <-(2*N_boy*p_boy+z^2+z*sqrt(4*N_boy*p_boy*(1-p_boy)+z^2))
    /(2*(N_boy+z^2))
> U_boy
[1] 0.9411586
>
> N_girl=20919+4680
> p_girl=20919/N_girl
> p_girl
[1] 0.8171804
> L_girl <-(2*N_girl*p_girl+z^2-z*sqrt(4*N_girl*p_girl*(1-p_girl)
    +z^2))/(2*(N_girl+z^2))
> L_girl
[1] 0.812398
> U_girl <-(2*N_girl*p_girl+z^2+z*sqrt(4*N_girl*p_girl*(1-p_girl)
    +z^2))/(2*(N_girl+z^2))
> U_girl
[1] 0.8218675
>
> RD<-p_boy-p_girl
> RD
```

```
[1] 0.1210297
> L_RD<-RD-sqrt((p_boy-L_boy)^2+(U_girl-p_girl)^2)
> L_RD
[1] 0.1154178
> U_RD<-RD+sqrt((U_boy-p_boy)^2+(p_girl-U_girl)^2)
> U_RD
[1] 0.1265671
```

Thus the difference in proportions of computer use between male and female students is estimated at 12.10%, with a 95% confidence interval equal to [11.54%, 12.66%].

If we would prefer to study the relative risk, the following codes can be applied. Here we compare boys versus girls (with girls the reference group).

```
> RR<- p_boy/p_girl
> RR
[1] 1.148106
> SE_RR<-sqrt((1-p_boy)/(N_boy*p_boy)+(1-p_girl)/(N_girl*p_girl))
> LL_RR<-log(RR)-z*SE_RR
> LU_RR<-log(RR)+z*SE_RR
> L_RR<-exp(LL_RR)
> L_RR
[1] 1.140524
> U_RR<-exp(LU_RR)
> U_RR
[1] 1.15574
```

Thus, the relative risk is estimated at 1.148 with 95% confidence interval [1.141, 1.156]. Although the name of the measure is referred to as relative risk, it is merely a ratio of probabilities. It indicates that 14.8% more boys use the computer than girls.

For the calculation of the odds ratio we may use the following R codes:

```
> NB0<-1512
> NB1<-22958
> NG0<-4680
> NG1<-20919
>
> OR<-(NB1/NB0)/(NG1/NG0)
> OR
[1] 3.396935
> SE_OR<-sqrt(1/NB1 + 1/NB0 +1/NG1 + 1/NG0)
> LL_OR<-log(OR)-z*SE_OR
> LU_OR<-log(OR)+z*SE_OR
> L_OR<-exp(LL_OR)
> L_OR
[1] 3.19614
> U_OR<-exp(LU_OR)
> U_OR
[1] 3.610344
```

The odds ratio for computer use of male students with respect to female students is estimated at 3.40 with a 95% confidence interval equal to [3.20, 3.61]. Recall that the odds ratio is agnostic to the subgroup-outcome setting. This means that the outcome and subgroup can be interchanged without changing the result. Thus the odds ratio could also have been used for the dependency between television watching and computer use, as it does not matter which variable creates the subgroup variable. Note that this argument is supported by the direct relation between the odds ratio and Yule's Q statistic in Eq. (6.13).

6.8.2.3   Measures of Similarity

The genetic data on the potatoes can be used to quantify how two potatoes are similar with respect to their genetic score. Here we will investigate Cohen's kappa statistic and several similarity measures.

To investigate how well the RNA sequencing method performs, we compared the score of `Bintje_1` with the score of `Bintje_2`. These scores represent two readings of the same potato. After reading in the data, the following R code provides us with the 2×2 contingency table:

```
> b1<-potato$Bintje_1
> b2<-potato$Bintje_2
> bb<-table(b1,b2)
> bb
   b2
b1      0     1
  0 22285  3726
  1   638 20933
```

The probability of agreement p_O is estimated at 90.83% $(= (22{,}285 + 20{,}933)/47{,}582)$. To eliminate the element of chance, Cohen's kappa statistic with its 95% confidence interval is estimated with the following R code:

```
> p_O<-(22285+20933)/47582
> p0_b1<-(22285+3726)/47582
> p1_b1<-(20933+638)/47582
> p0_b2<-(22285+638)/47582
> p1_b2<-(20933+3726)/47582
>
> p_E<-p0_b1*p0_b2+p1_b1*p1_b2
> p_E
[1] 0.4982978
> K<-(p_O-p_E)/(1-p_E)
> K
[1] 0.8171917
>
> SE_K<-sqrt(p_O*(1-p_O)/(47582*(1-p_E)^2))
> L_K<-K-z*SE_K
> L_K
[1] 0.8120226
> U_K<-K+z*SE_K
> U_K
[1] 0.8223607
```

Thus, the agreement between the two readings is estimated at $\hat{\kappa}_C = 0.817$ with 95% confidence interval equal to $[0.812, 0.822]$. According to the criteria in Sect. 6.5.4, the agreement between the two readings on a Bintje potato is substantial. Note that Cohen's kappa statistic is particularly useful for quantifying variability in binary (or categorical) variables, which should in principle be identical. If the sequencing method were perfect, the two readings for gene signals for Bintje potatoes should be identical. Thus Cohen's kappa statistic quantifies the amount of noise in the measurement system, which is less than 20% for detecting gene signals.

Calculating the agreement on gene signals between the first reading of Bintje and the experimental potato leads to an estimate of $\hat{\kappa}_C = 0.719[0.713, 0.725]$ and shows a lower agreement than the two readings on Bintje.[14] Instead of Cohen's kappa statistic, it may be more useful to apply another similarity measure, as we do not expect that both potatoes should provide identical gene scores. For such settings we may not want to correct for chance. It would be best to consider similarity measures that can also be interpreted as distance measures, since we would find it inappropriate if the similarity between the Bintje and experimental potatoes is closer than the two Bintje readings. Thus we will only illustrate the similarity measures S_θ and T_θ with $\theta \geq 1$. An alternative would be to study the dissimilarity measures $\sqrt{1 - S_\theta}$ or $\sqrt{1 - T_\theta}$ with $\theta \geq 1/3$ as distance measures, but here we want to study similarity.

Table 6.4 shows the similarity measures Sokal & Michner (S_1), Roger & Tanimoto (S_2), Jaccard (T_1), and Sokal & Sncath (T_2) for the similarity between the two Bintje readings and the first Bintje reading and the experimental potato.

The R code for the calculation of these similarity measures is only illustrated for the comparison of the first Bintje reading with the experimental potato, since simple adjustments are needed to calculate the similarity measures for the Bintje readings.

```
> b1<-potato$Bintje_1
> ex<-potato$Experimental
> be1<-table(b1,ex)
> be1
    ex
b1       0     1
  0 20448  5563
  1  1175 20396
>
> N00<-20448
> N11<-20396
> N01<-5563
> N10<-1175
>
> S1<-(N00+N11)/(N00+N11+1*(N01+N10))
> S1
[1] 0.8583918
> S2<-(N00+N11)/(N00+N11+2*(N01+N10))
> S2
[1] 0.7519146
> T1<-N11/(N11+1*(N01+N10))
> T1
[1] 0.7516769
> T2<-N11/(N11+2*(N01+N10))
```

---

[14] Calculations on the agreement between Bintje and the experimental potato are left to the reader.

Table 6.4 Similarity measures for the comparison of the genetic score of potatoes

	Bintje readings	Bintje and experimental
Sokal & Michner (S_1)	0.908	0.858
Roger & Tanimoto (S_2)	0.832	0.752
Jaccard (T_1)	0.827	0.752
Sokal & Sneath (T_2)	0.706	0.602

```
> T2
[1] 0.6021493
```

The results show that the similarity measures can be really different. S_1 seems to indicate a high similarity between the potatoes, while T_2 indicates almost a moderate similarity between Bintje and the experimental potato. This difference indicates that the similarity between potatoes reduces when we only focus on the presence of genes. Thus, part of the similarity between potatoes is coming from the absence of genes. On the other hand, the values of S_2 and T_1 are almost equal.

It is not straightforward to choose the most appropriate similarity measure. If we wish to compare the similarity of potatoes on the presence of genes, T_1 and T_2 seem most appropriate. Jaccard's measure is then more common than Sokal & Sneath and has an easy interpretation. It is the proportion of genes that are present among all active genes observed in both potatoes. If we also value the similarity of potatoes for the absence of genes we may want to use S_1 or S_2. They are both symmetric in the absence and presence of genes. Interchanging absence and presence does not make a difference in the calculation. Thus, they value similarity on presence and absence equally. Sokal & Michener has the easiest interpretation, being the proportion of genes that are present and absent in both potatoes.

6.8.3   Association Between Categorical Variables

To investigate the dependency between categorical variables we will use the data from the voting demo. Here we will study the variable voting choice and the variable on education. Both variables can be viewed as nominal or ordinal variables. The variable on education was already given in an ordinal format. Here the lowest level of education is given by "low" and the highest level of eduction is given by "high" and the numerical values list ordered levels of education between these extremes. This ordering is based on number of years studied, but this does not mean that the ordering of education is perfect. It does not take into account the type of education. Thus it merely shows some form of ordering between education. Political party can be (partially) ordered on a five-point scale as well, using for instance the categories strongly conservative, conservative, neutral, liberal, and strongly liberal. However, this ordering is imperfect since categorizing political parties into one of these five levels may be highly subjective and could be different depending on the political topic. We will first assume that the two variables are considered nominal and then change to ordinal variables.

The following code summarizes the two variables into a $K \times M$ contingency table (with $K = 4$ and $M = 3$). We also requested a chi-square statistic using an R package.

```
> x<- votedata$Educ
> y<- votedata$Choice
> xy<-table(x,y)
> xy
       y
x       CDU/CSU FDP SPD
  2          88  39  76
  3          36  16  27
  High       39  12  20
  Low       170  64 163
> chisq.test(xy, correct=FALSE)

    Pearson's Chi-squared test

data: xy
X-squared = 6.2894, df = 6, p-value = 0.3916
```

Thus Pearson's chi-square statistic is determined at 6.2894, which is not considered extremely large for a contingency table with six degrees of freedom ($df = (K - 1)(M - 1)$). To obtain a normalized value of Pearson's chi-square statistic, we may use Cramér's V statistic. The statistic divides the chi-square value by the number $n \min\{K - 1, M - 1\}$, as Pearson's chi-square can not exceed this number. Cramér's V can now be determined by the following R code:

```
> n<-dim(votedata)
> K<-4
> M<-3
> chi2<-6.2894
> V2<-chi2/(n[1]*min(K-1,M-1))
> V<-sqrt(V2)
> V
[1] 0.06475286
```

Cramér's V is determined at 0.065, which indicates that there is a negligible association between voting choice and education.

If we now assume that both variables are ordinal, we may calculate Goodman and Kruskal's gamma statistic. To do this we first need both variables as two ordinal variables where the levels can be identified as being in a particular order. In the education variable the level 2 is higher than the level "low", but R considers level 2 before the level "low", because R works with alphabetical order. Thus we must have a variable that is ordered properly. We will create a variable with the numbers 1, 2, 3, and 4, where the number 1 indicates the lowest level of education and 4 the highest level of education. For the voting variable we also need to create a variable with ordered levels. We will create a variable with levels 1, 2, and 3. The SPD is considered center-left, the FDP is considered center to center-right, and the CDU is considered center-right. Thus the levels 1, 2, and 3 will be given to the parties SPD, FDP, and CDU/CSU. The following R code provides the two new variables.

```
> Edu <- ifelse(x=='Low',1,as.character(x))
> Edu <- ifelse(Edu=='High',4,as.character(Edu))
> Edu <- as.numeric(Edu)
>
> Party <- ifelse(y=='SPD',1,as.character(y))
> Party <- ifelse(Party=='FDP',2,as.character(Party))
> Party <- ifelse(Party=='CDU/CSU',3,as.character(Party))
> Party <- as.numeric(Party)
```

Goodman and Kruskal's G estimator can be determined with an R function "gkgamma". To do this, we need to install the package "MESS". The following R code shows the steps that are needed to conduct the calculations. The first step is the installation of the package and the second step is the creation of the contingency table for education and voting choice, where the levels are now put in the correct order. The third and final step calculates the statistic and its 95% confidence interval.

```
> install.packages("MESS")
> library(MESS)
> EP<-table(Edu,Party)
> EP
   Party
Edu   1    2    3
  1  163   64  170
  2   76   39   88
  3   27   16   36
  4   20   12   39
> gkgamma(EP,conf.level=0.95)

    Goodman-Kruskal's gamma for ordinal categorical data

data: EP
Z = 1.8486, p-value = 0.06452
95 percent confidence interval:
 -0.005589579  0.196988457
sample estimates:
Goodman-Kruskal's gamma
            0.09569944
```

Thus Goodman and Kruskal's estimator G is determined at 0.096 with 95% confidence interval $[-0.006, 0.197]$. The association between education and political party, when they are considered ordinal, shows a negligible to low association, similar to the calculation of Cramér's V.

6.9 Conclusions

We have covered different topics for bivariate variables. We provided the theory of joint distribution functions to formalize dependency between two variables and made a distinction between discrete and continuous variables. The joint distribution function describes dependency in an abstract way. We tried to provide different ways of constructing bivariate distribution functions, to demonstrate that dependency can come in all kinds of forms. We also connected the bivariate distribution function to population characteristics, using expectations, and we discussed some calculation rules.

Another large part is the many different measures of association that exist to quantify the dependency between variables. These measures are described for continuous, categorical, and binary variables. For each type of variable several measures have been developed and studied. We described what aspect of the population these measures capture, i.e., formulated dependency parameters. These population parameters depend on how the dependency is constructed. They depend on the measure and the bivariate distribution. We also discussed estimators for these measures of association and we provided confidence intervals on many of these measures of association.

In the final section we illustrated many of the measures of associations on real data and discussed when certain estimators are more realistic than other measures of association. This section also provided the R code to help you calculate the measures on real data.

Although we have provided quite a range of measures of association, we have not been complete. There are still many more, and new measures are being developed. The reason is that these measures of associations may play a role in machine learning techniques and they are being studied for their performance on certain data science tasks.

Problems

6.1 Using the dataset `voting-demo.csv`, do the following:

1. Make a contingency table of `Vote` by `Choice`.
2. Compute the χ^2, ϕ, and Cramer's V value for this table. Do so using R and do so by hand.
3. Think of a way to visualize the relationship.
4. Create a new variable called `Age2`, which is the age in months, using `Age2 <- 12*Age`. What is the correlation between `Age` and `Age2`?
5. Now, let us add some noise to `Age2` using `Age2 <- Age2 + rnorm(length(Age2), mean=0, sd=x)` where you choose different values for x. Each time plot the relationship and compute the correlation coefficient.

6.2 Consider the following joint PMF for random variables X and Y where $x \in \{0, 1\}$ and $y \in \{0, 1, 2\}$:

$$
P_{XY}(x, y) = \begin{cases}
\frac{2}{12} & \text{if } x = 0 \text{ and } y = 0 \\
\frac{3}{12} & \text{if } x = 0 \text{ and } y = 1 \\
\frac{1}{12} & \text{if } x = 0 \text{ and } y = 2 \\
\frac{1}{12} & \text{if } x = 1 \text{ and } y = 0 \\
\frac{1}{12} & \text{if } x = 1 \text{ and } y = 1 \\
\frac{4}{12} & \text{if } x = 1 \text{ and } y = 2
\end{cases}
$$

1. What is the (marginal) expectation of X?
2. What is the conditional expectation of $Y|X = 1$?
3. Compute $\mathbb{E}(XY)$.
4. Are X and Y independent?

6.3 Consider the following questions:

1. Suppose $X \sim \mathcal{N}(0, 1)$ is a random variable and $Y = X^2$. What is the covariance $\text{COV}(X, Y)$ between X and Y? And ρ_{XY}? If you are unsure try it using simulations.
2. Can you think of another example where two variables are dependent but their correlation is zero?
3. How would you quantify the association between a variable of ordinal measurement level and one of interval measurement level? And nominal and interval? (Note that in different fields people use different methods. Try searching online for some proposed solutions and explore them.)

6.4 In this exercise we will calculate Pearson's rho, Spearman's rho, and Kendall's tau using the face data. We will use the variable on the first dimension and the rating variable.

1. Calculate the three correlation coefficients for the first dimension with the rating variable.
2. Provide a 95% confidence interval on the correlation coefficients. Based on the estimates what do you conclude on the association?
3. Produce a scatter plot between the first dimension and the rating. Based on this scatter plot, which correlation coefficient would you recommend for these two variables?

6.5 In this exercise we will calculate the measures of association between binary variables from the voting data. We are interested in the question of whether there is a relationship between religion and voting behavior. The binary variables of interest are "Church" and "Vote".

1. Create the 2×2 contingency table.

2. Calculate Pearson's chi-square statistic. Does this statistic show a dependency between religion and voting?
3. Calculate Pearson's ϕ coefficient between voting and religion. Do you think there is a dependency between the two variables?
4. Calculate Yule's Q statistic between voting and religion. Do you think there is a dependency between the two variables?
5. Do you think there is a difference in interpretation between Pearson's ϕ coefficient and Yule's Q statistic? If so, which one would you trust more in this setting?

6.6 In this exercise we will calculate the measures of risk for the voting data. We are interested in the question of whether religious people will vote more than non-religious people. The binary variables of interest are "Church" and "Vote".

1. Create the 2×2 contingency table
2. Calculate the proportion of voters for religious and non-religious people with their 95% confidence intervals
3. Calculate the risk difference on voting for religious people against non-religious people. Provide a 95% confidence interval on the risk difference.
4. Calculate the relative risk on voting for religious people against non-religious people. Provide a 95% confidence interval on the risk difference.
5. Calculate the odds ratio on voting for religious people against non-religious people. Provide a 95% confidence interval on the risk difference.
6. Based on the measures of risk, what is your conclusion about the dependence between voting and religion?

6.7 In this exercise we will calculate similarity measures between GMO1 and GMO2 potatoes. We would like to know whether the two GMO potatoes are similar on the active gene profile.

1. Create the 2×2 contingency table between the GMO1 and GMO2 potato.
2. Calculate Cohen's kappa statistic on the GMO1 and GMO2 data and provide the 95% confidence interval. What do you conclude on the agreement of the genetic profile?
3. Do you think that Cohen's Kappa statistic is suitable measure to quantify the similarity between the two GMO potatoes?
4. Calculate the similarity measures Sokal & Michner (S_1), Roger & Tanimoto (S_2), Jaccard (T_1), and Sokal & Sneath (T_2) for the GMO1 and GMO2 potatoes. What do you conclude on the similarity? Explain your answer.
5. Describe which similarity measure you would use to quantify the similarity on the active gene profile. Explain your answer.

6.8 In this exercise we will calculate measures of association between the nominal variables breakfast and favorite school subject using the data of the high-school students. We wish to answer the question of whether there is an association between school subjects and the eating of breakfast before school.

1. Create the 2×2 contingency table between breakfast and subject. What is the number of degrees of freedom in this contingency table?

2. Calculate Pearson's chi-square statistic. Do you think this indicates a strong association between breakfast and subject? Explain your answer.
3. Calculate Cramér's V statistic. Do you think there is a dependency between breakfast and school subject? Explain your answer.

6.9 In this exercise we will calculate measures of association between the ordinal variables age and allowance using the high-school data. Here the variable allowance will be changed into a categorical variable with just three levels: 0, (0, 10], and $(10, \infty)$. We are only interested in high-school students with an age in the interval $[10, 14]$. We wish to answer the question of whether there is an association between age and allowance categories.

1. Create the new ordinal variable for allowance and report the number of students in each of the three levels.
2. Create the 2×2 contingency table between age and the ordinal allowance variable. What is the number of degrees of freedom in this contingency table?
3. Calculate Pearson's chi-square statistic. Do you think this indicates a strong association between age and allowance? Explain your answer.
4. Calculate Goodman and Kruskal's G statistic. Do you think there is a dependency between age and allowance? Explain your answer.

References

A.N. Albatineh, Means and variances for a family of similarity indices used in cluster analysis. J. Stat. Plan. Inference **140**(10), 2828–2838 (2010)

N. Balakrishnan, C.D. Lai, *Continuous Bivariate Distributions* (Springer Science & Business Media, 2009)

D.G. Bonett, T.A. Wright, Sample size requirements for estimating Pearson, Kendall and Spearman correlations. Psychometrika **65**(1), 23–28 (2000)

C.B. Borkowf, Computing the nonnull asymptotic variance and the asymptotic relative efficiency of Spearman's rank correlation. Comput. Stat. Data Anal. **39**(3), 271–286 (2002)

J.C. Caruso, N. Cliff, Empirical size, coverage, and power of confidence intervals for Spearman's rho. Educ. Psychol. Measur. **57**(4), 637–654 (1997)

S.-S. Choi, S.-H. Cha, C.C. Tappert, A survey of binary similarity and distance measures. J. Syst., Cybern. Inform. **8**(1), 43–48 (2010)

J. Cohen, A coefficient of agreement for nominal scales. Educ. Psychol. Measur. **20**(1), 37–46 (1960)

A.A. Fatahi, R. Noorossana, P. Dokouhaki, B.F. Moghaddam, Copula-based bivariate ZIP control chart for monitoring rare events. Commun. Stat.-Theory Methods **41**(15), 2699–2716 (2012)

E.C. Fieller, H.O. Hartley, E.S. Pearson, Tests for rank correlation coefficients. I. Biometrika **44**(3/4), 470–481 (1957)

E.C. Fieller, E.S. Pearson, Tests for rank correlation coefficients: II. Biometrika, pp. 29–40 (1961)

R.A. Fisher, On the "probable error" of a coefficient of correlation deduced from a small sample. Metron **1**, 1–32 (1921)

R.A. Fisher, Frequency distribution of the values of the correlation coefficient in samples from an indefinitely large population. Biometrika **10**(4), 507–521 (1915)

S. Fuchs, Y. McCord, K.D. Schmidt, Characterizations of copulas attaining the bounds of multivariate Kendall's tau. J. Optim. Theory Appl. **178**(2), 424–438 (2018)

F. Galton, I. Co-relations and their measurement, chiefly from anthropometric data. Proc. R. Soc. Lond. **45**(273-279), 135–145 (1889)

L.A. Goodman, W.H. Kruskal, Measures of association for cross classifications, in *Measures of Association for Cross Classifications* (Springer, Berlin, 1979), pp. 2–34

J.C. Gower, P. Legendre, Metric and Euclidean properties of dissimilarity coefficients. J. Classif. **3**(1), 5–48 (1986)

D.E. Hinkle, W. Wiersma, S.G. Jurs, *Applied Statistics for the Behavioral Sciences*, vol. 663 (Houghton Mifflin College Division, Boston, 2003)

P. Jaccard, Distribution de la flore alpine dans le bassin des dranses et dans quelques régions voisines. Bull. Soc. Vaudoise Sci. Nat. **37**, 241–272 (1901)

N.L. Johnson, S. Kotz, N. Balakrishnan, *Discrete Multivariate Distributions*, vol. 165 (Wiley, New York, 1997)

M.G. Kendall, A new measure of rank correlation. Biometrika **30**(1/2), 81–93 (1938)

J.D. Long, N. Cliff, Confidence intervals for Kendall's tau. Br. J. Math. Stat. Psychol. **50**(1), 31–41 (1997)

A.W. Marshall, I. Olkin, Families of multivariate distributions. J. Am. Stat. Assoc. **83**(403), 834–841 (1988)

P. Moran, Rank correlation and product-moment correlation. Biometrika **35**(1/2), 203–206 (1948)

R.B. Nelsen, *An Introduction to Copulas* (Springer Science & Business Media, 2007)

R.G. Newcombe, Interval estimation for the difference between independent proportions: comparison of eleven methods. Stat. Med. **17**(8), 873–890 (1998)

R.L. Plackett, A class of bivariate distributions. J. Am. Stat. Assoc. **60**(310), 516–522 (1965)

M.-T. Puth, M. Neuhäuser, G.D. Ruxton, Effective use of Spearman's and Kendall's correlation coefficients for association between two measured traits. Anim. Behav. **102**, 77–84 (2015)

S. Ross, *A First Course in Probability* (Pearson, London, 2014)

K. Samuel, B. Narayanswamy, L. Johnson Norman, *Continuous Multivariate Distributions, Models and Applications* (2004)

W.R. Schucany, W.C. Parr, J.E. Boyer, Correlation structure in Farlie-Gumbel-Morgenstern distributions. Biometrika **65**(3), 650–653 (1978)

M.J. Warrens, On similarity coefficients for 2×2 tables and correction for chance. Psychometrika **73**(3), 487 (2008)

J. Yang, Y. Qi, R. Wang, A class of multivariate copulas with bivariate Fréchet marginal copulas. Insur.: Math. Econ. **45**(1), 139–147 (2009)

Chapter 7
Making Decisions in Uncertainty

7.1 Introduction

Up till now we have covered ways of summarizing data, and we have paid a lot of attention to understanding how summaries computed on sample data (sample statistics) vary as a function of the sampling plan and the population characteristics. In Chap. 5 we also covered the idea that we can use our sample to estimate population parameters; in this case we are basically making a decision—our best guess—regarding the population parameter given the sample data. The distribution function of the estimators—which are themselves sample statistics—that we studied in Chap. 5 gives us some feel for the precision of our inferences. However, what if your estimate of the population parameter is 10, and someone asks whether you are sure that it is not 10.2; what would your answer then be? In this Chapter we examine multiple approaches to answering this seemingly simple question.

A lot of the practical use of statistical analysis is to make decisions based on data. In this chapter we will cover some approaches to this end, although admittedly we will not provide a thorough overview of all the methods for decision-making under uncertainty that people have come up with over the years. We will focus on two different methods:

1. **Bootstrapping:** We will first focus on making decisions regarding population parameters based on a relatively simple procedure that is called bootstrapping. The bootstrap provides a very general way to obtain a quantification of the uncertainty of an estimator. We have already seen that obtaining a larger sample decreases the variance of an estimator (i.e., the estimator becomes more precise). Furthermore, when estimating a population mean or difference in population means, we find that a smaller population variance leads to a smaller variance of the estimator. The bootstrap has these exact same properties and is easy to carry out for many sample statistics; it provides a first entry into making decisions regarding populations based on sample data. We will introduce the bootstrap—and different variants thereof—in Sect. 7.2.

© Springer Nature Switzerland AG 2022
M. Kaptein and E. van den Heuvel, *Statistics for Data Scientists*, Undergraduate Topics in Computer Science, https://doi.org/10.1007/978-3-030-10531-0_7

2. **Hypothesis testing:** We often want to make binary decisions: does this medication have an effect—yes or no? Are the means of these two groups in the population equivalent yes or no? There is a large body of work in statistics that focusses on making such binary decisions; we will discuss this approach in some detail. In Sect. 7.3 we introduce the basic setup and discuss the errors that can occur when making binary decisions. Next we discuss hypothesis testing and provide a number of examples of different significance tests. In Sect. 7.3.3 we relate null hypothesis testing to confidence intervals and subsequently discuss *equivalence* testing. In Sect. 7.3.9 we discuss how we can make decisions regarding outliers and in Sect. 7.3.8 we discuss hypothesis testing for normality.

Note that in this chapter we omit a formal treatment of decision theory, which provides a formalized mathematical view on decision-making in uncertainty that extends to all kinds of decisions. We cover decision theory briefly in Chap. 8 when we discuss Bayesian methods, but we refer the interested student to Robert (2007). Furthermore, we do not go into much depth regarding hypothesis testing: we merely provide the main intuition and discuss a number of commonly used tests. We refer the interested reader to classical works like Lehmann and Romano (2006) for a much more detailed discussion of statistical hypothesis testing.

In this chapter you will learn:

- The use of non-parametric, parametric, and online bootstrap to quantify uncertainty and support decision making
- The errors involved in binary decision-making
- The rationale behind hypothesis testing
- The p-value
- The relationship between p-values and confidence intervals
- How to do a number of standard test statistics (e.g., t-tests, χ^2-tests).
- How to test for normality
- How to identify outliers in datasets
- How to conduct equivalence tests

7.2 Bootstrapping

To make our first steps into the area of decision-making, we will first provide a conceptually simple method (which can be made formal in many situations, see e.g., (Efron 1992), to address the following problem: Given an estimated sample statistic $\hat{\theta}$, can we say that the population value of that statistic is $\hat{\theta} \pm \delta$? Or, can we say something about how likely it is that the population value is larger than $\hat{\theta} + \delta$? Note that without making any assumptions regarding the sampling process and/or the population distributions involved, it is practically impossible to say anything about the population based on sample data with full certainty. However, informally, it is quite clear that a large random sample and a relatively small variance of the estimator $\hat{\theta}$ should both increase our confidence regarding statements we can make about the population.

The *bootstrap* provides a very simple way—at least when you have access to modern-day computers—to quantify the uncertainty of virtually any estimator given sample data. Once you have a quantification of this uncertainty—in the form of a distribution function—you can use this distribution function to make decisions. Note that the distribution of the estimator $\hat{\theta}$—either over repeated sampling or otherwise—is key to many statistical decision procedures; this will be true also for the hypothesis testing we discuss later in this chapter.

7.2.1 The Basic Idea Behind the Bootstrap

Given a (random) sample of size n from some population with distribution function F_X, we frequently set out to obtain an estimate of a population parameter $\theta = T(x)$.[1] Our point estimate of the parameter of interest is often what is called the "plug-in estimator" for θ: $\hat{\theta} = T(x_1, \ldots, x_n)$; i.e., it is the statistic of interest calculated on the sample data $x_1, \ldots, x_n$. Note that we have routinely used plug-in estimators for population statistics earlier in the book: e.g., we used $\bar{x} = \hat{\mu} = \sum_{x=1}^{n} x_n/n$ as a plug-in estimate for the population mean μ.

Next to our point estimate, we are often also interested in the distribution function of $\hat{\theta}$ over repeated samples: i.e., if we repeatedly obtain a (random) sample of size n from the population of interest, what would $F_{\hat{\theta}}$ look like? We are interested in $F_{\hat{\theta}}$ as it gives us information about the variability of our estimate over repeated samples: a key ingredient we can use to quantify the certainty surrounding our point estimate $\hat{\theta}$. Again, we have already discussed such properties for specific estimators. For instance, we investigated $F_{\hat{\mu}}$, with $\hat{\mu}$ the sample mean, when the density f_X is a normal density. The distribution function function $F_{\hat{\mu}}$ will be a normal distribution function (see Chap. 5). More generally, without making any assumptions on f_X, we calculated the standard error of the sample mean (see, e.g., Chaps. 2 and 5), which is simply the standard deviation of a random variable having CDF $F_{\hat{\mu}}$. Thus, depending on the statistic involved, the sampling plan, and the assumptions one is willing to make about F_X or f_X, we might be able to analytically derive $F_{\hat{\theta}}$. However, obtaining $F_{\hat{\theta}}$ in general (or properties thereof) can be challenging.

The bootstrap addresses the problem of deriving $F_{\hat{\theta}}$ using the power of computer simulation. It is appealing because it replaces the analytical approach that we have explored up till now by a simulation approach. Thus, the bootstrap can in principle approximate $F_{\hat{\theta}}$ for arbitrary statistics $\hat{\theta} = T(x_1, \ldots, x_n)$. The logic is quite simple: if we have a good estimate of the population distribution function F_X, i.e., $\hat{F}_X$, we can simply program a computer to obtain M samples (by simulation) of size n from

---

[1] If the population is finite, $T(x)$ represents a calculation on the values from all population units, as we discussed in Chap. 2. If the population is infinite, the $T(x)$ represents a characteristic of the random variable X having distribution function function F_X or density f_X, as we discussed in Chaps. 4 and 5. For instance, we may be interested in the expected value $\mathbb{E}(X)$, which would become a function of the parameters of the density f_X.

$\hat{F}_X$ using the same sampling plan that we have used to collect our initial sample. If $\hat{F}_X$ is close to F_X it does not really matter if we draw from $\hat{F}_X$ or F_X. On each sample $m = 1, \ldots, M$ we can subsequently compute the statistic of interest, $\hat{\theta}^{(1)}, \ldots, \hat{\theta}^{(M)}$ which themselves serve as approximate samples from $F_{\hat{\theta}}$ (approximate as we are using $\hat{F}_X$, and thus we obtain samples from $\hat{F}_{\hat{\theta}}$). As long as our estimate of $\hat{F}_X$ is close to the true F_X, our samples of $\hat{F}_{\hat{\theta}}$ can be used to approximate properties of $F_{\hat{\theta}}$: e.g., the standard error of a statistic can simply be computed by computing the standard deviation of the M samples of the statistic of interest:

$$\hat{SE}(\hat{\theta}) = \sqrt{\frac{\sum_{m=1}^{M}(\hat{\theta}^{(m)} - \bar{\theta}^2}{M - 1}}, \qquad (7.1)$$

where $\bar{\theta} = \sum_{m=1}^{M} \hat{\theta}^{(m)}/M$.

This leaves open the question of obtaining $\hat{F}_X$: how do we obtain an estimate of the distribution function of the variable of interest in the population? The simplest bootstrap approach—although others exist—is to simply use the *empirical distribution function*: the original samples $x_1, \ldots, x_n$ in our sample can be used to construct a discrete approximation of F_X by simply giving each unique value v_i in $x_1, \ldots, x_n$ probability $\frac{1}{n}$. Thus, we use the observed distribution function of X directly as our estimate for F_X. As a simple example, consider a sample of $n = 6$ binary observations $0, 1, 0, 1, 1, 1$. In this case we obtain

$$\hat{F}_X(x) = \begin{cases} 0 & \text{if } x < 0 \\ p_0 & \text{if } x \le 0 \\ p_0 + p_1 & \text{if } x \le 1 \end{cases} \qquad (7.2)$$

where $p_0 = \frac{2}{6}$ and $p_1 = \frac{4}{6}$. We can now use $\hat{F}_X$, in combination with computer simulation, to generate bootstrap samples $m = 1, \ldots, M$ and compute $\hat{\theta}^{(m)}$ for arbitrary statistics T. Note that the empirical distribution function will provide a reasonable approximation for F_X if the sample size n is large, and the sample has been obtained under *simple random sampling*. If the latter is not the case, we would obviously need to correct for the sampling plan when trying to construct $\hat{F}_X$.

The empirical distribution solution described above is often referred to as the *non-parametric bootstrap*: no parametric assumptions regarding F_X are made in this procedure. While appealing because of its generality, the non-parametric bootstrap might sometimes be outperformed (in terms of, e.g., estimation precision) by the so-called *parametric bootstrap*: in the parametric bootstrap $\hat{F}_X$ is assumed to be of a certain form (e.g., it is assumed to be normal), and plug-in estimates for its parameters (e.g., $\hat{\mu}$ and $\hat{\sigma}^2$ in the normal case) are used to estimate F_X. If the assumptions are correct, the parametric bootstrap is preferable over the non-parametric bootstrap. In the sections below we describe how the non-parametric and the parametric bootstrap can be carried out in R. Note that in this book we will not cover the theoretical properties of the bootstrap; we refer the interested reader to Tibshirani and Efron (1993) for more details.

Fig. 7.1 Histogram of
means computed on 10,000
bootstrap replicates of our
sample data

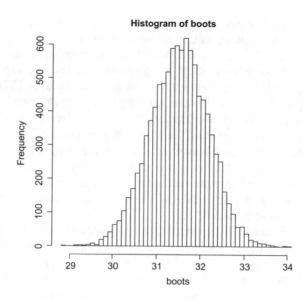

7.2.2 Applying the Bootstrap: The Non-parametric Bootstrap

The following code computes bootstrap replicates $\hat{\theta}^{(1)}, \ldots, \hat{\theta}^{(M)}$ for the sample mean
(thus $\hat{\theta} = \sum_{i=1}^{n} x_i/n$) by explicitly resampling $M - 10{,}000$ samples of size n. Here,
the sample is drawn from a known normal population with $\mu = 32$, and $\sigma = 5$. Note
that to generate the M bootstrap replicates, we (re-)sample the original sample *with
replacement*. Resampling the original sample with replacement effectively generates
a simple random sample from a population with distribution function F_X where the
empirical distribution function in the sample data is used to construct $\hat{F}_X$.

Figure 7.1 provides a histogram of these M so-called bootstrapped means. The
final line of R code was used to create the histogram (note that we first obtain a
sample of 50 observations from a normal population with mean 32 and standard
deviation 5, thus, in this simulation $F_X(x) = \Phi((x - 32)/5)$:

```
> n <- 50
> sample_data <- rnorm(n, mean=32, sd=5)
> M <- 10000 # Number of bootstrap replicates.
> boots <- rep(NA, times=M)
> for (m in 1:M) {
+    resamp <- sample(sample_data, size=n, replace=TRUE) # Note
     the sampling WITH replacement.
+    boots[m] <- mean(resamp)
+ }
> hist(boots, breaks=50)
```

The histogram seems to closely resemble a normal PDF: as we know from our
earlier analytical treatments, for normal populations the distribution function $F_{\hat{\mu}}$ is
indeed normal with $\mu = 32$ and $\sigma^2 = 25/50 = 0.5$. The histogram however gives us

a quantification of the uncertainty of the sample mean purely based on the sample data without any additional assumptions regarding the population.

The distribution function of bootstrap replicates can easily aid us in our subsequent decision-making: a population mean between approximately 30 and 33 seem plausible (which includes the true population mean of 32), while values smaller than 29 or larger than 34 seem very unlikely. If we want to make a binary decision—e.g., decide whether our population mean is larger or equal to 33, we can compute the proportion p of the bootstrap replicates that satisfy this criterion and, after setting some confidence bound p^{cut} we accept the proposed idea if $p > p^{cut}$.[2] Note that for such a procedure, as a rule of thumb, often a minimum of $M = 10,000$ bootstrap samples is recommended and samples sizes (of the original sample that is) of $n = 20$ are often considered the minimum (see Davidson and MacKinnon 2000) for a more formal discussion).

Interestingly, we can bootstrap not only the mean, but all kinds of sample statistics. Again, this is one of the great appeals of the bootstrap: for some estimators, deriving analytical results, such as an analytical statement of the standard error, might be hugely complex. In such cases in particular, the bootstrap procedure will give you an easy method of getting some idea of the uncertainty of your estimates. Here is the R code to carry out the bootstrapping procedure for the variance:

```
> # But, we can bootstrap all kinds of things:
> M <- 10000 # Number of bootstrap replicates.
> boots.var <- rep(NA, times=M)
> for (m in 1:M) {
+     resamp <- sample(sample_data, size=n, replace=TRUE)
+     boots.var[m] <- var(resamp)
+ }
> hist(boots.var, breaks=50)
```

The results are displayed in Fig. 7.2.

The histogram shows that the distribution function of the variance is somewhat skewed to the right. The right tail of the histogram is longer than the left tail of the histogram. We know this from analytical derivations when the distribution function of the original data F_X is normal. The distribution function of the sample variance follows a chi-square distribution function: $(n - 1)s^2/\sigma^2 \sim \chi^2_{n-1}$. The histogram shows that the true variance of the population is somewhere between 10 and 35, with a high probability. This contains the variance of 25 that we used for the simulation.

---

[2] The proportion p is, in some cases when M is large, indeed a good estimate of the probability that the population parameter falls within a certain range. Theoretical results that support this motivate the appeal of the bootstrap; however, for these to hold we need to make assumptions regarding the population. Here we discuss the bootstrap informally and do not examine these theoretical results.

Fig. 7.2 Histogram of bootstrap replicates of the sample variance

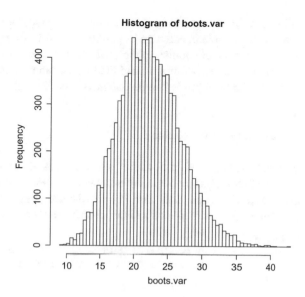

7.2.3 Applying the Bootstrap: The Parametric Bootstrap

What we have just covered is what is called the "non-parametric" bootstrap. As stated, other version(s) of the bootstrap exist, a very common one being called the *parametric* bootstrap. In this version of the bootstrap we do not resample the actual data, but rather create new datasets (as we have been doing ourselves earlier) based on properties of the observed sample (such as the mean and the standard deviation). We will not dig into the details of the parametric bootstrap, but it's good to know that it exists, and that—while introducing additional assumptions that might be erroneous— it sometimes provides better performance than the non-parametric bootstrap (again, see (Tibshirani and Efron 1993) for theoretical details). Here is the R code to generate a *parametric* bootstrap for a sample mean when we assume a normal population:

```
> K <- 10000 # Number of bootstrap replicates.
> par.boots <- rep(NA, times=K)
> xbar <- mean(sample_data)
> s <- sd(sample_data)
> for (k in 1:K) {
+   resamp <- rnorm(n, mean=xbar, sd=s)
+   par.boots[k] <- mean(resamp)
+ }
> summary(par.boots)
  Min. 1st Qu. Median  Mean 3rd Qu.  Max.
 28.48   30.83  31.30 31.30   31.76 33.93
```

In this specific case we generate a new sample each time from a normal distribution (using the `rnorm` function) where the mean and standard deviation are estimated using the mean and standard deviation of the sample. This illustrates the basic idea of the parametric bootstrap: if we assume some population distribution function, we can use the sample to estimate its parameters, and subsequently use this distribution function to generate bootstrap replicates. In the exercises we will further explore the results of both the parametric and the non-parametric bootstrap and compare these results to the asymptotic results obtained in earlier chapters.

7.2.3.1   The boot package

As you have seen before, R comes with all kinds of handy extensions (packages). A package for computing bootstrap estimates is called boot. It can be used to compute all kinds of bootstraps (see ?boot to read more), but a simple bootstrap of the mean (the non-parametric one we just discussed) can be computed like this:

```
> library(boot)
> bmean <- function(data, indices) {mean(data[indices])}
> boots <- boot(sample_data, statistic=bmean, R=10000)
> boots

ORDINARY NONPARAMETRIC BOOTSTRAP


Call:
boot(data = sample_data, statistic = bmean, R = 10000)


Bootstrap Statistics :
    original        bias    std. error
t1* 31.31461 -0.003718269 0.6822525
```

Here, the boots object will contain the bootstrap samples (and more). Once you generate the bootstrap samples, print(boots) and plot(boots) can be used to examine the results. Furthermore, you can use the boot.ci() function to obtain confidence intervals for the statistic(s). Note that you will have to pass a function to the boot() function that takes as arguments the data and a so-called index (the selected data); hence we had to write a new function to compute the mean ourselves, as the standard mean function does not accept the indices argument. Always look at ?boot if you are unsure.

7.2.4   Applying the Bootstrap: Bootstrapping Massive Datasets

The bootstrap is interesting because it allows us to gain quick insight into the variability of many different estimators (and even functions of multiple estimators, etc.). However, it can be computationally demanding. In particular, when you are working with extremely large datasets, or with data that arrives continuously in a so-called "data stream" (i.e., the data points are observed one by one, in sequence), boot-strapping might be computationally too demanding. A way to solve this is to use a so-called "online" (or "streaming" or "row-by-row") bootstrap.

The online bootstrap (see, e.g., Eckles and Kaptein 2019) was originally designed to deal with situations in which data points $x_1, \ldots, x_t$ arrive sequentially, over time. Thus, the sample $x_1, \ldots, x_t$ is continuously augmented (i.e., x_{t+1} gets added). In such cases, generating M bootstrap samples anew each time a new data point is added is computationally very demanding and, if the data points are observed in rapid sequence, perhaps even impossible. This is for example true when we consider data of the behavior of users of a website that is being logged to the web-servers: every user visiting the website will create a new data point and as such she will augment the sample. In such a case we can, instead of generating M bootstrap samples each time the sample is augmented, maintain M different estimators and "update" these estimators each time a new data point is observed. If we add some randomness to our updating mechanism by which, when a new data point arrives, some of the M estimators are updated while some are not, we—over time as the data are continuously augmented—end up with M different estimators that quantify the variability of interest.

To illustrate, suppose we are interested in quantifying the variability in an esti-mated proportion. Using—for illustration purposes—$M = 4$ and further suppos-ing that we initialize our bootstrapped estimates $p_1 = p_2 = p_3 = p_4 = 0.5$ using an effective sample size of $n_1 = n_2 = n_3 = n_4 = 2$, we can examine how the first datapoint would affect the M estimates. Suppose the first datapoint $x_1 = 1$, and we simply flip a fair coin M times to generate our randomness. If the sequence of coin tosses is $1, 1, 0, 1$, this implies that we update p_1, p_2, and p_3 as follows: $p_m = (p_m n_m + x_1)/(n_m + 1)$ and we subsequently update $n_m = n_m + 1$ (again, only for 1, 2, and 4). We thus end up with $p_1 = 2/3, p_2 = 2/3, p_3 = 0.5, p_4 = 2/3$. Next, if $x_2 = 0$, and our sequence of coin tosses is $0, 0, 1, 1$, we end up with $p_1 = 2/3, p_2 = 2/3, p_3 = 1/3, p_4 = 2/4$, etc. It is clear that as new datapoints are observed, this procedure will lead to M different estimates—each based on a slightly different dataset selected by virtue of our introduced randomness. A histogram over the M estimates can now again be used to examine the variability of the estimator of interest.

The online bootstrap for a sample mean can be implemented in R using the following code:

```
> # Online bootstrap:
> K <- 10000
> online.means <- rep(0, times=M)
> online.counts <- rep(0, times=M) # This could have been a
    matrix.
>
> # Function for online update of mean:
> update.mean <- function(current, n, x) {
+    return(current+(x-current)/n)
+ }
>
```

Fig. 7.3 Bootstrapping using the online bootstrap

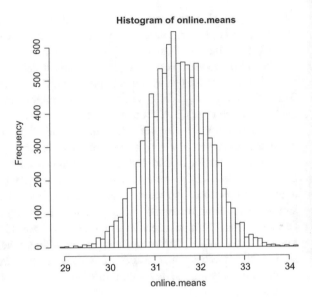

> # Run through the data in sequence:
> for (i in 1:length(sample_data)) {
+   # Sample / generate the randomness to determine which of the
    M samples to update:
+   update <- sample(c(TRUE, FALSE), size=K, replace=TRUE)
+   # Update the selected samples:
+   online.counts[update] <- online.counts[update] + 1
+   online.means[update] <- update.mean(current=online.means[
    update], n=online.counts[update], x=sample_data[i])
+ }
>
> summary(online.means)
  Min. 1st Qu. Median   Mean 3rd Qu.   Max.
 29.09 31.16 31.60 31.59 32.02 34.21
> hist(online.means, breaks=50)

This code produces Fig. 7.3. Note that instead of resampling the data multiple times, we actually maintain multiple (in this case $J = 10{,}000$) estimates, and for each datapoint that we encounter in the data we "update" our estimate with probability 0.5. Note that we could do this on much larger streams of data and that this is much more feasible computationally than the original non-parametric bootstrap; we will examine this in more detail in the assignments.

7.2.5 A Critical Discussion of the Bootstrap

The bootstrap provides an appealing, and very general, method of the quantifying uncertainty of any statistic $T(x)$. However, as is often the case with statistical methods, the approach also has its own caveats. It is relatively easy to see when the bootstrap procedure will provide us with poor results (although it is not always easy to solve). Recall that effectively the bootstrap procedure consists of two parts:

1. First, we estimate F_X based using our sample $x_1, \ldots, x_n$. This gives us $\hat{F}_X$.
2. Second, we obtain M random samples from $\hat{F}_X$ (each of size n), on which we computed our bootstrap estimates of the statistic of interest $\hat{\theta}^{(1)}, \ldots \hat{\theta}^{(M)}$ which we regard as (approximate) samples from $F_{\hat{\theta}}$.

Each of these parts can wreak havoc. First, it might be the case that $\hat{F}_X$ is a very poor estimate of F_X. This is often the case when n is small, but it might also be caused by the fact that the original sample $x_1, \ldots, x_n$ is not obtained through simple random sampling. If the latter is the case, the sampling scheme that was used should be taken into consideration when computing $\hat{F}_X$; this can be challenging. For well-known random sampling procedures such as cluster sampling, variations to the bootstrap that still provide accurate result are known (see (Tibshirani and Efron 1993) for details), but for more challenging—or unknown—sampling schemes assessing the accuracy of $\hat{F}_X$ is hard. Next, obviously, the sampling scheme implemented in the second step presented above should mimic the sampling scheme that was originally used: if the M bootstrap samples are generated using a different sampling scheme than the sampling scheme of interest, $F_{\hat{\theta}}$ might not be properly approximated by the M bootstrap samples. In our discussion above we assumed simple random sampling; resampling the sample with replacement as we discussed for the non-parametric bootstrap generates simple random samples from a population with distribution function $\hat{F}_X$. If, however, we are interested in the variability of our estimator over differently obtained samples, we should adopt our bootstrapping procedure accordingly. Thus, although the bootstrap is appealing as it allows one to quantify the uncertainty for virtually any statistic—by simply replacing tedious analytical work with simple computer operations—one should always be careful: for complex sampling schemes and complex population distributions, $\hat{F}_X$, or the resulting M bootstrap samples of the statistic of interest, might not provide a good quantification of the uncertainty associated with $\hat{\theta}$.

7.3 Hypothesis Testing

While the bootstrap provides an easy way of quantifying uncertainty that we can use to make decisions, it is hard in general to make statements about the *quality* of these decisions. Hypothesis testing provides a method for making binary decisions that, in many instances, does give us clear quantitative statements about the quality of the decisions we make.

Within hypothesis testing the general setup is as follows: we state our decision problem as a choice between two competing hypotheses regarding the population, often called the *null hypothesis* H_0, and the *alternative hypothesis* H_a. To provide a concrete example, we might be interested in testing whether the population mean $\mu(f)$ is equal to or below a certain known value μ_0. The respective hypotheses can in this case be formulated as follows:

$$H_0 : \mu(f) \leq \mu_0$$
$$H_a : \mu(f) > \mu_0$$

In this section we will discuss several different null hypotheses that would be useful in many different applications.

The subsequent rationale of hypothesis testing is that we assume that the null hypothesis is true and that we gather *sufficient* evidence to demonstrate that it is not true. Thus the goal of hypothesis testing is to reject the null hypothesis on the basis of sufficient and well-collected data. We will make the notion of "sufficient evidence" more precise below. However, before we do so it is good to think about the possible errors involved when making a binary decision between two hypotheses. Given that H_0 and H_a are complementary, one of the two *must* be true in the population. We will be making a decision in favor of one of the two hypotheses based on (random) sample data. We will decide that either H_0 is rejected (thus H_a must be true) or is not rejected (thus there is no or not enough evidence to demonstrate that H_0 is false). This setup gives rise to four different situations as depicted in Fig. 7.4.

In two of these situations the decision matches with the population: we do not reject H_0 when H_0 is true in the population (the upper left corner in Fig. 7.4) and we reject H_0, and subsequently accept H_a, when H_a is true in the population (the lower right corner in Fig. 7.4). We can also err on two sides. First of all, we can make a false positive or *type 1* error. In this case we reject H_0 while in reality it is true (bottom left corner in Fig. 7.4). The probability of a type 1 error is associated with the level α. The α is used as a maximal allowable type 1 error for a decision rule. Finally, we can also make a *type 2* error. In this case we do not reject H_0, while in actuality H_a is true (the top right corner in Fig. 7.4). The probability of a type 2 error is associated with the level β. The value β is used a maximal allowable type 2 error.[3] One minus the type 2 error is called the *power* of the binary decision rule. It indicates how likely the null hypothesis is rejected when the alternative hypothesis is true.

Note that it is easy to create a decision procedure that has a type 1 error probability equal to zero: if we simply *never* reject H_0—in this case basically we state that there is never sufficient evidence to reject H_0—we will never make a type 1 error. While this decision procedure does control the type 1 error, it is clearly not very useful, as the power of this decision rule is zero: we never accept the alternative hypothesis

---

[3] These two errors generally exist for any binary decision. Suppose we need to make a choice between A and B: we can choose A while B is true, or choose B when A is true. In different contexts these two types of errors have different names, such as sensitivity and specificity in diagnostic testing, as discussed in Chap. 3.

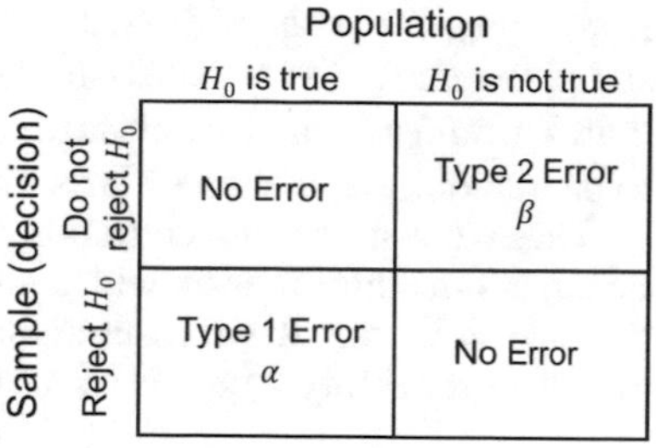

Fig. 7.4 Types of errors in
hypothesis testing

		Population	
		H_0 is true	H_0 is not true
Sample (decision)	Do not reject H_0	No Error	Type 2 Error β
	Reject H_0	Type 1 Error α	No Error

when it is true. To alleviate this problem, the procedure of hypothesis testing aims
to be less conservative (e.g., it will reject the null hypothesis sometimes but not too
often when it would be true). It defines *sufficient evidence* such that the probability
of making a type 1 error is *at most* α.[4] The level α is called the *significance level* and,
as we have already mentioned, it is the maximal allowable probability of rejecting
the null hypothesis when the null hypothesis is actually true. It is often set equal to
a value of $\alpha = 0.05$ or $\alpha = 0.01$.

7.3.1 The One-Sided z-Test for a Single Mean

To illustrate the concept of hypothesis testing, now that we understand the type
of mistakes we can make, we will return to the hypothesis on the population mean
above: $H_0 : \mu(f) \leq \mu_0$ versus $H_a : \mu(f) > \mu_0$. We will illustrate how it works, using
asymptotic theory, i.e., we assume that the sample is large enough to be able to use
the normal distribution function as an approximation to the distribution function of
the statistic that we are using to make a decision about the null hypothesis.

Let's assume that we have collected a random sample $Y_1, Y_2, \ldots, Y_n$ from the
population. We may estimate the population mean $\mu(f)$ with the sample average $\bar{Y}$.
Clearly, when $\bar{Y}$ is smaller or equal to μ_0 the random variable $\bar{Y}$ seems to be in line
with the null hypothesis $H_0 : \mu(f) \leq \mu_0$. In other words, there is no evidence that the
null hypothesis is false. Although the random variable does not suggest any conflict
with the null hypothesis $H_0 : \mu(f) \leq \mu_0$, it does not guarantee that $\mu(f) \leq \mu_0$ either.
Indeed, if the population mean $\mu(f)$ were somewhat larger than μ_0, it might not be
completely unlikely to observe a sample average still below μ_0 due to the sampling
(a type 2 error). When $\bar{Y}$ is larger than μ_0, we might start to believe that the null
hypothesis is incorrect. However, when $\bar{Y}$ is just a little higher than μ_0 this might not
be very unlikely either, even when $\mu(f) \leq \mu_0$. For instance, for any symmetric f at

---

[4] Here we discuss merely bounds on the type 1 error for a given sample size n. Subsequent theory
exists to also bound the type 2 error with the level β, from which the required sample size n would
then follow. This latter theory is of interest when planning (e.g.) experiments and determining their
sample size. Here we consider situations in which the sample data of size n is given and we have
no real control over the power.

$\mu(f) = \mu_0$, the probability that $\bar{Y}$ is larger than μ_0 is equal to 0.5. Only when the sample average $\bar{Y}$ is substantially larger—thus when there is sufficient evidence—than $\bar{Y}$ would we start to indicate that the null hypothesis $H_0 : \mu(f) \leq \mu_0$ is unlikely to be true (although a type 1 error could be made here).

Thus we want to find a criterion for the average $\bar{Y}$ such that the average can only be larger than this criterion with a probability that is at most equal to α when the null hypothesis is true. If we assume that this criterion is equal to $\mu_0 + \delta$, with $\delta > 0$, then the probability that $\bar{Y}$ is larger than $\mu_0 + \delta$ is given by

$$\Pr\left(\bar{Y} > \mu_0 + \delta\right) = \Pr\left(\frac{(\bar{Y} - \mu(f))}{\sigma(f)/\sqrt{n}} > \frac{\mu_0 - \mu(f) + \delta}{\sigma/\sqrt{n}}\right) \approx 1 - \Phi\left(\frac{\mu_0 - \mu(f) + \delta}{\sigma(f)/\sqrt{n}}\right)$$

where Φ is the standard normal PDF. Note that we have made use of asymptotic theory here to approximate the probability. Under the null hypothesis $H_0 : \mu(f) \leq \mu_0$ this probability is the type 1 error and it is *maximized* when $\mu(f) = \mu_0$. Hence, if we deliberately set $\mu(f) = \mu_0$ to maximize the type 1 error, the probability $\Pr\left(\bar{Y} > \mu_0 + \delta\right)$ is given by $1 - \Phi(\delta\sqrt{n}/\sigma)$. When we choose δ equal to $\delta = z_{1-\alpha}\sigma(f)/\sqrt{n}$, the probability becomes $\Pr\left(\bar{Y} > \mu_0 + \delta\right) \approx \alpha$. Thus, when $\bar{Y} > \mu_0 + z_{1-\alpha}\sigma(f)/\sqrt{n}$ the probability of rejecting the null hypothesis is at most α—and hence we have defined *sufficient evidence* by the criterion $\mu_0 + z_{1-\alpha}\sigma(f)/\sqrt{n}$ for the null hypothesis $H_0 : \mu(f) \leq \mu_0$ using the statistic $\bar{Y}$. Note that the null hypothesis could still be true, but that it is just bad luck to have observed such an unlikely large average under the null hypothesis. However, we know that this probability of having bad luck is less than or equal to α and therefore we accept making this potential type 1 error.

In practice, we cannot use the criterion $\bar{Y} > \mu_0 + z_{1-\alpha}\sigma(f)/\sqrt{n}$ directly, as it depends on the population standard deviation $\sigma(f)$, which is generally not known. We could estimate it from the data and the most natural candidate would be to take the sample standard deviation $S = [\frac{1}{n-1}\sum_{i=1}^{n}(Y_i - \bar{Y})^2]^{1/2}$. If the sample size is large enough, the probability $\Pr(\bar{Y} > \mu_0 + z_{1-\alpha}S/\sqrt{n})$ would still be close to α. Thus when the sample size is large enough, we may reject the null hypothesis in practice when $\bar{Y} > \mu_0 + z_{1-\alpha}S/\sqrt{n}$ or in other words when the *asymptotic test statistic* $T_n = (\bar{Y} - \mu_0)/(S/\sqrt{n})$ is larger than $z_{1-\alpha}$. Thus for large sample sizes n, the null hypothesis $H_0 : \mu(f) \leq \mu_0$ is tested with test statistic $T_n = (\bar{Y} - \mu_0)/(S/\sqrt{n})$ and the null hypothesis is rejected with significance level α when the test statistic is larger than the *critical value* $z_{1-\alpha}$. The null hypothesis is not rejected when $T_n = (\bar{Y} - \mu_0)/(S/\sqrt{n}) \leq z_{1-\alpha}$, since there would not be enough evidence to reject the null hypothesis with significance level α (i.e., the observed result is not unlikely enough under H_0). This does not mean that the null hypothesis is true. Note that the asymptotic approach for testing $H_0 : \mu(f) \leq \mu_0$ will work for most population densities f_X whenever the sample size is large enough.

Instead of using the critical value $z_{1-\alpha}$ for the asymptotic test statistic $T_n = (\bar{Y} - \mu_0)/(S/\sqrt{n})$ to reject a null hypothesis, we often compute a so-called p-value. As the p-value is very often reported and used, it is good to provide a definition first:

The p-value is the probability that the observed test statistic t_n and more extreme observations would occur under the null hypothesis.[5]

After computing a p-value, we can reject H_0 when $p < \alpha$. In the case of the asymptotic test statistic for H_0, this probability is equal to $p = 1 - \Phi(t_n)$, with Φ the standard normal distribution function.

To summarize, the basic rationale of null hypothesis testing is that it provides a procedure to bound the errors made when making binary decisions (rejecting H_0 or not) based on sample data. To do so, we compute a test statistic T_n, and we compute the distribution function of this test statistic given the assumption that H_0 is true. When we have the distribution function of our test statistic under H_0, we can then see whether the observed test statistic t_n (or a more extreme value of t_n) is sufficient to reject H_0. When the observed test statistic t_n is unlikely to occur under the null hypothesis and we reject H_0 the result is often said to be *statistically significant*.

7.3.1.1 Example: Watching Television

To provide a simple example of the one-sided null hypothesis test we just explained, consider the data on approximately 50,000 children at high schools in the Netherlands with an age of 11 to 13 years and focus on the amount of time spent behind the TV. The population density is highly skewed, since this is suggested by the histogram in Fig. 7.5 and the sample skewness of 1.568 (use the function call `skewness(high_school$TV, type=3)` after you have loaded the package `e1071`). However, the sample size is $n = 50{,}069$, which may be considered large enough to apply the asymptotic test statistic.

In this case we would like to know if children spend less than 14 hours per week in front of the television on average, since we have reason to believe that they spend (much) more. Thus the null hypothesis is $H_0 : \mu(f) \leq \mu_0 = 14$ and the alternative hypothesis is $H_a : \mu(f) > 14$, since we would like to show the alternative. The following R code provides the observed test statistic $t_n = (\bar{y} - \mu_0)/(s/\sqrt{n})$, with $\bar{y}$ the sample average and s the sample standard deviation.

```
> mu0 <- 14
> ybar <- mean(high_school$TV)
> s <- sd(high_school$TV)
> n <- nrow(high_school)
> tn <- (ybar-mu0)/(s/sqrt(n))
> ybar
[1] 14.22914
> s
[1] 10.43579
> n
[1] 50069
> tn
```

---

[5] Here t_n is considered the realization of the test statistic T_n and referred to as the observed test statistic.

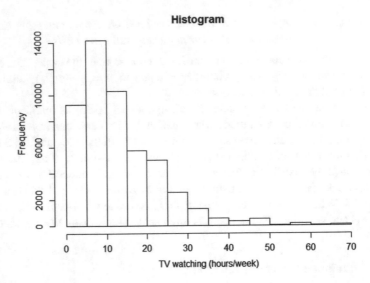

Fig. 7.5 Histogram of TV-watching of children at age 11 to 13 years

[1] 4.913231

If we take as the significance level $\alpha = 0.05$, this test statistic is substantially larger than the critical value $z_{1-\alpha} = z_{0.95} = 1.644854$ (to see this use R and use `qnorm(0.95, mean=0, sd=1)`), which means that we will reject the null hypothesis $H_0 : \mu(f) \leq 14$ for the hours of watching TV.

7.3.2 The Two-Sided z-Test for a Single Mean

Testing $H_0 : \mu(f) \leq \mu_0$ against $H_a : \mu(f) > \mu_0$ (or the other form $H_0 : \mu(f) \geq \mu_0$ against $H_a : \mu(f) < \mu_0$) is called *one-sided hypothesis testing*. We are only interested in rejecting the null hypothesis in one direction, when the test statistic $T_n = (\bar{Y} - \mu_0)/(S/\sqrt{n})$ is large positive (or negative).

In other applications we might be interested in *two-sided hypothesis testing*. The null hypothesis is formulated as $H_0 : \mu(f) = \mu_0$ and the alternative hypothesis is $H_a : \mu(f) \neq \mu_0$. We would reject the null hypothesis when either $T_n = (\bar{Y} - \mu_0)/(S/\sqrt{n})$ is large positive *or* large negative. If we still keep our significance level at α, we will reject the null hypothesis $H_0 : \mu(f) \neq \mu_0$ when $(\bar{Y} - \mu_0)/(S/\sqrt{n}) > z_{1-\alpha/2}$, with $z_{1-\alpha/2}$ the $1 - \alpha/2$ quantile of the standard normal distribution (when using a normal approximation), or when $(\bar{Y} - \mu_0)/(S/\sqrt{n}) < -z_{1-\alpha/2}$. Thus we would reject when $|\bar{Y} - \mu_0|/(S/\sqrt{n}) > z_{1-\alpha/2}$, with $|x|$ the absolute value of x. Figure 7.6 illustrates the difference between one- and two-sided tests by indicating the rejection region under the null hypothesis.

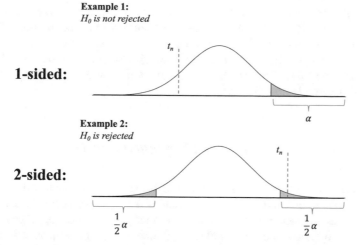

Fig. 7.6 The difference between one-sided and two-sided tests. In both cases H_0 is not rejected if the observed test statistic t_n falls outside of the rejection region α. However, for a one-sided test (top) the rejection region is located on one tail of the distribution (the blue area denoted α). Hence, only if $t_n > z_{1-\alpha}$ (in this case), and sufficiently large to fall within the rejection region, is H_0 rejected. For a two-sided test the rejection region is split over the two tails of the distribution: hence H_0 can be rejected if t_n is either sufficiently small or sufficiently large. Finally, note that if $t_n > z_{1-\alpha}$—if this is the direction of the one-sided test—a one-sided test might reject H_0, whereas the same t_n might not lead to a rejection in the two-sided case, because the critical value for the two-sided test is larger than for the one-sided test

Two-sided tests are very often used; they are the default testing method reported in many studies in the social sciences (and beyond). In practice, when moving from a one-sided test to a two-sided test we merely change our rejection region: we still derive the distribution function of the test statistic under the null hypothesis, but now we reject this hypothesis when the test statistic is sufficiently small *or* sufficiently large. To control our type 1 error, we therefore split up our rejection region in two: this is why we work with $z_{1-\alpha/2}$ as opposed to $z_{1-\alpha}$ as we did before for one-sided hypothesis testing. Note that two-sided tests, while used very often in practice, are less useful when sample sizes are very large: with a very large n, the standard error of a test statistic will become small, and we eventually will reject the null hypothesis in most cases.[6]

Although we have shown how to compute the values of the test statistics for one-sided and two-sided tests by writing our own custom R code, R also contains a number of default functions to perform statistical tests. The R function `z.test (x, y = NULL, alternative="two.sided", mu = 0, sigma.x= NULL, sigma.y = NULL, conf.level = 0.95)` takes a vector x and

---

[6] One way to see why this is true is to consider testing the two-sided null hypothesis that the difference in means of two independent normal populations is zero. As this concerns a continuous random variable, the probability that this difference is *exactly* zero is itself zero. Thus, as long as we gather a sufficient number of observations, we will eventually reject the null hypothesis.

optionally a vector y to test whether the mean of x is equal to μ (the default value
of which is 0, but it can be set using the mu argument) or whether the *difference*
in means of x and y is equal to μ. The argument `alternative` can be used to
specify the alternative hypothesis and thus specify whether we are considering a
one- or two-sided test. The function `z.test` will provide a p-value using a(n)
(asymptotic) normal approximation.

7.3.3 Confidence Intervals and Hypothesis Testing

Both in Chap. 5 and when we covered the bootstrap method we discussed confidence
intervals. A confidence interval quantifies that, if the same population is sampled on
numerous occasions and interval estimates are made on each occasion, the resulting
intervals would include the true population parameter in (approximately) 95% of the
cases. We showed how to compute (asymptotic) confidence intervals for specific test
statistics. For example, we showed that a $1 - \alpha$ asymptotic confidence interval for
$\mu(f)$ could be calculated by $[\bar{Y} - z_{1-\alpha/2}S/\sqrt{n}, \bar{Y} + z_{1-\alpha/2}S/\sqrt{n}]$.

Interestingly, this confidence interval contains the statistic we just discussed for
hypothesis testing. In the case that μ_0 is not contained in the confidence interval, either
$(\bar{Y} - \mu_0)/(S/\sqrt{n}) > z_{1-\alpha/2}$ or $(\bar{Y} - \mu_0)/(S/\sqrt{n}) < -z_{1-\alpha/2}$ would have occurred.
Thus the null hypothesis $H_0 : \mu(f) = \mu_0$ would be rejected if μ_0 is not contained
in $[\bar{Y} - z_{1-\alpha/2}S/\sqrt{n}, \bar{Y} - z_{1-\alpha/2}S/\sqrt{n}]$. Thus confidence intervals relate directly to
two-sided hypothesis tests.

This example demonstrates that, if we can compute confidence intervals, we can
directly use the confidence intervals to test a null hypothesis. If the null hypothesis
is included in the confidence interval, we do not have sufficient evidence to reject it.
If the confidence interval lies fully outside of the null hypothesis, there is sufficient
evidence to reject the null hypothesis. Theoretically (and asymptotically) using confidence intervals for testing will lead to the same type 1 error probabilities. However,
whether the asymptotic theory holds depends quite heavily on the test statistic of
interest, the population distribution, and the sample size: as we saw in the assignments in Chap. 5, confidence intervals do not always have their desired coverage
probability for small sample sizes. The bootstrap approach we discussed above as
an analytical approach to obtain confidence intervals for certain statistics may be an
alternative to generate confidence intervals (instead of using asymptotic theory).

We would like to encourage the reporting of confidence intervals, as opposed to
the common practice of reporting only p-values. We feel that the current emphasis on
p-values in scientific reporting is unwarranted: the p-value is useful for controlling
type 1 errors, but its use should not be overstated. This has a number of reasons:
first of all, the p-value states whether it is unlikely to see a value of t_n under H_0 to
declare "significance"; this, however, does not at all imply that the value of t_n actually
relates to a practically significant or important finding. Second, not rejecting H_0 is
often considered evidence in favor of H_0, but this is erroneous: for very small n we
would hardly ever reject H_0, but this is just due to a lack of sufficient evidence, it
is not due to the fact that H_0 is likely to be true. A confidence interval would then

show a very wide interval, indicating that the test statistic is not very precise. The R functions we will discuss in the next sections for testing null hypotheses frequently provide (by default 95%) confidence intervals for the test statistic of interest.

7.3.4 The t-Tests for Means

In the following sections we discuss a number of test statistics for testing population means assuming that the data comes from a normal distribution. In this case we do not have to make use of large sample sizes and the asymptotic theory. The test statistics will work for any sample size n and not just for large sample sizes.

7.3.4.1 The t-Test for a Single Sample

Under the assumption that the random variables $Y_1, Y_2, \ldots, Y_n$ are i.i.d. normally $N\left(\mu_0, \sigma^2\right)$ distributed, we would know that the test statistic $T_n = (\bar{Y} - \mu_0)/(S/\sqrt{n})$ has a t-distribution with $n - 1$ degrees of freedom. Thus, instead of using the $1 - \alpha$ quantile $z_{1-\alpha}$ of the normal distribution, we may better use the $1 - \alpha$ quantile $x_{1-\alpha}(f_t)$ of the t-distribution for one-sided testing. Indeed, we would obtain that $\Pr((\bar{Y} - \mu_0)/(S/\sqrt{n}) > x_{1-\alpha}(f_t)) = \alpha$ for any sample size n. The test statistic $T_n = (\bar{Y} - \mu_0)/(S/\sqrt{n})$ is called the *one-sample t-test* and we would reject the one-sided null hypothesis $H_0 : \mu(f) \leq \mu_0$ in favor of $H_a : \mu(f) > \mu_0$ when the observed value $t_n > x_{1-\alpha}(f_t)$ and do not reject the null hypothesis when $t_n \leq x_{1-\alpha}(f_t)$. On the other hand, we would reject the one-sided null hypothesis $H_0 : \mu(f) \geq \mu_0$ in favor of $H_0 : \mu(f) < \mu_0$ when $t_n < -x_{1-\alpha}(f_t)$ and do not reject the null hypothesis when $t_n \geq -x_{1-\alpha}(f_t)$. Finally, we would reject the two-sided null hypothesis $H_0 : \mu(f) = \mu_0$ in favor of $H_a : \mu(f) \neq \mu_0$ when $|t_n| > x_{1-\alpha/2}(f_t)$ and do not reject the null hypothesis when $|t_n| \leq x_{1-\alpha/2}(f_t)$.

Again, we could also calculate a p-value instead of using quantiles as critical values. In this case, the probability for $H_0 : \mu(f) \leq \mu_0$ against $H_a : \mu(f) > \mu_0$ is equal to $p = 1 - F_t(t_n)$, with F_t the t-distribution function with $n - 1$ degrees of freedom and t_n is calculated from the data. If this p-value is below α, we believe that it is unlikely to obtain the result t_n or larger results under the null hypothesis $H_0 : \mu(f) \leq \mu_0$. The p-value would be exactly equal to α if t_n equals the critical value $x_{1-\alpha}(f_t)$. Indeed, $\alpha = 1 - F_t(x_{1-\alpha}(f_t)) = \Pr(T_n > x_{1-\alpha}(f_t))$. For the null hypothesis $H_0 : \mu(f) \geq \mu_0$ against $H_a : \mu(f) < \mu_0$ the p-value is calculated as $p = F_t(t_n)$ and for the null hypothesis $H_0 : \mu(f) = \mu_0$ against $H_a : \mu(f) \neq \mu_0$, the p-value is calculated as $p = 2[1 - F_t(|t_n|)]$.

The one-sample t-test is the optimal test statistic under the assumption of $Y_1, Y_2, \ldots, Y_n$ are i.i.d. $N\left(\mu_0, \sigma^2\right)$: There is no other test statistic with the same type 1 error α that would reject the null hypothesis quicker than the t-test when the alternative hypothesis is true, i.e., it has the highest power compared to any other test statistic. The one-sample t-test is illustrated with R in the following subsection.

7.3.4.2   The t-Test for Two Independent Samples

We often want to compare the means of two independent samples with each other. In this case, we essentially have one sample from population $h = 1$ with sample size n_1 and one sample from population $h = 2$ with sample size n_2. The random variables are denoted by $Y_{h,1}, Y_{h,2}, \ldots, Y_{h,n_h}$ for population h. If we assume that $Y_{h,1}, Y_{h,2}, \ldots, Y_{h,n_h}$ are i.i.d. $N(\mu_h, \sigma_h^2)$ and we are interested in a testing hypothesis regarding the difference $\mu_1 - \mu_2$ (i.e., the difference in population means), a natural estimator for this difference is $\bar{Y}_1 - \bar{Y}_2$, and the standard error of this estimator is $\sqrt{\sigma_1^2/n_1 + \sigma_2^2/n_2}$.[7] This standard error can be estimated by substituting the sample variance S_h^2 for σ_h^2, $h = 1, 2$.

If the standard deviations of the two populations are equal ($\sigma = \sigma_1 = \sigma_2$; see below), the standard error $\sqrt{\sigma_1^2/n_1 + \sigma_2^2/n_2}$ becomes equal to $\sigma\sqrt{1/n_1 + 1/n_2}$. In the case of equal variances, both sample variances S_1^2 and S_2^2 provide information on the variance σ^2. The variance σ^2 can now be estimated by the *pooled sample variance* S_p^2 given by

$$S_p^2 = \frac{(n_1 - 1)S_1^2 + (n_2 - 1)S_2^2}{n_1 + n_2 - 2}.$$

S_p^2 is a weighted average of the sample variances where the weights are based on the degrees of freedom. The random variable $(\bar{Y}_1 - \bar{Y}_2)/[S_p\sqrt{1/n_1 + 1/n_2}]$ has a t-distribution function with $n_1 + n_2 - 2$ degrees of freedom. It can be used to test a one-sided or two-sided null-hypothesis on the mean difference $\mu_1 - \mu_2$.

When the two variances are unequal, the statistic $(\bar{Y}_1 - \bar{Y}_2)/\sqrt{S_1^2/n_1 + S_2^2/n_2}$ does not follow a t-distribution and the distribution function of the statistic is not free from the ratio σ_1/σ_2. This does not mean that we should always use asymptotic theory for normally distributed data, since the t-distribution still provides a better approximation when the degrees of freedom are estimated from the data. This approximation is often referred to as the Satterthwaite or Welch approximation and can easily be computed using R.

The R function `t.test(x, y = NULL, alternative = "two.sided", mu = 0, paired = FALSE, var.equal = FALSE, conf.level = 0.95, ...)` can be used similarly to the function `z.test` but uses the t-distribution as opposed to the normal distribution function. Hence, this function can be used to compute both the one-sample t-test and the two independent samples t-test (with the assumption of equal and unequal variances). The code below provides an example of the one-sample and two independent samples tests based on simulated data:

---

[7] The variance of the difference of two independent random variables is the sum of the two variances of the random variables (see the calculation rules in Chap. 4). Indeed, $\mathsf{VAR}(X - Y) = \mathsf{VAR}(X) + \mathsf{VAR}(Y)$, with $\mathsf{VAR}(X - Y) = \mathbb{E}(X - Y - \mathbb{E}(X - Y))^2$, $\mathsf{VAR}(X) = \mathbb{E}(X - \mathbb{E}(X))^2$, and $\mathsf{VAR}(Y) = \mathbb{E}(Y - \mathbb{E}(Y))^2$. The variances of $\bar{Y}_1$ and $\bar{Y}_2$ are equal to σ_1^2/n_1 and σ_2^2/n_2, respectively.

```
> # Simulating 20 observations
> set.seed(21349)
> x <- rnorm(20, 5, 1)
> # t.test one mean, H0 : mu=0
> t.test(x)

	One Sample t-test

data: x
t = 25.697, df = 19, p-value = 3.206e-16
alternative hypothesis: true mean is not equal to 0
95 percent confidence interval:
 4.547128 5.353552
sample estimates:
mean of x
  4.95034


>
> # Adding a set of observations
> y <- rnorm(20, 3, 3)
> # t.test comparing means
> t.test(x, y)

	Welch Two Sample t-test

data: x and y
t = 3.5683, df = 26.042, p-value = 0.001423
alternative hypothesis: true difference in means is not equal to
    0
95 percent confidence interval:
 0.7259575 2.6984199
sample estimates:
mean of x mean of y
 4.950340 3.238152
```

The code above shows that in this case the null hypothesis for the one-sample t-test (first example, $H_0 : \mu = 0$ by default), using $\alpha = 0.05$, is rejected as $p << 0.05$. Note that the 95% confidence interval is provided and it does not contain the value 0. For the second example, using $\alpha = 0.05$ the null hypothesis $H_0 : \mu_1 = \mu_2$ is also rejected. Again, a confidence interval for the mean difference is provided and does not contain the value 0. As we did not assume equal variances, the degrees of freedom for the test statistic were calculated at $df = 26.042$, which is less than the degrees of freedom of $df = 38 (= n_1 + n_2 - 2 = 20 + 20 - 2)$ that we would have used when assuming equal variances.

Deciding between the two t-tests with equal or unequal variances for testing the means from two independent samples has been a long-standing debate among statisticians. Some statisticians believe that we should always use the t-test with unequal variances, while others first want to investigate whether the variances between the two populations are different. If they are different, then a t-test with unequal variances is used, otherwise the t-test with equal variances is used. These statisticians have recommended using a significance level of $\alpha = 0.25$ for testing equality of the two variances: $H_0 : \sigma_1^2 = \sigma_2^2$ or equivalently $H_0 : \sigma_1 = \sigma_2$. Testing this null-hypothesis under the assumption of normality is discussed in Sect. 7.3.6.1.

7.3.4.3   The t-Test for Two Dependent Samples

We sometimes want to compare the means of two dependent samples: e.g., the means of two related groups. This occurs frequently when we want to compare a property that we measure regarding people (e.g., their happiness) at one point in time with that same property at a later point in time. In this case, contrary to the independent sample case discussed above, the data are paired (as we have discussed in Chap. 6). The null hypothesis is still formulated on the difference in means $\mu_1 - \mu_2$ for the two samples, like we did for the two independent samples, but now we must take into account that the data may be dependent.

We can consider hypothesis testing regarding the difference in means by calculating difference scores. Thus, we can consider the difference scores $D_1 = Y_{1,1} - Y_{2,1}$, $D_2 = Y_{1,2} - Y_{2,2}$, ... $D_n = Y_{1,n} - Y_{2,n}$. By considering the difference score we have brought the two samples back to one sample of difference scores and addressed the dependency between the two samples. Thus, the test static $T_n = \bar{D}/\hat{SE}(\bar{D})$, where $\bar{D}$ is the average of the difference scores and $\hat{SE}(\bar{D}) = s_D/\sqrt{n}$ with s_D the sample standard deviation of the difference scores. The test statistic T_n follows a t-distribution with $n - 1$ degrees of freedom (when the difference scores are normally distributed). This test statistic can be used to test the null hypotheses $H_0 : \mu_D = \mu_1 - \mu_2 \leq 0$, $H_0 : \mu_D = \mu_1 - \mu_2 \geq 0$ or $H_0 : \mu_D = \mu_1 - \mu_2 = 0$.

The R function `t.test()` discussed above can also be used to conduct a dependent samples t-test: using the argument `paired` we can differentiate between an independent and a dependent samples t-test. The code below shows the results for a paired samples (i.e., dependent samples) t-test. We use a slightly different set-up for simulating data, since we want to simulate paired data. Here we generate a random variable Z from a normal distribution that will be used for the X and Y observation on one unit. We add a little bit of normal noise and shift the mean for the Y variable. Using this simulation setup, the null hypothesis $H_0 : \mu_D = 0$ is rejected. The 95% confidence interval of the difference in means is provided.

```
> set.seed(21349)
> z <- rnorm(20, 3, 5)
> e1 <- rnorm(20, 0, 1)
> e2 <- rnorm(20, 0, 1)
> x <- z + e1
> y <- z + 1 + e2
> t.test(x, y, paired=TRUE)

        Paired t-test

data:  x and y
t = -3.4888, df = 19, p-value = 0.002457
alternative hypothesis: true difference in means is not equal to
    0
95 percent confidence interval:
```

```
-1.4061194 -0.3516121
sample estimates:
mean of the differences
            -0.8788657
```

If we had ignored the dependency between the two samples and performed a two independent samples t-test, we would not have rejected the null hypothesis of $H_0 : \mu_D = \mu_1 - \mu_2 = 0$. The reason is that the two independent samples t-test does not eliminate the common part Z in the two paired observations X and Y. Thus when the data are paired, you should use the paired t-test and when they are not paired you should use the two independent samples t-test.

7.3.5 Non-parametric Tests for Medians

The z-tests and the t-tests have been developed under certain assumptions. The z-test heavily depends on asymptotic theory and therefore requires large sample sizes, while the t-test can be used for small sample sizes but under the strict assumption of having collected data from a normal distribution. Alternative approaches for the z-test and t-test have been developed that require fewer or no assumptions. These tests are referred to as *non-parametric tests*. We will discuss the Mann–Whitney U test for the comparison of two population medians when two independent samples are collected and the Wilcoxon's signed rank test and the sign test for the comparison of two medians from two dependent samples.

7.3.5.1 Mann–Whitney U Test for Two Independent Samples

The t-tests are based on the assumption of normality. This assumption does not always hold, so we may be in need of alternative approaches that do not assume normality of the data. One such approach is the Mann–Whitney U test for two independent samples. It tests the null hypothesis $H_0 : Pr(Y_1 > Y_2) = Pr(Y_1 < Y_2)$, with Y_1 a random draw from the first population and Y_2 a random draw from the second population. If the null hypothesis is false, it is more likely to observe larger values in one of the populations with respect to the other population. Thus one population is *stochastically greater* than the other population. Note that the null hypothesis is equivalent to $H_0 : \Pr(Y_1 > Y_2) = 0.5$.

If we now assume that the distribution function F_2 for the second population is given by $F_2(y) = F_1(y + \delta)$, with F_1 the distribution function of the first population and δ a (shift) parameter, the null hypothesis $H_0 : \Pr(Y_1 > Y_2) = \Pr(Y_1 < Y_2)$ implies that the medians of the two populations must be equal (i.e., $\delta = 0$). Thus, only in this somewhat restrictive formulation of population distributions, the Mann–Whitney U test investigates whether the two populations have equal medians.

The Mann–Whitney U test statistic is given by

$$U = \sum_{i=1}^{n_1} \sum_{j=1}^{n_2} \left[1_{(Y_{2,j} < Y_{1,i})} + 0.5 \cdot 1_{(Y_{1,i} = Y_{2,j})} \right],$$

with 1_A the indicator function equal to 1 if A is true and zero otherwise. It represents the number of pairs $(Y_{1,i}, Y_{2,j})$ of observations from the two populations for which the first population provides a larger value than the second population. If there are ties (i.e., $Y_{1,i} = Y_{2,j}$), we cannot determine which population provides the larger value; thus this pair only contributes half to the total number of observations for which the first population is larger than the second population. If the variable Y_h is continuous, we should not observe any ties if we use enough decimal places in the recording of the values.

The Mann–Whitney U test only makes use of the ordering of the observations. The total number of pairs for which the comparison between the two populations is made is equal to $n_1 n_2$. Indeed, each observation of the first sample is compared to each observation in the second sample. Thus the statistic $U/(n_1 n_2)$ is an estimator of the probability $\Pr(Y_1 > Y_2)$. When $U/(n_1 n_2)$ is away from 0.5 or in other words, when U is away from $0.5 n_1 n_2$, the null hypothesis $H_0 : \Pr(Y_1 > Y_2) = 0.5$ is rejected and one population is considered stochastically greater than the other population.

To determine whether the U statistic is away from $0.5 n_1 n_2$, we will make use of asymptotic theory, as we did with the z-test for means. If the sample sizes n_1 and n_2 are large enough, the Mann–Whitney U statistic is approximately normally distributed with mean μ_U and variance σ_U^2. Under the null hypothesis $H_0 : \Pr(Y_1 > Y_2) = 0.5$, the mean μ_U is of course equal to $\mu_U = 0.5 n_1 n_2$. The variance is then equal to $\sigma_U^2 = n_1 n_2 (n_1 + n_2 + 1)/12$, but only when there are no ties. If there are ties, a correction to this variance is required to make the variance smaller. The ties reduce the variability in U. To test the null hypothesis $H_0 : \Pr(Y_1 > Y_2) = 0.5$ we calculate the standardized value $Z = [U - 0.5 n_1 n_2]/\sqrt{n_1 n_2 (n_1 + n_2 + 1)/12}$ and compare this with the quantiles $z_{\alpha/2}$ and $z_{1-\alpha/2}$ of the standard normal distribution. If $|Z| > z_{1-\alpha/2}$ we reject the null hypothesis.

To illustrate the Mann–Whitney U test, we will use the simulated data we have already used for the t-test for two independent samples.

```
> set.seed(21349)
> x <- rnorm(20, 5, 1)
> y <- rnorm(20, 3, 3)
> sort(x)
 [1] 3.641566 3.710570 4.037056 4.059152 4.392723 4.447961
     4.498769 4.508335 4.585216 4.774473 4.827863 4.893561
     5.182205
[14] 5.417988 5.568407 5.647138 5.666598 5.956765 6.562305
     6.628153
> sort(y)
 [1] -0.3466269 0.2560427 0.8378082 1.4280921 1.6181263 1.6484644
      2.0823331 2.1738134 2.9864973 3.3260533
[11]  3.8210475 3.8994514 4.2691191 4.3280371 4.5659308 4.9015815
      4.9347916 5.1430118 6.0666133 6.8228439
```

Calculating the number of pairs from the sorted values can now be quickly conducted manually. The first value 3.641566 of the first population is larger than the first 10 observations of the second population. If we do this for each of the observations in the first population we obtain: 10, 10, 12, 12, 14, 14, 14, 14, 15, 15, 15, 15, 18, 18, 18, 18, 18, 18, 19, and 19. Thus the Mann–Whitney U statistic is equal to $U = 306$ by summing up all these numbers. The mean and variance of the Mann–Whitney U test (under the null hypothesis) are equal to $\mu_U = 200$ and $\sigma_U^2 = 4,100/3$. Thus the standardized statistic is calculated at $2.867 = [306 - 200]/\sqrt{4,100/3}$, which indicates that the null hypothesis $H_0 : \Pr(Y_1 > Y_2) = 0.5$ is rejected at significance level $\alpha = 0.05$, since the normal quantiles are equal to $z_{1-\alpha/2} = -z_{\alpha/2} = 1.96$. It seems that the first population is stochastically greater than the second population.

In R we can carry out the Mann–Whitney U test as follows:

```
> wilcox.test(x,y,correct=FALSE,exact=FALSE)

	Wilcoxon rank sum test

data: x and y
W = 306, p-value = 0.00414
alternative hypothesis: true location shift is not equal to 0
```

We need to eliminate the "exact" calculation of the p-value if we wish to use the asymptotic test statistic. We also need to eliminate a continuity correction, which is built in as the default calculation. The asymptotic test statistic is in most cases appropriate, unless the sample sizes are small.[8]

The function `wilcox.test` is named after Frank Wilcoxon, because it calculates the Wilcoxon rank sum test. Frank Wilcoxon formulated a test statistic based on the ranks of the observations, which later turned out to be identical to the Mann–Whitney U test. They calculated different statistics, but they were not so different after all.

7.3.5.2 The Sign Test for Two Related Samples

For two related datasets we observe the pairs of data $(Y_{1,i}, Y_{2,i})$ for the units $i = 1, 2, \ldots, n$. A relevant question for paired data is whether one component is stochastically greater than the other component: $H_0 : \Pr(Y_1 > Y_2) = 0.5$. For instance, if we test a new dermatological treatment against an existing treatment, we could possibly use both arms on each patient, where one arm receives the new treatment and the other arm receives the existing treatment. If the number of patients for which the new treatment scores better than the old treatment is much larger than

---

[8] The exact test makes use of permutations, where the index of the populations are permuted but the observations are not. Then for each permutation, the U test can be calculated and the position of the observed U from the original (non-permuted) data among all permuted values can be determined. If the observed U is in the tail of the permuted values, the null hypothesis is rejected. For small sample sizes the exact test is more appropriate, but for larger samples sizes the calculation of the exact test can take a long time.

$0.5n$, it becomes reasonable to think that the new treatment is better than the existing treatment. The sign test makes this procedure or intuition precise.

The sign test is mathematically given by $S = \sum_{i=1}^{n} 1_{(Y_{2,i} < Y_{1,i})}$, with 1_A the indicator variable that is equal to one if A is true and zero otherwise. It represents the number of units for which the first observation is larger than the second observation. Under the null hypothesis (and when there are no ties $Y_{1,i} = Y_{2,i}$) the statistic follows a binomial distribution with proportion $p = 0.5$ and n trials (i.e., comparisons between Y_1 and Y_2). Thus the binomial distribution function can be used to determine when S becomes too large or too small. The sign test can be easily executed with R using the function `binom.test`. We have used the simulated data on pairs that we used for the paired t-test.

```
> set.seed(21349)
> z <- rnorm(20, 3, 5)
> e1 <- rnorm(20, 0, 1)
> e2 <- rnorm(20, 0, 1)
> x <- z + e1
> y <- z + 1 + e2
>
> z<-ifelse(x-y>0,1,0)
> count <- sum(z)
> count
[1] 5
> binom.test(count,20)

        Exact binomial test

data:  count and 20
number of successes = 5, number of trials = 20, p-value = 0.04139
alternative hypothesis: true probability of success is not equal
        to 0.5
95 percent confidence interval:
 0.08657147 0.49104587
sample estimates:
probability of success
                0.25
```

Thus from the 20 comparisons, only five resulted in a larger value of the first component (x) compared to the second component (y). We had to create this count by ourselves before we could put it into the binomial test function. The output shows that the p-value, which was calculated from the binomial distribution, is just below $\alpha = 0.05$, which means that the null hypothesis $H_0 : \Pr(Y_1 > Y_2) = 0.5$ is rejected at significance level $\alpha = 0.05$. The estimated probability for $\Pr(Y_1 > Y_2)$ is equal to 0.25 with 95% confidence interval [0.087, 0.491]. This confidence interval is not calculated from asymptotic theory and also illustrates that the null hypothesis should be rejected, since the value 0.5 is not contained in the interval.

The sign test needs a small adjustment when ties ($Y_{1,i} = Y_{2,i}$) are present in the data. In the case of ties we cannot judge which component of the pair is larger than the other component. This means that these pairs cannot be used. These pairs should be excluded from the calculations. The test statistic remains unchanged, but the number of trials n should be reduced by the number of ties. To illustrate this, assume that we had two ties in the simulated pairs. These pairs with ties must have occurred in the 15 pairs that did not demonstrate $Y_{1,i} > Y_{2,i}$. Thus the test statistic remains $S = 5$, but now the number of trials is 18, since two pairs did not give us a decision on $Y_{1,i} > Y_{2,i}$ or $Y_{1,i} < Y_{2,i}$. In this case, we should have used the following R code

```
> binom.test(count,18)

    Exact binomial test

data: count and 18
number of successes = 5, number of trials = 18, p-value = 0.09625
alternative hypothesis: true probability of success is not equal
    to 0.5
95 percent confidence interval:
 0.09694921 0.53480197
sample estimates:
probability of success
          0.2777778
```

Thus, if there had been two ties, we could not have rejected the null hypothesis. It is therefore important to pay attention to ties if you want to use the sign test.

7.3.5.3   Wilcoxon's Signed Rank Test for Two Related Samples

The advantage of the sign test is that we did not have to assume anything about the distribution function of the pairs $(Y_{1,i}, Y_{2,i})$. Under the null hypothesis we could determine the distribution function of our test statistic. Therefore, we could determine when the test statistic would result in unlikely results if the null hypothesis is true. The disadvantage of the sign test is that it does not consider the size of the distances between the two components.

To illustrate this disadvantage, assume again that we are comparing two treatments, where we observe the paired results $(Y_{1,i}, Y_{2,i})$ on treatment one and two. If the number of pairs for which the first treatment performs better than the second treatment $(Y_{1,i} > Y_{2,i})$ is small, the sign test may conclude that the second treatment outperforms the first treatment. However, if the difference $Y_{1,i} - Y_{2,i}$ for the two treatments with $Y_{1,i} > Y_{2,i}$ is very large and the difference $Y_{2,i} - Y_{1,i}$ for the treatments $Y_{1,i} < Y_{2,i}$ is very small, we may argue that the first treatment does better, because the difference between the two treatments is negligible when the second treatment shows better results, but the first treatment shows much better results when it does better than the second treatment (although it does not occur frequently).

The Wilcoxon signed rank test takes into account these differences for the two groups of pairs with $Y_{1,i} > Y_{2,i}$ and $Y_{1,i} < Y_{2,i}$. It can be executed in the following steps:

1. Calculate for each pair the difference $D_i = Y_{1,i} - Y_{2,i}$.
2. Create for each pair an indicator $1_{(Y_{1,i} > Y_{2,i})}$.

3. Calculate for each pair the rank R_i of the absolute differences $|D_i|$.
4. Calculate the sum of ranks for the positive differences: $W^+ = \sum_{i=1}^{n} R_i 1_{(Y_{1,i} > Y_{2,i})}$.

If the distribution function of the positive differences D_i is equivalent to the distribution function of the negative differences D_i, we would expect that the *average* rank for the positive differences is equal to the *average* rank of the negative differences. This translates to a null hypothesis on the median of the difference: $H_0 : m_D = 0$, with m_D the median of the distribution of all differences D_i. If the average rank for the positive differences is really different from the average rank of the negative differences, the median of all differences can no longer be zero.

For the Wilcoxon signed rank test, we will make use of asymptotic theory again. If the sample sizes are large enough and the null hypothesis is true, the sum of the ranks for the positive differences W^+ is approximately normal with mean μ_W and variance σ_W^2. The mean and variance are determined by $\mu_W = n(n + 1)/4$ and the variance is $\sigma_W^2 = n(n + 1)(2n + 1)/24$, respectively. Note that the mean is just half of the sum of all ranks, as the sum of all ranks is equal to $n(n + 1)/2$. Similar to the Mann–Whitney U test, we can standardize the Wilcoxon signed rank test to $Z = [W^+ - \mu_W]/\sigma_W$ and compare this standardized statistic with the quantiles of the standard normal distribution to reject the null hypothesis $H_0 : m_D = 0$ or not.

To illustrate the Wilcoxon signed rank test, consider the simulated paired data we used for the paired t-test and the sign test. The following R code helps us understand the calculations.

```
> set.seed(21349)
> z <- rnorm(20, 3, 5)
> e1 <- rnorm(20, 0, 1)
> e2 <- rnorm(20, 0, 1)
> x <- z + e1
> y <- z + 1 + e2
>
> d<-x-y
> ad<-abs(d)
> sgn<-sign(d)
> ad
 [1]  0.83246900 1.17190367 0.36570753 0.97984688 0.34540792
      1.99806325 2.81096726 0.06015728 0.94722108 0.27203432
[11]  1.49564909 1.15050241 0.57035769 1.00001388 0.92194825
      2.09065266 2.46242541 0.44392413 2.42612844 1.07738070
> rank(ad)
 [1]  7 14  4 10  3 16 20  1  9  2 15 13  6 11  8 17 19  5 18 12
> sgn
 [1] -1 -1  1 -1  1 -1 -1 -1 -1  1  1 -1 -1 -1 -1 -1 -1  1 -1 -1
```

The sum of the ranks for the positive differences is now calculated as $W^+ = 4 + 3 + 2 + 15 + 5 = 29$. The standardized statistic is then equal to $[29 - 105]/\sqrt{717.5} = -2.837$. This value is clearly outside the interval $[-1.96, 1.96]$, which means that the null hypothesis $H_0 : m_D = 0$ is rejected. The differences D_i for the positive differences $D_i > 0$ are ranked lower than the differences D_i for the negative differences D_i. An average rank of 5.8 ($= 29/5$) for positive differences compares with an average rank of 12.1 ($= (210 - 29)/15$) for negative difference. Thus we not only obtain a significantly low number of positive differences D_i, we also see that these differences are smaller than the differences for $D_i < 0$.

Instead of doing these calculations manually, we could have calculated the results with R using the function `wilcox.test`.

```
> wilcox.test(x,y,paired=TRUE, correct=FALSE, exact=FALSE)

    Wilcoxon signed rank test

data: x and y
V = 29, p-value = 0.00455
alternative hypothesis: true location shift is not equal to 0
```

The output shows that W^+ is equal to 29 (although they used the notation V) and that the likelihood of this result (or lower values) under the null hypothesis is determined as $p = 0.005$. Thus the null hypothesis should be rejected at significance level $\alpha = 0.05$.

7.3.6 Tests for Equality of Variation from Two Independent Samples

The term *heteroskedasticity* refers to differences in variation or variability. In hypothesis testing this is often translated to a hypothesis on the variances or standard deviations from two different populations, as we described for the two samples t-test. Under the assumption of normality the most efficient test statistic is based on a ratio of the two sample variances, but under non-normal data an alternative approach has been suggested. Here we discuss the F-test for normal data and Levene's test for non-normal data.

7.3.6.1 The F-Test for Equal Variances

Under the two-sided null hypothesis $H_0 : \sigma_1 = \sigma_2$, the ratio S_1^2/S_2^2 has an F-distribution with $n_1 - 1$ and $n_2 - 1$ degrees of freedom and the ratio S_2^2/S_1^2 has an F-distribution function with $n_2 - 1$ and $n_1 - 1$ degrees of freedom. If S_1^2/S_2^2 or S_2^2/S_1^2 is large, then the null hypothesis $H_0 : \sigma_1 = \sigma_2$ is rejected. We may choose just one of the two ratios, since there is symmetry. If one ratio is large, the other ratio must be small, or the other way around. Thus if we choose one of the ratios, the null hypothesis is rejected when this ratio is large or when this ratio is small. The critical value can be determined by using the function $\mathtt{qf}(1 - \alpha/2, d_1, d_2)$ for large ratios or $\mathtt{qf}(\alpha/2, d_1, d_2)$ for small ratios, with d_1 the degrees of freedom for the sample variance in the numerator and d_2 the degrees of freedom of the sample variance in the denominator. The choice of degrees of freedom depends on which ratio S_1^2/S_2^2 or S_2^2/S_1^2 is selected, but one of them is equal to $n_1 - 1$ and the other to $n_2 - 1$. Note that we must use $\alpha/2$ since we are using the two-sided test. The test statistic is referred to as the F-test. The function var.test(x, y, ratio = 1, alternative ="two.sided", conf.level = 0.95) can be used to test equality of variances using R and is demonstrated in the code below:

```
> # Generating two samples with unequal variances:
> set.seed(21349)
> x <- rnorm(20, 0, 1)
> y <- rnorm(20, 0, 3)
> # Test for variances equal:
> var.test(x, y)

    F test to compare two variances

data: x and y
F = 0.19217, num df = 19, denom df = 19, p-value = 0.0007381
alternative hypothesis: true ratio of variances is not equal to 1
95 percent confidence interval:
 0.07606166 0.48549812
sample estimates:
ratio of variances
        0.1921661
```

Note that R has chosen the ratio S_1^2/S_2^2 for the analysis. In this case, the null hypothesis that the two variances σ_1^2 and σ_2^2 are equal is rejected, since the ratio of 0.19217 is too unlikely to be generated under the null hypothesis. Having this ratio or smaller ratios only occurs with probability 0.00074 under the null hypothesis. Note that an observed F value larger than 3 or 4 (or smaller than $1/3$ or $1/4$) typically indicates that the variances between the two populations are likely to be unequal (even if the data in the two samples are not from a normal distribution). Thus the observed data suggest that we should accept the alternative hypothesis that the variances in the two populations are not equal. A 95% confidence interval for the ratio of the two variances is also provided and is equal to [0.076, 0.4855]. If sample sizes gets large, rejection of the null hypothesis can happen at values closer to one.

7.3.6.2 Levene's Test for Equal Variation

For non-normal data the variability or variation around the mean or median is not fully described by the standard deviation alone, as is the case for normal data. Instead of comparing the two sample variances, an alternative measure of variation is created. Here the distance of each observation from its group mean or median is calculated first. Thus for sample h we calculate distances $Z_{h,i} = |Y_{h,i} - m_h|$, with $i = 1, 2, \ldots, n_h$. The value m_h represents some kind of location of the sample h, like the average or median. Initially, Howard Levene suggested using the means for the location m_h in the calculation of the distances, while Morton Brown and Alan Forsythe later suggested using the median.

  If the distribution functions F_1 and F_2 for the two populations ($h = 1, 2$) are the same, except for a difference in the population mean or median, the distances $Z_{1,i}$ and $Z_{2,i}$ may be viewed as two samples from one and the same population of distances, i.e., the distances $Z_{1,i}$ and $Z_{2,i}$ would be drawn from the same distribution of distances. To investigate if these samples of distances come from one distribution, we may study a difference in means. If the samples on distances $Z_{1,i}$ and $Z_{2,i}$ suggest that they have different means, they do not come from the same distribution of distances and there must exist differences between the two distribution functions F_1 and F_2 that are not induced by a shift in mean or median alone. This would imply that a difference in variation in the two populations $h = 1$ and $h = 2$ is warranted, since differences in the means between the two samples of distances $Z_{1,i}$ and $Z_{2,i}$ indicate that one set of distances are on average larger than the other set of distances.

Levene's test statistic just uses the t-test with equal variances on the two independent samples of distances $Z_{1,i}$ and $Z_{2,i}$ to determine if the means of the population of distances that they may represent are different. A two-sided t-test is used, since there is no preference in knowing which variable X or Y has a higher or lower variability. Brown & Forsythe demonstrated that the use of the median for m_h gives a somewhat more robust statistic for many different distribution functions F_1 and F_2 than Levene's choice of means. Thus the Brown & Forsythe version of Levene's test is often recommended over Levene's test.

7.3.7 Tests for Independence Between Two Variables

In Chap. 6 we discussed several measures of association, all quantifying some form of dependency between two variables. Hypothesis tests are often carried out to test the null hypothesis of independence, with the goal of rejecting this null hypothesis. Demonstrating a dependency between two variables is sometimes all we need to know. For instance, knowing that there exists a positive relation between eating breakfast and performances in high school, may be enough to initiate advice to high-school students to eat breakfast before they come to school (although it may be better to also know the strength of this relation). A demonstrated dependency between eating breakfast and school performance could motivate high-school students to eat breakfast more regularly.

7.3.7.1 Correlation Tests for Numerical Variables

To test the dependency between two normally distributed variables X and Y, Pearson's correlation coefficient can be applied. The null hypothesis of independence is then formulated as $H_0 : \rho_P = 0$ against the alternative hypothesis $H_a : \rho_P \neq 0$, with ρ_P the correlation coefficient of the bivariate normal distribution function (which equals Pearson's definition of correlation coefficient). Under normality, independence is the same as being uncorrelated, but this may not be true when X and Y do not follow a normal distribution.

The test statistic for the null hypothesis $H_0 : \rho_P = 0$ is

$$T_P = \frac{r_P \sqrt{n-2}}{\sqrt{1 - r_P^2}}$$

with $r_P = S_{XY}/[S_X S_Y]$ Pearson's product moment estimator given in Eq. (6.10). Under the assumption of normality and under the null hypothesis $H_0 : \rho_P = 0$ the test statistic T_P has a t-distribution with $n - 2$ degrees of freedom. Thus when the observed value t_P is larger than the upper $\alpha/2$-quantile of the t-distribution with $n - 2$ degrees of freedom or smaller than the lower $\alpha/2$-quantile of the t-distribution with $n - 2$ degrees of freedom, the null hypothesis $H_0 : \rho_P = 0$ is rejected. If the observed value t_P does not deviate enough from zero we cannot reject the null hypothesis (but this means that there is no evidence that the null hypothesis is correct).

The R function cor.test(x, y, alternative = "two.sided", conf.level = 0.95) can be used to test the null hypothesis $H_0 : \rho_P = 0$ when the data on the X and Y variables are normally distributed. Here we simulate X and Y as we have done earlier for testing two means from independent samples. Thus we simulated data knowing that $H_0 : \rho_P = 0$ is correct.

```
> set.seed(21349)
> x <- rnorm(20, 3, 1)
> y <- rnorm(20, 2, 1)
> cor.test(x,y)

        Pearson's product-moment correlation

data: x and y
t = -0.40839, df = 18, p-value = 0.6878
alternative hypothesis: true correlation is not equal to 0
95 percent confidence interval:
 -0.5164388  0.3620570
sample estimates:
        cor
-0.09581508
```

Note that in the simulation example provided above, the null hypothesis $H_0 : \rho_P = 0$ is not rejected. This is reasonable in this case as there is no relationship between X and Y in the simulated data generated. R directly provides the 95% confidence interval for the correlation (which includes 0 in this case). Note that this confidence interval is different from the calculation we proposed in Chap. 6, where we used Fisher's z-transformation on Pearson's product moment estimator. Clearly, this alternative confidence interval could also have been used to test $H_0 : \rho_P = 0$ and is often recommended above T_P. The reason is that the Fisher z-transformed Pearson's product moment estimator would more quickly reject the null hypothesis than T_P when the alternative hypothesis is true. Thus the Fisher z-transformed Pearson's product moment estimator has a slightly higher power than the test statistic T_P, while they both have the same type 1 error.

As an alternative approach we could also have used another correlation estimator, like Kendall's tau or Spearman's rho estimators. Then the null hypothesis changes (of course) to $H_0 : \tau_K = 0$ and $H_0 : \rho_S = 0$ for Kendall's tau and Spearman's rho, respectively. If such a null hypothesis were true, the two variables X and Y may still be dependent. It merely says that the two variables X and Y are uncorrelated. The 95% confidence intervals based on the Fisher z-transformation, which were discussed in Chap. 6, can then be used to test the corresponding null hypothesis. These alternative correlation coefficients are typically used when X and/or Y are continuous but not normally distributed. If the null hypothesis is rejected, i.e., the value zero is not contained in the 95% confidence interval, the variables X and Y are considered dependent.

If one or both variables X and Y are discrete, care should be taken in using any of the described methods. In this case, there will most likely be ties, i.e., there exists pairs of data for which the pairs cannot be ordered, either by the first dimension or by the second dimension (or both). For such variables X and Y, the distribution function of the test statistics just described does not has the same distribution function that was used for the test statistic on continuous variables X and Y. Alternative distribution functions for the test statistics have been developed, but this topic is outside the scope of this book. One approach to handle this situation is to use bootstrap and calculate the bootstrap quantiles of one of these correlation coefficients. If zero is not contained within the lower and upper 2.5% bootstrap quantiles, the null hypothesis of independence is rejected.

7.3.7.2 The χ^2 Test for Categorical Variables

In Chap. 6 we also discussed dependency or associations between two categorical (nominal and ordinal) variables X and Y. Thus the null hypothesis is formulated as $H_0 : \Pr(X = x, Y = y) = \Pr(X = x)\Pr(Y = y)$ against the alternative hypothesis of $H_a : \Pr(X = x, Y = y) \neq \Pr(X = x)\Pr(Y = y)$. The collected data are typically presented by a contingency table and Pearson's chi-square statistic is a proper statistic for testing the independence between X and Y. The statistic was designed to compare the observed cell counts for the contingency table with the expected cell counts for the contingency table under independence.

Using K for the number of rows in the contingency table, M for the number of columns, Pearson's chi-square statistic is given by (see also Chap. 6):

$$\chi_P^2 = \sum_{x=1}^{K} \sum_{y=1}^{M} (N_{xy} - (N_{x.}N_{.y}/N))^2/(N_{x.}N_{.y}/N),$$

with N_{xy} the observed count in cell (x, y) of the contingency table, $N_{x.}$ the row total for $X = x$, $N_{.y}$ the column total for $Y = y$, and N the total count in the contingency table. The product $N_{x.}N_{.y}/N$ is the expected count for cell (x, y) under independence. Under the null hypothesis of independence, Pearson's chi-square statistic follows the χ^2-distribution with $(K - 1)(M - 1)$ degrees of freedom. Thus, the p-value for testing the null hypothesis is found by calculating $\Pr(\chi^2 > \chi_P^2)$, the probability that a chi-square distributed random variable χ^2 exceeds the observed value χ_P^2 for χ_P^2.

The R function `chisq.test(x)`, where x is a contingency table (a `matrix`) can also be used to directly test $H_0 : \Pr(X = x, Y = y) = \Pr(X = x) \Pr(Y = y)$. The following code generates data for a 2×2 contingency table and conducts Pearson's χ^2 test (as we already illustrated in Chap. 6):

```
> # Generate a 2x2 table
> Table <- as.table(rbind(c(762, 327), c(484, 289)))
> dimnames(Table) <- list(study = c("Yes", "No"), exam = c("Pass"
  , "Fail"))
> Table
      exam
study Pass Fail
  Yes  762  327
  No   484  289
> chisq.test(Table)

    Pearson's Chi-squared test with Yates' continuity correction

data: Table
X-squared = 10.73, df = 1, p-value = 0.001054
```

In this case the null hypothesis of independence is rejected.[9]

Note that the list of statistical tests introduced in this subsection is not at all complete; many more test statistics for specific circumstances (e.g., for proportions or for multiple groups) are known. Books like Field (2013) will list many of the well-known test statistics and procedures for computing them. However, knowing the appropriate test statistic by heart for each situation does not seem very useful; it is much more useful to understand the basic principles behind hypothesis testing.

7.3.8 Tests for Normality

Many statistical techniques "require" that the data come from a normal distribution (e.g., the t-test). Thus it becomes important to verify or evaluate whether the data has come from a normal distribution. In the past many different solutions have been proposed. We will describe the graphical approach and the Shapiro–Wilk test. Assume that we have selected a sample of data $Y_1, Y_2, \ldots, Y_n$.

An often used approach, which is somewhat subjective, is to visualize the data and make a judgment. A common visualization method for this purpose is the quantile-quantile or q-q plot. In this case, the ordered observations $Y_{(1)}, Y_{(2)}, \ldots, Y_{(n)}$ are plotted against the quantiles of the normal distribution. The quantile q_k that belongs to $Y_{(k)}$ is given by $q_k = \Phi^{-1}((k - 0.375)/(n + 0.125))$, with Φ the standard normal distribution function. This is close to the quantile $\Phi^{-1}(k/n)$, but the constants 0.375

---

[9] Note that the χ^2 value is slightly different than what you would obtain by computing $\chi_P^2 = \sum_{x=1}^{K} \sum_{y=1}^{M} (N_{xy} - (N_x.N._y/N))^2 / (N_x.N._y/N)$; this is due to the fact that R automatically uses Yate's "continuity" correction: *always* inspect the help files for an R function to know exactly how the results are computed. In this case the help can be accessed using `?chisq.test`.

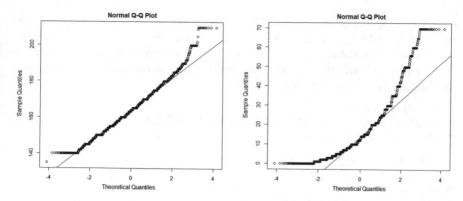

Fig. 7.7 q-q plots on height (left) and watching television (right) for high-school boys

and 0.125 are used to make the graphical approach somewhat less biased. Depending on the software package, the q-q plot reports the theoretical quantiles q_k on the horizontal axis and the order statistics $Y_{(k)}$ (or the quantiles of the sample) on the vertical axis. The function qqnorm in R plots the theoretical quantiles on the horizontal axis. If the data $Y_1, Y_2, \ldots, Y_n$ come from a normal distribution, the graph $(q_k, Y_{(k)})$ would form approximately a straight line. The intercept of this line is an estimate of the population mean and the slope of the line is then an estimate of the population standard deviation.

To illustrate the q-q plot we use the data of the high-school children. The function in R that provides a q-q plot is qqnorm. If you also want to have a straight line through the data in the plot you have to run qqline after qqnorm. We applied the two functions to the height measurements of the school children and to the amount of television watching. We restricted the data to boys. The following code was used:

```
> boys <- high_school[high_school$GENDER=="Boy", ]
> qqnorm(boys$HEIGHT)
> qqline(boys$HEIGHT)
> qqnorm(boys$TV)
> qqline(boys$TV)
```

The two q-q plots are provided in Fig. 7.7.

For the variable height, the data seem to fit closer to the straight line than the data on watching television. Thus the height measure seems to be more normal than watching television. Actually, the non-normality for watching television is quite obvious, but the judgment for height is somewhat arbitrary. Indeed, the height measurements also seem to deviate from the line in the tails of the distribution function (the lower and upper ends of the data). Thus it is not immediately obvious if the height is or is not normally distributed.

To make the evaluation of normality less subjective, formal test statistics have been developed. One such test statistic actually investigates how well the line goes through the points $(q_k, Y_{(k)})$. This is the Shapiro–Wilk test for normality. The test statistic is of the form $W = \sum_{k=1}^{n}[a_{n,k}Y_{(k)}]/[S\sqrt{n-1}]$, with $a_{n,k}$ constants that depend on the sample size and on the normal quantiles and S the sample standard deviation. Large values of W indicate that it is unlikely that the data was collected from a normal distribution, but intuition on this value is missing. Instead we may use the p-value that is calculated with the statistic. The function `shapiro.test` in R calculates the Shapiro–Wilk normality test.

```
> set.seed(21349)
> sample<-sample(boys$HEIGHT,5000,replace=FALSE)
> shapiro.test(sample)

    Shapiro-Wilk normality test

data: sample
W = 0.99288, p-value = 4.097e-15
```

The test does not allow for more than 5,000 observations, so that is why we draw a random sample of 5,000 high-school students before testing for normality of their height. The p-value suggests that the height of (this sample of) students is deviating from normality. Together with the q-q plot, we conclude that height may not follow a normal distribution.

The q-q plot and Shapiro–Wilk test together may help in assessing normality. When the (sample) data deviate greatly from the normal distribution the properties of test statistics that rely on normality assumptions will not hold. This means that their p-values will need to be interpreted with more caution. There are many other formal test statistics of normality, but we do not discuss these in this book. The Shapiro–Wilk test is often considered the most appropriate normality test. It is, however, sensitive to ties. Under normality ties cannot occur. Thus when there are many ties, the Shapiro–Wilk test may show that the data is not normal. If these ties are caused by rounding issues and the sampled data actually come from a normal distribution, the Shapiro–Wilk test may incorrectly reject normality. For that reason it is important to always look at the q-q plot and never just trust the p-value.

7.3.9   Tests for Outliers

An outlier is an observation that seems to deviate from the other observations such that it arouses suspicion that some unintended mechanism has interfered with the random sampling. This means that the observation is not drawn from the same population distribution as the other observations.

Outliers are a nuisance for many calculations since they may highly affect the estimator or test statistic, and therefore influence the conclusions strongly. To limit their effect, one may be inclined to remove the outliers and continue with the remaining observations. But this is really dangerous if there is no clear argument for why the observation is an outlier (other than a statistical argument), since it could truly belong to the population distribution (of which we did not know that it had such a long tail). Outliers should only be removed when a reason or cause has been established; in all other siutations the outlier should not be removed. It is often considered good practice to report the most conservative statistical analysis of the data. Thus, if removal of the outlier results in the least favorable analysis, the outlier is removed from the data, otherwise it remains part of the data.

In some practical cases, an outlier or a group of outliers are the main study of interest. For instance, in fraud detection, fraudulent applications for credit cards or fraudulent bank transactions may appear as outliers in the data. The outliers may help find criminal activity and are thus the study of interest. In health sciences an outlier observation may indicate a signal for a (symptom of a) disease, in particular if the observation is (far) outside the normal range of values.[10]

Outlier detection is not an easy task. First of all, it is difficult to distinguish whether the outlier observation is truly caused by some kind of unintended mechanism or whether is belongs to a population distribution that is just different than what we anticipated. Secondly, statistical methods that were developed for checking only one outlier observation are diminished in their detection capacity if multiple outliers are present. This is called the *masking effect* and relates to the type 2 error rate in hypothesis testing, since a real outlier is not being detected. Additionally, outlier detection requires certain assumptions on the underlying population distribution. Without such assumptions we can never make confident statements like we try to do with hypothesis testing. If we are not willing to make such assumption, we all have to agree to the same definition that an observation is called an outlier whenever the observation satisfies some kind of predefined criterion.

There are many outlier detection methods, but here we will discuss the Grubbs test and Tukey's method. The Grubbs test was developed for normally distributed data and it can be formulated in terms of hypothesis testing. Tukey's method has been implemented in many software packages for visualization of the data with a box plot. John Tukey simply provided a definition for outliers, in the sense that an observation is called an outlier when the observation is further away from the median value than a predefined distance. In the box plot they are typically indicated by dots or stars.

7.3.9.1 Grubbs Test for Outliers

The Grubbs test is sometimes called the *extreme studentized deviate test* or the *maximum normed residual test*. The null hypothesis is that no observation is an outlier, while the alternative hypothesis is that there is one observation that is an outlier. Thus, the Grubbs test is only searching for *one* outlier observation. The calculation of the Grubbs test is quite easy, since it is equal to $G = \max\{|Z_1|, |Z_2|, \ldots, |Z_n|\}$, with $Z_i = (Y_i - \bar{Y})/S$ the standardized observations (with $\bar{Y}$ the average and S the standard deviation).

The probability that G is larger than some critical value c_α is equal to

---

[10] The normal range of values represents an interval or range of values that seems common or likely in a healthy population.

Fig. 7.8 Critical values for the Grubbs test for outliers: dark gray curve:$\alpha = 0.05$ and light gray curve:$\alpha = 0.01$

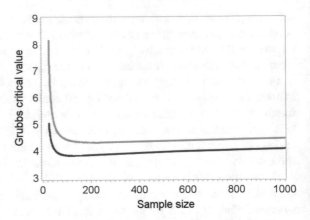

$$\Pr\left(G > c_\alpha\right) \approx 1 - \prod_{i=1}^{n} \Pr\left(|Z_i| \leq c_\alpha\right).$$

The larger the sample size, the better the approximation. This critical value can be visualized for different values of n; see Fig. 7.8.

For large sample sizes the critical value seems to stabilize or becomes almost constant. For $\alpha = 0.05$ the critical value becomes approximately equal to $c_\alpha = 4$ and for $\alpha = 0.01$ the critical value becomes approximately $c_\alpha = 4.5$. It is not uncommon to qualify an observation as an outlier when the absolute standardized value is larger than 4. Some software packages would even signal absolute standardized values larger than 3 as potential outliers.

In R one can execute the Grubbs test with the function grubbs.test. This requires the installation of the package outliers. The following code investigates if there is one outlier in the height variable of the children at high school:

```
> library(outliers)
> grubbs.test(high_school$HEIGHT)

    Grubbs test for one outlier

data: high_school$HEIGHT
G = 5.4735, U = 0.9994, p-value = 0.001099
alternative hypothesis: highest value 210 is an outlier
```

Based on the output, the null hypothesis that no observation is an outlier is rejected, and we would accept the alternative hypothesis that there is an outlier.

7.3.9.2 Tukey's Method

Tukey suggested that an observation is an outlier whenever the observation Y_k is 1.5 times the interquartile range below the first quartile or 1.5 times the interquartile range

Fig. 7.9 Expected
proportion of population
units outside Tukey's outlier
criteria for the lognormal
distribution function

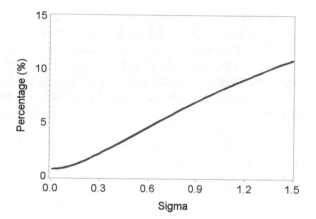

above the third quartile. To write this in mathematical terms, an observation is called
a lower-tail outlier when $Y_k < Q_1 - 1.5\text{IQR}$ and it is called an upper-tail outlier
when $Y_k > Q_3 + 1.5\text{IQR}$, with $\text{IQR} = Q_3 - Q_1$ the interquartile range. Clearly, the
first observation that may satisfy the lower tail criterion would be the minimum,
while for the upper tail it is the maximum value.

Tukey's method is often referred to as a method that does not depend on any
assumptions of the population distribution, since it is only based on quartiles. However, the shape of the population distribution does determine how likely an observation will satisfy the criterion and thus will be declared an outlier. If we consider the full population, we may substitute the expected or true values for Q_1 and
Q_3 in the criteria and determine the proportion of the population that would satisfy the lower and upper criteria. The lower criterion in the population is given
by $F^{-1}(0.25) - 1.5(F^{-1}(0.75) - F^{-1}(0.25))$ and the upper criterion is given by
$F^{-1}(0.75) + 1.5(F^{-1}(0.75) - F^{-1}(0.25))$.

For the normal distribution function $\Phi((x - \mu)/\sigma)$, with Φ the standard normal
distribution function, these criteria become $\mu + \Phi^{-1}(0.25)\sigma - 1.5(\Phi^{-1}(0.75) -
\Phi^{-1}(0.25))\sigma = \mu - 2.69796\sigma$ and $\mu + 2.69796\sigma$ for the lower and upper tail,
respectively. Thus the percentage of units from the population that fall outside these
criteria is approximately only 0.70%, which does not depend on the parameters μ
and σ. Thus under the assumption of normality, an outlier observation has 0.70%
probability to occur in the sample.

For the lognormal distribution function $\Phi((\ln(x) - \mu)/\sigma)$ though, the lower
and upper criteria now become $\exp(\mu - 0.67449\sigma) - 1.5(\exp(\mu + 0.67449\sigma) -
\exp(\mu - 0.67449\sigma))$ and $\exp(\mu + 0.67449\sigma) + 1.5(\exp(\mu + 0.67449\sigma) - \exp(\mu -
0.67449\sigma))$, respectively. Calculating the proportion of population units outside
Tukey's criteria depends only on σ, not on μ. This proportion is visualized in Fig. 7.9.

Figure 7.9 clearly indicates that for the lognormal distribution function the proportion outside Tukey's criteria can be as high as 10%. This would typically occur at
the upper tail and not on the lower tail, since the proportion below the lower tail criterion would become equal to zero. Thus fixing the criteria for an outlier observation

does not always imply that the occurrence of such observations in the sample will be unlikely. For normal distribution functions, Tukey's criteria for outlier observations is unlikely if no outliers are present, but using Tukey's criteria for other distribution functions should be implemented with care. The code below shows the identification of outliers using Tukey's criteria in R using the `boxplot.stats` function. Note that in the example the two added extreme values (-10 and 500) are indeed identified.

```
> # Test tukey
> x <- rnorm(100,10,5)
> x <- c(x, -10, 500)
> boxplot.stats(x)$out
[1] -10 500
```

7.3.10 Equivalence Testing

The common practice of testing the (two-sided) null hypothesis has a few drawbacks. As we have already noted, for (extremely) large samples we will almost always reject the null hypothesis. This might not be desirable, as rejecting the null hypothesis in such cases does not actually imply that the (e.g.,) difference in means of interest is indeed large. Furthermore, not rejecting the null hypothesis $H_0 : \mu(f) = \mu_0$ is no proof that the null hypothesis is true. Indeed, it is very easy not to reject the null hypothesis: you just need to collect as little information as possible. If we would collect only a few observations the confidence interval would be very wide and the value μ_0 is likely to fall inside this wide confidence interval, but this does not guarantee that $\mu(f) = \mu_0$ or even close to it.

One way of dealing with these two issues on hypothesis testing is to reduce the significance level α to a much lower value than the commonly used $\alpha = 0.05$. We may use $\alpha = 0.001$ or even $\alpha = 0.0001$ to make the hypothesis statements more confident than just 95%. An alternative approach, possibly in combination with lower levels of α, is to use *equivalence testing*. Equivalence testing would first formulate a margin Δ that would be used to form a range of values around the value μ_0 that we would see as equivalent settings. The null hypothesis is then formulated as $H_0 : |\mu(f) - \mu_0| > \Delta$ against the alternative hypothesis $H_a : |\mu(f) - \mu_0| \leq \Delta$. Thus, the null hypothesis specifies non-equivalence: the difference between $\mu(f)$ and μ_0 is larger than Δ. If the null hypothesis is rejected, we have demonstrated with sufficient confidence $(1 - \alpha)$ that the true population mean $\mu(f)$ is equivalent to the value μ_0. The level Δ is called the *equivalence margin*.

Using confidence intervals, a test for equivalence is reasonably simple. We just need to calculate a $1 - 2\alpha$ confidence interval on $\mu(f) - \mu_0$ and then compare it to the interval $[-\Delta, \Delta]$. If the confidence interval falls fully in the interval $[-\Delta, \Delta]$, the null hypothesis $H_0 : |\mu(f) - \mu_0| > \Delta$ is rejected at significance level α. To test at significance level α, we need to calculate a $1 - 2\alpha$ confidence interval. The reason is that the equivalence test is actually the application of two one-sided tests: one test for showing that $\mu(f) < \mu_0 + \Delta$ and another test for showing that $\mu(f) > \mu_0 - \Delta$. Each test is done at significance level α. This is similar to having the $1 - 2\alpha$ confidence interval on $\mu(f) - \mu_0$ fall inside $[-\Delta, \Delta]$.

In the case of equivalence testing, where the equivalence margin is given by $\Delta > 0$, we can formulate the following hypotheses:

1. Non $-$ inferiority : $H_0 : \mu_1 - \mu_2 \leq -\Delta$ versus $H_a : \mu_1 - \mu_2 > -\Delta$
2. Non $-$ inferiority : $H_0 : \mu_1 - \mu_2 \geq \Delta$ versus $H_a : \mu_1 - \mu_2 < \Delta$ (7.3)
3. Equivalence   : $H_0 : |\mu_1 - \mu_2| \geq \Delta$ versus $H_a : |\mu_1 - \mu_2| < \Delta$

The one-sided equivalence tests are referred to as *non-inferiority tests*.

Equivalence testing $H_0 : |\mu(f) - \mu_0| > \Delta$ against $H_a : |\mu(f) - \mu_0| \leq \Delta$ is really different in concept from traditional hypothesis testing $H_0 : \mu(f) = \mu_0$ against $H_a : \mu(f) \neq \mu_0$. To show this difference we have listed six different settings in Fig. 7.10 for which we will discuss the rejections or not. The vertical dotted lines represents the interval $[-\Delta, \Delta]$ and the vertical solid line represents the difference $\mu(f) - \mu_0 = 0$. The horizontal x-axis represents the difference $\mu(f) - \mu_0$ and each dot in the figure represents an estimate of this difference. The horizontal lines through the dots represent the $1 - 2\alpha$ confidence intervals.

1. The $1 - 2\alpha$ confidence interval fall completely within $[-\Delta, \Delta]$, which means that the null hypothesis $H_0 : |\mu(f) - \mu_0| > \Delta$ is being rejected. It is not completely clear whether $H_0 : \mu(f) = \mu_0$ is being rejected, since this can only be established if the $1 - \alpha$ confidence interval has been applied. If we assume that this $1 - \alpha$ interval would fit within $[-\Delta, 0]$, the null hypothesis $H_0 : \mu(f) = \mu_0$ would also be rejected. This is a setting that we described earlier for large datasets, where the traditional hypothesis can be rejected for irrelevant differences.

2. The null hypothesis $H_0 : |\mu(f) - \mu_0| > \Delta$ for equivalence is being rejected, indicating that the population mean $\mu(f)$ is equivalent to μ_0. The traditional hypothesis $H_0 : \mu(f) = \mu_0$ is not being rejected, which implies that there is not enough evidence to believe that the population mean $\mu(f)$ is different from μ_0. This means that both conclusions seem to coincide, although this does not imply that $\mu(f) = \mu_0$.

3. The traditional null hypothesis is being rejected (assuming that the 95% confidence interval would not contain zero) and equivalence cannot be claimed. Thus both approaches seem to have similar conclusions: $\mu(f)$ is different and not equivalent to μ_0.

4. This is the same as in the previous setting, although the results lie on the other side of the vertical line $\mu(f) = \mu_0$.

5. Again this is similar to the third setting. The fact that one side of the confidence interval is contained in the interval $[-\Delta, \Delta]$ does not change the conclusions.

6. Here we cannot demonstrate equivalence, but we cannot reject the traditional null hypothesis either. Thus on the one hand there is not enough evidence to reject $\mu(f) = \mu_0$, but there is not enough evidence either to reject that $|\mu(f) - \mu_0| > \Delta$. This seems contradictionary, but it is probably an issue of a lack of information, since the 90% confidence interval is too wide.

Fig. 7.10 Comparison
between equivalence testing
and traditional hypothesis
testing

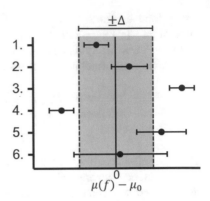

The non-inferiority hypotheses look somewhat similar to the one-sided hypotheses
we discussed earlier, but they are not. To illustrate this we consider the situation
where we want to compare a new treatment to an existing or an older treatment.
In the traditional hypothesis testing format we would investigate the null hypothesis
$H_0 : \mu_{New} \leq \mu_{Old}$ against $H_a : \mu_{New} > \mu_{Old}$, assuming that a higher mean is a better
clinical outcome. The collected data (typically through a randomized controlled
clinical trial) would then have to demonstrate that the null hypothesis is rejected.
Thus the new treatment must demonstrate *superiority*. Often new treatments are
not (much) better in their clinical outcome, but they bring a secondary benefit. For
instance, the new treatment is much less invasive. Superiority is then difficult to
demonstrate. Instead, we would like to show that the new treatment is not inferior
to the old treatment. Thus, the data must demonstrate that the new treatment is not
too much worse than the old treatment: $H_a : \mu_{New} > \mu_{Old} - \Delta$, with $\Delta > 0$ the *non-inferiority margin*. Thus, the null hypothesis becomes $H_0 : \mu_{New} \leq \mu_{Old} - \Delta$. If this
null hypothesis cannot be rejected, the new treatment is assumed inferior to the old
treatment, since non-inferiority could not be proven.

7.4 Conclusions

In this chapter we used the theory developed in the previous chapters to make decisions regarding a population. Such decisions will always contain uncertainty; however, in some cases we can make quantitative statements regarding the quality of our
decisions.

We first introduced the bootstrap as an informal, but very generally applicable,
method of making decisions. The bootstrap allows us to quantify our uncertainty
around any sample statistic $\hat{\theta}$, and we can use its bootstrapped distribution function
to make decisions: if the value of $\hat{\theta}$ seems unlikely, we might reject it as a good
estimate for our population. Note that in some situations the bootstrap can be made

precise (e.g., in terms of controlling type 1 and 2 errors), but we do not study these further in this book.

After presenting the bootstrap, we discussed hypothesis testing: this often used procedure for binary decision making controls the type 1 and type 2 errors involved. We have discussed the rationale of null hypothesis testing, and we discussed the distributions of several test statistics under the assumption that the null-hypothesis was true. We presented a number of well-known methods, but in any applied situation you should always look up the state-of-art for the problem you are facing. We discussed t- and F-tests, non-parametric tests for medians and variation, tests for associations, normality tests and outlier tests. In practice, you would use R if you want to carry out a hypothesis test: we discussed several R functions that allow you to conduct these tests. We also discussed equivalence testing and non-inferiority testing, to be able to demonstrate that the parameter of interest is equal to a value within a certain margin.

Finally, we would like to note that the methods for decision-making we presented here only "scratch the surface": statistical decision theory is a large and active research field. In Chap. 8 we will briefly encounter an alternative approach to decision making. However, for now we have covered both a method that is extremely widely applicable and easily available (the bootstrap), and a method that is very frequently used (hypothesis testing). These methods allow you to make your first steps into making decisions regarding a population based on sample data.

Problems

7.1 This assignment uses the dataset `voting-demo.csv`.

1. Look at the frequency distribution of the variable `Choice`. Use a non-parametric bootstrap (with the number of bootstrap replicates $M = 10,000$) to create a histogram of the distribution of the proportion of people who vote for the SPD. Does this follow a bell curve?
2. We now want to compare the average age of those who did vote and those who did not vote. Note that in the sample the mean age of voters is 45.46, and the mean age of non-voters is 39.18. Hence, we would estimate the *difference* to be 6.28. We now want to say something about whether this is "significant". Try to compute the difference between voters and non-voters based on $K = 10,000$ non-parametric bootstrap replicates and make a histogram of the differences. In which proportion of cases are the voters on average older than the non-voters?
3. Compute an *online* bootstrap distribution of the variance of the variable `Age` in the dataset. Use the code we used in Sect. 7.2.4. However, now (a) run through the dataset `voting-demo.csv` line by line, and (b) replace the `update.mean` function by a function that computes the variance in a data stream (or *online*, see https://www.researchgate.net/publication/306358951_Dealing_with_Data_Streams_an_Online_Row-by-Row_Estimation_Tutorial for examples).

7.2 This assignment uses the dataset `high-school.csv` and focusses on the variable TV (i.e., the number of hours a student watches television per week).

1. Use a non-parametric bootstrap (with the number of bootstrap replicates $M = 10{,}000$) to create a histogram of the distribution of the average number of hours students watch television.
2. Use a parametric (normal) bootstrap with the same number of replicates to create a second histogram; how do the two compare?
3. Use both the non-parametric bootstrap replicates and the parametric bootstrap replicates derived in the previous two questions to construct 95% confidence intervals. How do these compare?
4. Compute the asymptotic 95% confidence interval for the mean; how does this compare to those obtained using bootstrapping?
5. Repeat the above assignments for the sample variance (opposed to the mean).

7.3 This assignment uses the dataset `high-school.csv`. The variable of interest is the allowance that children receive per week in euros. We are only interested in those children that do receive an allowance (thus eliminate the children that do not get any allowance). Hypothesis testing should be used with significance level $\alpha = 0.05$.

1. Test whether the mean allowance is equal to 10 euros per month using a two-sided test.
2. Test whether the mean allowance is equivalent to 10 euros with equivalence margin $\Delta = 2$ euros.
3. Investigate whether the variable allowance is normally distributed. Explain what you did and what your conclusion is.
4. Create a box plot of the log transformed allowance. How many observations are considered outliers according to Tukey's criterion?
5. Use the Grubbs test on the log transformed allowance. Do you think there is an outlier?
6. Test whether the mean log transformed allowances of boys and girls are equal using a t-test. Would you use equal or unequal variances? Why? What do you conclude for the hypothesis test?
7. Test whether the allowances of girls are stochastically greater than the allowances of boys using the Mann–Whitney U test. Note that we are looking at a one-sided test. What is your conclusion?
8. Test whether girls receive a similar allowance to boys, at least not less than the non-inferiority margin of $\Delta = 2$ euros.

7.4 This assignment uses the dataset `high-school.csv`. This time the variable of interest is the amount of money that children earn per week, which is the variable WORK. We will focus on those children who earn money. A hypothesis test should be used with significance level $\alpha = 0.05$.

1. Investigate whether the log transformed amount of money for boys is normally distributed.
2. Investigate whether the log transformed amount of money for girls has an outlier.

3. Investigate with Levene's test whether the variability in the log transformed amount of money differs between boys and girls.
4. Is there a difference in the average log transformed amount of money between boys and girls? Which test statistic would you use? Explain.
5. Is the average log transformed amount of money of boys and girls equivalent within equivalence margin 0.10?
6. Do breakfast eaters earn more money than non-breakfast eaters? Formulate the null hypothesis and describe which test statistic would be best for this question.

7.5 In this exercise we will investigate dependency on variables of the face data. We will use the variable on the second dimension and the rating variable.

1. Test whether the two variables are normally distributed. What is your conclusion?
2. Test whether the two variables are independent. What is the null hypothesis and what test statistic did you use? Explain your choices.
3. We want to investigate whether there is a difference in correlation coefficient between rating and the second dimension for women and men. Use a non parametric bootstrap approach to answer this question. What is your conclusion? Explain why.

7.6 In this exercise we will investigate dependency between variables from the voting data.

1. Determine whether religion and voting behavior are independent. Formulate the null hypothesis and describe the test statistic. What do you conclude?
2. When the percentage of voting is within the margin of 10%, voting behavior between subgroups is considered equivalent. Test the equivalence behavior for religious and non-religious people. Describe what you did and describe your conclusion.
3. Determine whether voting behavior and voting choice are independent. Formulate the null hypothesis and describe the test statistic. What do you conclude?

References

R. Davidson, J.G. MacKinnon, Bootstrap tests: how many bootstraps? Economet. Rev. **19**(1), 55–68 (2000)

D. Eckles, M. Kaptein, Bootstrap Thompson sampling and sequential decision problems in the behavioral sciences. SAGE Open **9**(2), 2158244019851675 (2019)

B. Efron, Bootstrap methods: another look at the jackknife, in *Breakthroughs in Statistics* (Springer, Berlin, 1992), pp. 569–593

A. Field, *Discovering Statistics Using IBM SPSS Statistics* (SAGE, California, 2013)

E.L. Lehmann, J.P. Romano, *Testing Statistical Hypotheses* (Springer Science & Business Media, 2006)

C. Robert, *The Bayesian Choice: From Decision-Theoretic Foundations to Computational Implementation* (Springer Science & Business Media, 2007)

R.J. Tibshirani, B. Efron, An introduction to the bootstrap. Monogr. Stat. Appl. Probab. **57**, 1–436 (1993)

Chapter 8
Bayesian Statistics

8.1 Introduction

In this book we have introduced both the practice of analyzing data using R, and covered probability theory, including estimation and testing. For most of the text we have, however, considered what some would call *Frequentist* statistics (the name deriving from the notion of probability as a long-run frequency): in this school of thought regarding probability it is generally assumed that population values are fixed quantities (e.g., θ is, despite being unknown, theoretically knowable and has a fixed value). Any uncertainty (and hence our resort to probability theory) arises from our sampling procedure: because we use (ostensibly) random sampling we have access to only one of the many possible samples that we could have obtained from the population of interest, and when estimating θ (by, e.g., computing $\hat{\theta}$) we will need to consider the fact that another sample might have produced a different value.

There is, however, another school of thought, called *Bayesian* statistics. Its name is derived from Bayes' Theorem, as this theorem is used almost constantly in this latter stream of thought. We have introduced Bayes' Theorem in Chap. 3 when discussing conditional events:

$$\Pr(A|B) = \frac{\Pr(B|A)\Pr(A)}{\Pr(B)}. \tag{8.1}$$

Both the Frequentist and the Bayesian school of thought accept Bayes' rule as a logical consequence derived from the probability axioms. However, in Bayesian statistics the prominence of Bayes' rule is much greater than it is in Frequentist statistics. In Bayesian statistics probabilities are not solely regarded as long-run frequencies, but rather, they are regarded as a tool to quantify the "degree of belief" one holds in relation to a specific event. In this framework, probabilities can be regarded as an extension to boolean logic (in which statements/events are true or false, 0 or 1), to deal with events that have some probability (Jaynes 2003). Bayes rule is subsequently used to update the degree of belief regarding an event in the

© Springer Nature Switzerland AG 2022
M. Kaptein and E. van den Heuvel, *Statistics for Data Scientists*, Undergraduate Topics in Computer Science, https://doi.org/10.1007/978-3-030-10531-0_8

face of new evidence: i.e., it is used to update $\Pr(A)$ in the formula above to include the evidence provided by $\Pr(B)$, the result of which is expressed as $\Pr(A|B)$. As we will see below, this approach takes on a very powerful meaning when we replace $\Pr(A)$ by $\Pr(\theta)$ (i.e., the distribution of the parameters) and $\Pr(B)$ by $\Pr(D)$ where D relates to the data we observe in (e.g.,) a scientific study. In this interpretation, Bayes' rule provides us with a means of updating our belief regarding a population parameter $(\Pr(\theta))$ when we observe sample data $(\Pr(D))$. In this chapter we will introduce this "Bayesian" school off thought and discuss its relations to Frequentist thinking.

It is good to note that although philosophically the Bayesian and Frequentist schools of thought are very different, we will, in this book, not strongly argue in favor of one or the other. For a data scientists both approaches have their merits and we strongly encourage applied data scientists to be able to move from one school of thought to the other flexibly. In support of this flexible view it is good to know that in many practical situations (but certainly not all!) the conclusions one would draw from a Bayesian or a Frequentist analysis are similar (or even exactly the same). This chapter merely provides an introduction and short discussion with regard to Bayesian thinking; we could easily fill multiple books on Bayesian statistics alone (which is something others have done; see for example Gelman et al. (2014)).

We will cover (in brief, and here and there informally) the following:

- The extension of Bayes' Theorem for events to statements about population parameters: i.e., from $\Pr(A|B)$ to $f(\theta|D)$.
- Bayesian analysis by example: we will examine the Bayesian estimation of various population parameters for Bernoulli and Normally distributed populations.
- The general Bayesian approach: Bayesian decision-making.
- Bayesian hypothesis testing: the Bayes Factor.
- Choosing a prior distribution, i.e., choosing $f(\theta)$.
- Bayesian computation: the intractable (or simply difficult to compute) marginal likelihood.
- A discussion of the differences and commonalities of Bayesian and Frequentist methods.

8.2 Bayes' Theorem for Population Parameters

In this section we explore how we get from Bayes law for events, as presented in Chap. 3, to the much more general interpretation of Bayes' law as it is used in modern Bayesian statistical practice. Let us start by revisiting the derivation of Bayes' law itself: suppose we have two events of interest, A and B, that are not independent. In this case we know, based on the axioms presented in Chap. 3, that the conditional probability of A given B is given by $\Pr(A|B) = \Pr(A \cap B)\Pr(B)$. Or, in words, the conditional probability is given by dividing the joint probability of A and B by the (marginal) probability of B. Similarly, we can consider $\Pr(B|A) = \Pr(A \cap B)\Pr(A)$. Putting this together allows for the relatively simple derivation of Bayes' law:

$$\Pr(A \cap B) = \Pr(A|B) \Pr(B)$$
$$\Pr(B \cap A) = \Pr(B|A) \Pr(A)$$
$$\Pr(A|B) \Pr(B) = \Pr(B|A) \Pr(A)$$
$$\Pr(A|B) = \frac{\Pr(B|A) \Pr(A)}{\Pr(B)}$$

Here, the last line is the famous Bayes' law, and it can be considered a device for inverting conditional probabilities: it can be used to compute $\Pr(A|B)$ when $\Pr(B|A)$ is observed.

We have already seen a number of examples of the use of Bayes' law in Chap. 3, but it is useful to recall an example of its practical use. Suppose that in a population of data science students we know that about 2% are Bayesians whereas the remaining 98% identify as Frequentist. Thus, we have $\Pr(B) = 0.02$ and $\Pr(B^c) = 0.98$. However, the Bayesian data scientists do not always like to admit that they are indeed Bayesian: when we conduct a survey in which we ask students about their preference the answers to the question "Are you a Bayesian" may not always be fully accurate. Let us assume that the survey is 95% accurate on classifying Bayesians (thus $\Pr(C|B) = 0.95$), and let us further assume that the survey is 97% accurate on negative classifications; i.e., $\Pr(C^c|B^c) = 0.97$. We are now interested in learning the probability of a student truly being a Bayesian, after the student has been classified as a Bayesian by the survey; thus, we are interested in learning $\Pr(B|C)$. A few simple steps give us the answer. It is first useful to compute the marginal probability of classifying a respondent as a Bayesian, $\Pr(C)$ using the law of total probability:

$$\begin{aligned}
\Pr(C) &= \Pr(C \cap B) + \Pr(C \cap B^c) \\
&= \Pr(C|B) \Pr(B) + [1 - \Pr(C^c|B^c)] \Pr(B^c) \\
&= 0.95 * 0.02 + [1 - 0.97] * 0.98 \\
&\approx 0.05
\end{aligned}$$

after which applying Bayes' Theorem to compute the desired $\Pr(B|C)$ is straightforward:

$$\begin{aligned}
\Pr(B|C) &= \frac{\Pr(C|B) \Pr(B)}{\Pr(C)} \\
&= \frac{0.95 * 0.02}{0.05} \\
&\approx 0.38.
\end{aligned}$$

This computation shows that when a students is classified as a Bayesian in the survey this event increases our belief regarding whether or not the student truly is a Bayesian from 0.02 before the survey results came in (i.e., our prior belief was $\Pr(B) = 0.02$) to 0.38 after observing the survey result. This shows how Bayes' Theorem can be used to update our degree of belief regarding events when new evidence is observed.

8.2.1 Bayes' Law for Multiple Events

While Bayes' Theorem as we have seen it up till now already has many useful applications, its use can be greatly extended. Just as we extended our thinking about probability from events (in Chap. 3) to random variables and distributions (in Chaps. 4 and 5), so too Bayes' Theorem can be generalized and put to use in many situations. A first step towards such a generalization is to apply Bayes' law to multiple events. Suppose we consider three disjoint events A, B, and C, whose union comprises the entire sample space of interest (i.e., they are exhaustive). Next, we are interested in learning how our a priori belief regarding the probability of these events occurring (i.e., $\Pr(A)$, $\Pr(B)$, and $\Pr(C)$) changes when we observe new data D. Thus, we are interested in $\Pr(A|D)$, $\Pr(B|D)$, and $\Pr(C|D)$. Focussing on $\Pr(A)$, we know from Bayes' rule that:

$$\Pr(A|D) = \frac{\Pr(D|A)\Pr(A)}{\Pr(D)}.$$

Using the law of total probability and our assumptions regarding $\Pr(A)$, $\Pr(B)$, and $\Pr(C)$ we also know that

$$\Pr(D) = \Pr(A \cap D) + \Pr(B \cap D) + \Pr(C \cap D)$$
$$= \Pr(D|A)\Pr(A) + \Pr(D|B)\Pr(B) + \Pr(D|C)\Pr(C)$$

which we can substitute into our application of Bayes' Theorem to obtain:

$$\Pr(A|D) = \frac{\Pr(D|A)\Pr(A)}{\Pr(D|A)\Pr(A) + \Pr(D|B)\Pr(B) + \Pr(D|C)\Pr(C)}.$$

The derivations above demonstrate that we can compute the probability—conditional on the data—of each of the three events A, B, and C, of interest as long as we have access to the unconditional (or prior) probabilities $\Pr(A)$, $\Pr(B)$, and $\Pr(C)$ and the probability of the observed data given these events, $\Pr(D|A)$, $\Pr(D|B)$, and $\Pr(D|C)$.

8.2.2 Bayes' Law for Competing Hypotheses

Effectively, applying the above allows one to state how much more or less likely specific events (A, B, etc.) become due to observing the data D. This itself is particularly interesting when the events of interest concern rival (again disjoint and exhaustive) hypotheses regarding (e.g.) an unknown parameter of interest. Suppose we assume we are trying to learn about the value of a parameter of interest θ, and further suppose we have a number of specific hypotheses regarding its value, e.g., "Hypothesis A: $\theta < 0$" (although we could effectively make any statement we want regarding the parameter values of interest). Let us further introduce the notation θ_i to refer to the first hypothesis regarding θ (i.e., Hypothesis A). If we have k such hypotheses regarding θ (thus we have $\theta_1, \ldots, \theta_k$) we can now compute the probability of each of these hypothesis after observing some data by applying Bayes' rule as we did before:

$$\Pr(\theta_i|D) = \frac{\Pr(D|\theta_i)\,\Pr(\theta_i)}{\sum_{i=1}^{k} \Pr(D|\theta_i)\,\Pr(\theta_i)}.$$

Note that in the above we once again need both the marginal probabilities of the hypothesis prior to observing the data, i.e., $\Pr(\theta_i)$, and the probability of the data given a specific hypothesis, $\Pr(D|\theta_i)$. The marginal probability of the hypothesis is often referred to as the prior probability, whereas the probability of the data conditional on some hypothesis should be recognizable by now to readers as the *likelihood* that we have encountered in previous chapters. The resulting conditional probability of the hypothesis given the data is often called the *posterior* probability of the hypothesis.

8.2.3 Bayes' Law for Statistical Models

Given our treatment above, it is straightforward to extend Bayesian reasoning to statistical models, i.e., the types of models we have been studying in previous chapters. In previous chapters we were often interested in estimating (or otherwise making decisions about) the value of some population parameter θ by computing $\hat{\theta}$. One fairly general method of computing $\hat{\theta}$ was choosing the value of θ that maximized the likelihood $l(\theta)$ where, when considering a dataset of n i.i.d. observations, we had (see Chap. 5):

$$\begin{aligned} l(\theta) &= l(\theta|X_1, X_2, X_3, \ldots, X_n) \\ &= \Pr(X_1|\theta)\,\Pr(X_2|\theta) \ldots \Pr(X_n|\theta) \\ &= \Pr(D|\theta). \end{aligned}$$

Bayes theorem in a way provides an alternative to for example Maximum Likelihood Estimation by—opposed to picking the value of θ that maximizes $\Pr(D|\theta)$—allowing one to quantify the probability of a large number of possible values of θ_i after observing the data:

$$\begin{aligned} \Pr(\theta_i|D) &= \frac{\Pr(D|\theta_i)\,\Pr(\theta_i)}{\sum_{i=1}^{k} \Pr(D|\theta_i)\,\Pr(\theta_i)} \\ &= \frac{l(\theta_i|X_1, X_2, X_3, \ldots, X_n)\,\Pr(\theta_i)}{\sum_{i=1}^{k} l(\theta_i|X_1, X_2, X_3, \ldots, X_n)\,\Pr(\theta_i)}. \end{aligned}$$

At this point, it isn't a large stretch to specify the prior probability $\Pr(\theta_i)$ off each hypothesis i using a Probability Mass Function (PMF), as we have studied in Chap. 4. This highlights that Bayes' Theorem can be used not just for events, but for random variables. The general form of Bayes' Theorem that is most often encountered in Bayesian statistics generalizes the ideas presented above to random variables as these are described by their distribution functions (i.e., their PMFs or PDFs):

$$f(\theta|D) = \frac{l(\theta)f(\theta)}{\int l(\theta)f(\theta)d\theta}. \tag{8.2}$$

The expression above effectively describes that our posterior beliefs regarding the parameter of interest—i.e., our belief after observing the data—can be quantified using a (potentially continuous) PDF $f(\theta|D)$ which itself is obtained by multiplying the likelihood $l(\theta) = \Pr(D|\theta)$ by our prior belief regarding the parameters $f(\theta)$ and dividing by the so-called *marginal likelihood* $\int l(\theta)f(\theta)d\theta$. The latter simply acts as a Normalizing constant to make sure that $f(\theta|D)$ is a valid PDF.

8.2.4   The Fundamentals of Bayesian Data Analysis

Eq. 8.2 is the main work-horse of Bayesian data analysis. Contrary to Frequentist MLE estimation where a single point $\hat{\theta}$ is often the end-point of the analysis (i.e., her or his best guess regarding the true population value), to a Bayesian, $f(\theta|D)$ effectively quantifies all that can be learned based on the observed data: it quantifies, using a PDF, our belief regarding the population value after seeing the data. Generally, the steps involved in conducting a Bayesian analysis are thus:

- Create a sampling model, i.e., specify the likelihood of the observed data as a function of the parameters θ.
- Specify a prior distribution over the parameters $f(\theta)$. Note that we will often encounter sampling models that involve multiple parameters and hence θ itself is a vector and $f(\theta)$ is a multivariate distribution (PMF or PDF).
- Compute $f(\theta|D)$, i.e., compute the posterior distribution of θ after seeing the data D.

To a Bayesian, that's pretty much it; we have now updated our belief regarding the population parameters of interest based on the data. In practice, however, $f(\theta|D)$ is not always the final point of the analysis: often some summary of $f(\theta|D)$ is communicated (akin to the selection of the MLE estimate $\hat{\theta}$ that we have seen for frequentists analysis) such as it's expected value $\mathbb{E}(\theta|D) = \int \theta f(\theta|D)d\theta$. We will look at these summaries and the practice of Bayesian analysis in more detail below. First, however, we will examine a few concrete cases of a Bayesian data analysis.

8.3 Bayesian Data Analysis by Example

In the previous section we extended Bayes' Theorem from events to (distribution functions of) random variables and we inserted a particular meaning: using θ and the data D in our notation we showed how Bayes' Theorem can, in theory, be used to update one's belief regarding a population parameter θ in the face of the evidence provided by the data D. In this section we will examine two specific examples: first, we will focus on a Bernoulli population: assuming our observations $X_1, \ldots, X_n$ are i.i.d. Bernoulli p, what can we learn about the parameter of interest $\theta = p$ based on a set of realizations $x_1, \ldots, x_n$? Second, after examining the Bernoulli population case, we will examine the Normal population case in which we assume observations $X_1, \ldots X_n$ are i.i.d. $\mathcal{N}(\mu, \sigma^2)$.

8.3.1 Estimating the Parameter of a Bernoulli Population

Let θ denote the probability of heads when throwing a (possibly unfair) coin, and let us treat the observations $X_1, \ldots, X_n$ from the coin as i.i.d. Bernoulli(θ). In line with the steps regarding a Bayesian analysis detailed above we first need to specify the sampling model. Given our assumption of a Bernoulli population and using x to denote the vector of all x_i realizations, we have

$$l(\theta) = p(x|\theta)$$

$$= \prod_{i-1}^{n} \theta^{x_i} (1-\theta)^{1-x_i}$$

$$= \theta^{\sum_{i=1}^{n} x_i} (1-\theta)^{n-\sum_{i=1}^{n} x_i}$$

$$= \theta^{n\bar{x}_n} (1-\theta)^{n(1-\bar{x}_n)}.$$

Next, we need to specify our prior beliefs regarding the parameter values by specifying $f(\theta)$. Although prior choice is a heavily debated subject (see Sect. 8.5.1) we will simply choose a reasonable and easy to work with prior in this case. A reasonable prior has support (i.e., $f(\theta) > 0$) for all plausible values of θ. As in our case θ is a Bernoulli p for which $0 \le p \le 1$ we should choose $f(\theta)$ such that it has support on $[0, 1]$. Furthermore, it seems reasonable (again, more on this in Sect. 8.5.1), in the absence of any other information, to choose a prior distribution that gives equal likelihood to all plausible values of θ: the uniform distribution function on $[0, 1]$ thus seems a reasonable choice. The uniform on $[0, 1]$ is however a special case of a much more flexible distribution function called the beta distribution function. The beta(α, β) distribution function is a continuous distribution function with PDF

$$f(\theta) = \frac{\Gamma(\alpha + \beta)}{\Gamma(\alpha)\Gamma(\beta)} \theta^{\alpha-1} (1-\theta)^{\beta-1}.$$

Fig. 8.1 Three beta densities for different values of it's parameters α and β

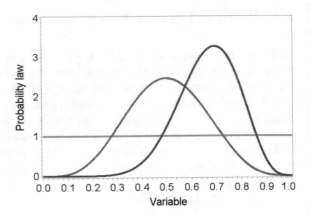

for $0 < \theta < 1$ and zero otherwise. Here $\Gamma()$ denotes the Gamma function $\Gamma(x) = \int_0^\infty s^{x-1}e^{-s}ds$. Note that $\alpha > 0$ and $\beta > 0$. When we choose $\alpha = 1$ and $\beta = 1$ the resulting beta distribution function is actually a uniform distribution function on $[0, 1]$. Figure 8.1 shows the PDF of the beta for various alternative choices of α and β. The horizontal line is given by $\alpha = \beta = 1$ and shows that the uniform distribution function on the unit interval is a special case of the beta distribution function. The beta$(1,1)$ is considered a relatively *uninformative prior* for the Bernoulli p. The density with its peak at 0.5 is given by $\alpha = \beta = 5$; as long as both parameters of the beta are equal and larger than 1 the maximum of the density will be 0.5. When the parameters are unequal this is no longer the case: the third density in the figure is produced using $\alpha = 10$ *and* $\beta = 5$.

Now that we have both our sampling distribution and our prior, we can compute the posterior. Here we use a common trick used in many Bayesian analysis attempts: we first compute the posterior up to some Normalizing constant (i.e., we ignore the marginal likelihood for now), and later see if we can find the correct Normalizing constant to ensure that $f(\theta|x)$ is indeed a valid probability distribution. We find:

$$f(\theta|x) \propto p(x|\theta) f(\theta)$$

$$= \theta^{\sum_{i=1}^n x_i}(1-\theta)^{n-\sum_{i=1}^n x_i} \frac{\Gamma(\alpha+\beta)}{\Gamma(\alpha)\Gamma(\beta)} \theta^{\alpha-1}(1-\theta)^{\beta-1}$$

$$\propto \theta^{(\alpha+\sum_{i=1}^n x_i)-1}(1-\theta)^{(\beta+n-\sum_{i=1}^n x_i)-1}.$$

Hence, we find that in this specific problem the posterior $f(\theta|D)$ is proportional to the density of a beta(α', β') distribution, with

- $\alpha' = \alpha + \sum_{i=1}^n x_i$,
- $\beta' = \beta + n - \sum_{i=1}^n x_i$.

Since the posterior is proportional to the density of the beta, and we are only missing the Normalizing constant, clearly, this will be the Normalizing constant of the beta. Thus, we know that our posterior belief about θ is just a new beta distribution with updated parameters. At this point the role of the prior in this model is also more easily understood: the beta prior can be regarded as adding additional observations where α specifies the a-priori number of successes, and β specifies the a-priori number of failures.

The beta distribution is considered a *conjugate prior distribution*. Formally, when $\mathscr{F}$ is a class of sampling distribution functions, and $\mathscr{P}$ is a class of prior distribution functions, then the class $\mathscr{P}$ is called conjugate for $\mathscr{F}$ if

$$p(\theta|x) \in \mathscr{P} \text{ for all } p(\cdot|\theta) \in \mathscr{F} \text{ and } p(\cdot) \in \mathscr{P}.$$

This definition is cumbersome, since if we choose $\mathscr{P}$ as the class of all distribution functions, then $\mathscr{P}$ is always conjugate. However, in practice we are most interested in *natural* conjugate prior families, which arise by taking $\mathscr{P}$ to be the set of all densities having the same functional form as the sampling distribution (i.e., the likelihood).

Practically, in R, you can use the `rbeta()` function to obtain samples from a beta distribution. If you start with a prior using $\alpha = \beta = 1$ and the vector x contains your observations, then you can obtain $n = 1,000$ draws from the posterior using `rbeta(1000, 1+sum(x), 1+length(x)-sum(x))`. Hence, we now have both an analytical description of our posterior belief regarding θ after seeing a dataset, and we can use R to obtain draws from this posterior distribution or to visualize it.

8.3.2 Estimating the Parameters of a Normal Population

Here we will examine the Bayesian analysis of data resulting from a Normal population. Thus, we will assume $X_1, \ldots X_n$ are i.i.d. $\mathscr{N}(\mu, \sigma^2)$. Although the general principles detailed above still hold—i.e., we need to specify a sampling model, and a prior distribution to compute the posterior distribution—the Normal case is quite a bit more complex, in part due to the fact that the Normal distribution function has two parameters that might be of interest. To simplify matters, and to provide an idea of the steps involved in computing the posterior distribution $f(\theta|D) = f(\mu, \sigma^2|D)$, we first consider the case where the population variance σ^2 is known, and we are only dealing with a single realization x_1. Next, we will extend this to including multiple observations. Finally, we will present the more general result for a Bayesian analysis of n realizations from a Normal population where both μ and σ^2 are unknown.

8.3.3 Bayesian Analysis for Normal Populations Based on Single Observation

To somewhat ease into the Bayesian analysis of Normal populations we will first consider the situation in which we assume $X_1 \sim \mathcal{N}(\mu, \sigma^2)$, we only observe a single realization x, and we assume σ^2 known; hence, we only derive the posterior distribution for μ. The sampling model is:

$$
\begin{aligned}
l(\theta) &= p(x|\mu) \\
&= \frac{1}{\sqrt{2\pi\sigma^2}} e^{\frac{-(x-\mu)^2}{2\sigma^2}},
\end{aligned}
$$

which is simply the Normal PDF we have seen in Chap. 4. Since $\mu \in (-\infty, \infty)$ a natural choice for a prior has support on the full real number line. It turns out that the Normal distribution provides a reasonable choice of prior for μ:

$$
f(\mu) = \frac{1}{\sqrt{2\pi\sigma_0^2}} e^{\frac{-(\mu-\mu_0)^2}{2\sigma_0^2}},
$$

where we use μ_0 and σ_0^2 to denote the parameters of this prior distribution (not to be confused with μ and σ^2 as they occur in the likelihood). We focus, as we did before, on computing the posterior distribution up to the Normalizing constant:

$$
f(\mu|x) \propto p(x|\mu)\, f(\mu)
$$

$$
= \frac{1}{\sqrt{2\pi\sigma^2}} \exp\left(-\frac{(x-\mu)^2}{2\sigma^2}\right) \times \frac{1}{\sqrt{2\pi\sigma_0^2}} \exp\left(-\frac{(\mu-\mu_0)^2}{2\sigma_0^2}\right)
$$

$$
= \frac{1}{2\pi\sqrt{\sigma^2\sigma_0^2}} \exp\left(-\frac{(x-\mu)^2}{2\sigma^2} - \frac{(\mu-\mu_0)^2}{2\sigma_0^2}\right)
$$

$$
= \frac{1}{2\pi\sqrt{\sigma^2\sigma_0^2}} \exp\left(-\frac{\mu^2 - 2\mu\mu_0 + \mu_0^2}{2\sigma_0^2} - \frac{x^2 - 2\mu x + \mu^2}{2\sigma^2}\right)
$$

$$
= c \times \exp\left(-\frac{\mu^2\sigma^2 - 2\mu\mu_0\sigma^2 + \mu_0^2\sigma^2 + \sigma_0^2 x^2 - 2\mu\sigma_0^2 x + \mu^2\sigma_0^2}{2\sigma_0^2\sigma^2}\right)
$$

$$
= c \times \exp\left(-\frac{\mu^2(\sigma^2 + \sigma_0^2) - 2\mu(\mu_0\sigma^2 + \sigma_0^2 x) + (\mu_0^2\sigma^2 + \sigma_0^2 x^2)}{2\sigma_0^2\sigma^2}\right)
$$

$$
= c \times \exp\left(-\frac{\mu^2 + 2\mu\frac{\mu_0\sigma^2 + \sigma_0^2 x}{\sigma^2 + \sigma_0^2} - \left(\frac{\mu_0\sigma^2 + \sigma_0^2 x}{\sigma^2 + \sigma_0^2}\right)^2 + -\left(\frac{\mu_0\sigma^2 + \sigma_0^2 x}{\sigma^2 + \sigma_0^2}\right)^2}{\frac{2\sigma_0^2\sigma^2}{\sigma^2 + \sigma_0^2}}\right)
$$

$$\times \exp\left(-\frac{\mu_0^2\sigma^2 + \sigma_0^2 x^2}{2\sigma_0^2\sigma^2}\right)$$

$$\propto \exp\left\{\frac{-\left(\mu - \frac{\mu_0\sigma^2 + x\sigma_0^2}{\sigma^2 + \sigma_0^2}\right)^2}{2\frac{\sigma^2\sigma_0^2}{\sigma^2 + \sigma_0^2}}\right\}$$

where c is a the Normalizing constant. Next, if we let

$$\sigma_1^2 = \frac{\sigma^2\sigma_0^2}{\sigma^2 + \sigma_0^2} = \frac{1}{\sigma^{-2} + \sigma_0^{-2}},$$

and

$$\mu_1 = \frac{\mu_0\sigma^2 + x\sigma_0^2}{\sigma^2 + \sigma_0^2}$$

$$= \frac{\mu_0\sigma^{-2} + x\sigma_0^{-2}}{\sigma^{-2} + \sigma_0^{-2}}$$

$$= \sigma_1^2(\mu_0\sigma_0^{-2} + x\sigma^{-2}),$$

we obtain

$$\sigma_1^{-2} = \sigma^{-2} + \sigma_0^{-2}$$

and

$$\mu_1\sigma_1^{-2} = \mu_0\sigma_0^{-2} + x\sigma^{-2}.$$

After these derivations we see that

$$f(\mu|x) \propto e^{\frac{(\mu - \mu_1)^2}{2\sigma_1^2}}$$

which is recognizable as the Normal PDF and hence the Normalizing constant that we are after is $\frac{1}{\sqrt{2\pi\sigma_1^2}}$ and we find that

$$f(\mu|x) = \frac{1}{\sqrt{2\pi\sigma_1^2}} e^{\frac{(\mu - \mu_1)^2}{2\sigma_1^2}}.$$

This derivation shows that if σ^2 is assumed known, and we use a Normal prior for μ, we find that the posterior for μ is itself a Normal: $\mu|X_1 = x \sim \mathcal{N}(\mu_1, \sigma_1^2)$. It is worthwhile to have another look at the definition of μ_1 above to see that the posterior mean μ_1 is effectively a weighted average of the prior mean μ and the observation x.

8.3.4 Bayesian Analysis for Normal Populations Based on Multiple Observations

The analysis in the previous section gave an idea of the Bayesian analysis of data originating from a Normal population; however, we considered σ^2 known, and we only considered a single data point x. It is much more realistic that we would be dealing with a much larger set of realizations $x_1, \ldots, x_n$. Here, we are assuming that effectively these are realizations from n identically distributed, independent, random variables $X_i \sim \mathcal{N}(\mu, \sigma^2)$. Adding this step to our analysis is conceptually not hard (although it takes some algebra). The only real changes affects the sampling model: the likelihood is now specified as:

$$l(\theta) = p(x_1, \ldots, x_n | \mu)$$
$$= \prod_{i=1}^{n} \frac{1}{\sqrt{2\pi\sigma^2}} e^{\frac{-(x_i - \mu)^2}{2\sigma^2}},$$

Multiplying the likelihood $l(\theta)$ of this dataset by the (Normal) prior $f(\mu)$ we defined in the previous section again gives us the posterior up to the Normalizing constant. Working through all the algebra (see Gill 2014 for an example), we find that

$$f(\mu | x_1, \ldots, x_n) = \frac{1}{\sqrt{2\pi\sigma_1^2}} e^{\frac{(\mu - \mu_1)^2}{2\sigma_1^2}}.$$

where this time

$$\sigma_1^2 = \left(\frac{1}{\sigma_0^2} + \frac{1}{\sigma^2/n} \right)^2$$

and

$$\mu_1 = \sigma_1^2 \left(\frac{\mu_0}{\sigma_0^2} + \frac{\bar{x}}{\sigma^2/n} \right).$$

Thus, after following through the derivation, we find that the posterior distribution for μ given the realizations $x_1, \ldots, x_n$ is itself again Normal as we found before. Additionally, also in this case the posterior mean μ_1 is a weighted average of the prior mean μ_0 and the data: it is simply a weighted average of the prior mean and the mean of the observations $\bar{x} = \sum_{i=1}^{n} x_i/n$. Note that in this case the variance of the posterior is affected by the number of observations; the term σ^2/n clearly shrinks as the number of observations increases leading to a decrease in the variance of the posterior. If $n \to \infty$, then μ_1 will converge to the sample mean, and σ_1^2 will converge to zero.

8.3.5 Bayesian Analysis for Normal Populations with Unknown Mean and Variance

To complete our Bayesian analysis of the Normal population we will also consider the case in which both σ^2 and μ are unknown. The sampling model does not change much, although we do add σ^2 explicitly as an unknown:

$$l(\theta) = p(x_1, \ldots, x_n | \mu, \sigma^2)$$

$$= \prod_{i=1}^{n} \frac{1}{\sqrt{2\pi\sigma^2}} e^{\frac{-(x_i - \mu)^2}{2\sigma^2}}.$$

The prior choice, however, is more involved than it is in the examples we have seen hitherto; this time we need a prior for the joint distribution function of μ and σ^2; hence, we are looking for a multivariate PDF $f(\mu, \sigma^2)$ with the correct support (note that $\sigma^2 \in (0, \infty]$). Many possible choices are available, but one often used (conjugate) prior is the Normal-inverse-χ^2 distribution. It's PDF is:

$$\begin{aligned} f(\mu, \sigma) &= NI\chi^2(\mu_0, \kappa_0, \nu_0, \sigma_0^2) \\ &= p(\mu, \sigma^2) \\ &= p(\mu | \sigma^2) p(\sigma^2) \\ &= \mathcal{N}(\mu_0, \sigma^2/\kappa_0) \times \chi^{-2}(\nu_0, \sigma_0^2) \\ &= \frac{1}{Z(\mu_0, \kappa_0, \nu_0, \sigma_0^2)} (\sigma^2)^{-(\nu_0/2+1)/2} \exp\left(-\frac{1}{2\sigma^2}[\nu_0\sigma_0^2 + \kappa_0(\mu_0 - \mu)^2]\right) \end{aligned}$$

where

$$Z(\mu_0, \kappa_0, \nu_0, \sigma_0^2) = \frac{\sqrt{2\pi}}{\sqrt{\kappa_0}} \Gamma(\nu_0/2) \left(\frac{2}{\nu_0\sigma_0^2}\right)^{\nu_0/2}.$$

The above shows that our prior for σ^2 is a scaled inverse-χ^2 distribution and the prior for μ conditional on σ is a Normal distribution; both of these are known and well-understood univariate distribution functions. While the Normal-inverse-χ^2 PDF may look unwieldy—see Gill (2014) for step-by-step derivations—the interesting result for us is that the posterior $f(\mu, \sigma^2 | x_1, \ldots, x_n)$ is itself again a Normal-inverse-χ^2-PDF:

$$f(\mu, \sigma^2 | x_1, \ldots, x_n) = NI\chi^2(\mu_n, \kappa_n, \nu_n, \sigma_n^2)$$

with

$$\mu_n = \frac{\kappa_0 \mu_0 + n\bar{x}}{\kappa_n}$$

$$\kappa_n = \kappa_0 + n$$

$$\nu_n = \nu_0 + n$$

$$\sigma_n^2 = \frac{1}{\nu_n}\left(\nu_0 \sigma_0^2 + \sum_{i=1}^{n}(x_i - \bar{x})^2 + \frac{n\kappa_0}{\kappa_0 + n}(\mu_0 - \bar{x})^2\right).$$

As we have learned in Chap. 6, when working with bivariate distribution functions it is often interesting to look at the marginal distribution functions of the variables involved. We will skip the derivations, but for the Normal-inverse-χ^2 distribution it is possible to derive both the marginal PDF of μ and σ^2 in closed form. For the posterior Normal-inverse-χ^2 distribution provided above we find that

$$f(\sigma^2|x_1, \ldots, x_n) = \chi^{-2}(\nu_n, \sigma_n^2) \tag{8.3}$$

and

$$f(\mu|x_1, \ldots, x_n) = t(\mu_n, \sigma_n^2/\kappa_n). \tag{8.4}$$

Thus, the marginal posterior PDF of σ^2 is a scaled inverse χ^2 PDF with scale ν_n and degrees of freedom σ_n^2, while the marginal posterior distribution for μ is the familiar t-distribution with mean μ_n and σ_n^2/κ_n degrees of freedom.

In R it is easy to generate samples from the posterior for μ using the `rt()` function. Sampling from the bivariate Normal-inverse-χ^2 distribution is a bit more involved. The following code defines three functions that jointly (together with the functions available in core R, allow you to obtain samples from the Normal-inverse-χ^2 distribution:

```
# Sample from inverse gamma
> rinvgamma = function(N,a,b){
+   return (1/rgamma(N,a,rate = b))
+ }


# Sample from scaled inverse chi-squared
> rinvchisq = function(N,df,s2=1){
+   return (rinvgamma(N,a = df/2,b = df*s2/2))
+ }


# Sample from bivariate Normal Inverse Chi Squared
> rnorminvchisq = function(N,mu0,kappa0,nu0,sigma02){
+   sigmas_square <- rinvchisq(N, nu0, sigma02) + mus <-
sapply(sigmas_square, function(s2) rnorm(1,mu0,sqrt(s2/kappa0)))
    +
+   return (rbind(mus, sigmas_square))
+ }
```

8.4 Bayesian Decision-Making in Uncertainty

In the previous sections we first extended Bayes' law from events to distributions, and subsequently we showed how we can use these results to quantify our beliefs regarding population parameters conditional on observed data. Next, we provided two specific examples: we have demonstrated how to compute our posterior belief about the parameter p for observations from a Bernoulli population, and similarly we have shown how to update our beliefs regarding μ and σ^2 when faced with observations from a Normal population. More generally, a Bayesian analysis will, after choosing a prior (more on that subject in the next section), allow you to compute (or at least sample from) the posterior $f(\theta|D)$. However, simply providing the parameters of the posterior or a batch of samples from the posterior often does not answer the questions that a statistical analysis is supposed to answer: as we have seen in previous chapters we would often like to provide a specific estimate for a population value, we would like to make probabilistic statements regarding populations values, or we would like to test specific hypotheses and make decisions. We have seen how many of these tasks are accomplished in the Frequentist framework in prior chapters; here we briefly highlight approaches that are often used in Bayesian analysis for similar purposes.

8.4.1 Providing Point Estimates of Parameters

After obtaining either an analytical expression for $f(\theta|D)$, or, as is more often the case in Bayesian analysis in practice, a way to obtain samples from this posterior PDF, we might want to provide a so-called point estimate of the population value in question. A point estimate is simply a single-number summary of the results obtained regarding a single parameter. The Frequentist analogy in this case would be providing (e.g.) a maximum likelihood estimate of a population value.

In the Bayesian case providing point-estimates is conceptually very simple: we have already seen various ways of summarizing distributions using (e.g.) the expected value, the median, or the mode. When an analytical expression for $f(\theta|D)$ is available these can often readily be computed. The expected value is

$$\mathbb{E}(\theta) = \int_{\mathbb{R}} \theta f(\theta)\, d\theta, \tag{8.5}$$

whereas the median is the value of m for which

$$\int_{\infty}^{m} f(\theta)\, d\theta = \frac{1}{2}, \tag{8.6}$$

and the mode is the (potentially local) maximum of the distribution function, i.e., where $f'(\theta) = 0$. Note that the mode is, in the literature on Bayesian analysis, often referred to as the *Maximum A Posteriori* (or MAP) estimate.

In some sense, which we will make more precise in the next subsection, the expected value, the median, or the mode provide a Bayesian's "best guess" towards the population value if she or he is forced to make a single number statement. Note that very often in practice $f(\theta|D)$ is not analytically available; rather, we are able to use Markov Chain Monte-Carlo methods (see details below) to generate samples from $f(\theta|D)$. Given m samples of the posterior (i.e., given $\theta_1, \ldots, \theta_m$), computation of the various point estimates is straight forward (and effectively brings us right back to the definitions of the mean, median, and mode of a sample, as we discussed in Chap. 1): the core R functions `mean()` and `median()`, together with the function `get_mode()` we defined in Chap. 1 suffice.

8.4.1.1   A Formal Justification for Specific Point Estimates: Bayesian Decision Theory

Which exact point estimate should be reported can be made (much) more formal by considering the reporting of such a point estimate a (Bayesian) statistical decision problem. Here we briefly introduce some of the basic ideas behind Bayesian decision theory, and in doing so provide a rationale for reporting the expected value of the posterior distribution of θ. This section is by no means an exhaustive discussion of Bayesian decision theory; it is merely intended to be a short introduction to the topic and a view on things to come in future courses in Bayesian statistics that you might follow.[1]

Suppose we want to make a decision d based on the data x—note that we change notation here slightly and use x for data as opposed to D—where $d(x) \in \mathcal{D}$; we assume that the decision is an element of the decision space $\mathcal{D}$ (i.e., the space of all possible decisions) and furthermore we assume that the decision will be a function of the data. Subsequently, we stipulate that every time we make a decision, we can *loose* something by making a "wrong" decision. Next, we define *risk* as the expected loss (taking the expectation over the paremeter θ that we estimate in the process), and we state that our goal when making decisions is to *minimize risk*. We can set this up mathematically by quantifying the loss involved, which is done using a loss function:

$$L = L(d(x), \theta) \geq 0$$

One interpretation of the loss function is that it gives a numerical (positive) value for any decision $d(x)$ given a value of a parameter θ. A specific example of a loss function in a parameter estimation context is $L(d, \theta) = (d - \theta)^2$, the so-called squared error loss. This loss function basically states that if we make a specific decision $d(x)$ regarding the parameter θ based on the observed data x, we incur no loss if $d = \theta$

---

[1] For those interested we would recommend the book Robert (2007).

and the further away from the actual parameter our choice is, the more loss we incur. Note that L itself is a random variable (since it depends on the data) and thus we can take (e.g.) its expectation or compute its variance.

Now, risk, as we stated above, is the *posterior expected loss*, which is a function of the data due to the $\Pr(\theta|x)$ term:

$$\mathbb{E}\left(L(d(x), \theta)|X = x\right) = \int_{\Theta} L\left(d(x), \theta\right) p(\theta|x)d\theta$$

Any decision $d(x)$ that minimizes the posterior expected loss—given some loss function—is called a *Bayes optimal decision*.

If we use the *squared error* loss function $L(d(x), \theta) = (d(x) - \theta)^2$ introduced above and denote $\mathbb{E}(\theta|X = x)$ by $\mu(x)$, then (without detailing the proof) we can state

$$\mathbb{E}\left(L(d(x), \theta)|X = x\right) = \mathbb{E}\left((d(x) - \theta)^2|X = x\right)$$

$$= \dots$$

$$= (d(x) - \mu)^2 + \mathsf{VAR}(\theta|X = x)$$

which clearly is minimized when $d(x) = \mu(x) = \mathbb{E}(\theta|X = x)$. Thus, the posterior *mean* minimizes the squared error loss, providing a decision-theoretic justification for the reporting of the posterior mean as a point-estimate.[2]

The above derivation of the decision that minimizes the expected squared error loss was general (e.g., we did not make any additional assumptions regarding the priors or the likelihood), and hence we have just shown that the Bayes estimate under squared error loss is always the posterior mean. Alternatively, if the loss function is the absolute value function, $L(d(x), \theta) = |d(x) - \theta|$, we would obtain the posterior median as the decision that minimizes the risk. In this way, a Bayesian analysis, in combination with formal statistical decision theory, motivates specific decisions regarding the parameters of interest.

8.4.2 Providing Interval Estimates of the Parameters

Next to point estimates, we often like to quantify our uncertainty regarding a parameter. In previous chapters we have frequently encountered the (Frequentist) confidence interval as a tool for quantifying the uncertainty regarding an estimate. In Bayesian data analysis it is common to consider *credible intervals*. As usual, no new analysis is required: the Bayesian credible interval is merely a summary of the posterior PDF $f(\theta|D)$. A $100\%(1 - \alpha)$ credible interval gives the region in parameter space in which the (true) parameter value falls with probability $(1 - \alpha)$. Often, the $q\%$ credible interval identifies the bounds of the middle q percent of posterior mass

---

[2] For interested students we refer to Robert (2007) for a step-by-step derivation.

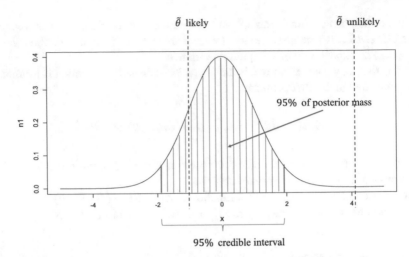

Fig. 8.2 Posterior PDF of a parameter, likely and unlikely true population values $\tilde{\theta}$, and the 95% credible interval

(i.e., the mass outside of the interval is $\frac{1}{2}(1 - q)$ on each side). An example of this is given in Fig. 8.2. It has to be noted that credible intervals are not uniquely defined for arbitrary posterior distribution functions and several definitions exist:

- Choosing the narrowest interval. For a *unimodal* distribution function this will involve choosing those values of highest probability density including the mode.
- Choosing the interval where the probability of being below the interval is as likely as being above it. This interval will include the median. This is sometimes called the equal-tailed interval.
- Assuming that the mean exists, choosing the interval for which the mean is the central point.

Also, credible intervals need not be a connected set. The shortest possible interval with a given probability is called the *Highest Posterior Density Interval* (HPDI).

8.4.2.1   Comparing Confidence Intervals and Credible Intervals

For students first encountering credible intervals it is often unclear how these summaries of the posterior PDF compare to confidence intervals (please do revisit Chap. 5 for our initial discussion of confidence intervals). Adding to this confusion is the fact that in a number of situations the numerical values of a $100\%(1 - \alpha)$ credible interval might coincide exactly with the numerical values of the $100\%(1 - \alpha)$ confidence interval. This latter is, for example, true for the credible/confidence interval for the mean parameter μ when dealing with Normal populations: the confidence classical interval around $\bar{x}$ is numerically exactly the same as the credible interval around the posterior for μ (given a suitable choice of prior).

However, this occasional numerical similarity does not imply that both intervals quantify the same thing; far from it. The true meaning of the confidence interval is very different from the true meaning of the credible interval, which lays bare a big philosophical distinction between Bayesian and Frequentist analysis (which we will discuss in more detail below). Recall the interpretation of the confidence interval provided in Chap. 5:

- *The confidence interval quantifies that, if the same population is sampled on numerous occasions and interval estimates are made on each occasion, the resulting intervals would include the true population parameter in (approximately) $100\%(1 - \alpha)$ of the cases.*

In this definition we assume the true population value to be a fixed quantity; the uncertainty captured by the interval is due to our sampling from the population. The confidence interval is a probabilistic statement regarding the provided bounds: it will include the true (fixed) population value in $100\%(1 - \alpha)$ of the cases it is computed on a (similarly collected) sample. The interpretation of the credible interval is different:

- *Given our prior and our observed data, there is a $100\%(1 - \alpha)$ probability that the (unobserved) value of θ is within the interval.*

The credible interval is thus a statement about the population parameter taking on specific values with a specific probability. For a more extensive discussion of credible intervals and confidence intervals see Hespanhol et al. (2019).

8.4.3 Testing Hypotheses

As we encountered in previous chapters, specifically Chap. 7, we would often like to not only provide point or interval estimates, but would like to test specific hypotheses. While there is a large literature on Bayesian hypothesis testing (see, e.g., Gu et al. 2014; Wagenmakers et al. 2017), the simplest approach to Bayesian hypothesis testing follows directly from our earlier discussion of extending Bayes' Theorem from events to multiple specific assumptions/hypothesis regarding parameter values: Bayes' Theorem can be used to directly quantify the probability of a parameter having a specific value (or being in a range of values). This idea can be extended to quantify probability in favor of a specific hypothesis.

However, an alternative approach that is fairly popular for hypothesis testing in a Bayesian framework is the use of so-called *Bayes factors*. The Bayes factor relies on the idea that the posterior probability $\Pr(M|D)$ of a model M given data D is also given by Bayes' Theorem:

$$\Pr(M|D) = \frac{\Pr(D|M)\Pr(M)}{\Pr(D)}$$

and the subsequent idea that we can compare the posterior probability of different models M_1, M_2 and make a choice in favor of one or the other. Note that a *model* here often refers to a specific value or range of the parameters of a parametric statistical model. Given this setup the term $\Pr(D|M)$ is effectively the likelihood of the data under this specific model, and it thus represents the probability that the data are produced under the assumptions encoded in the model. Evaluating this likelihood is key to Bayesian model comparison. Given a model selection problem, in which we have to choose between two models on the basis of observed data D, the plausibility of the two different models M_1 and M_2, parametrized by model parameter vectors θ_1 and θ_2, is assessed by the *Bayes factor K* given by:

$$K = \frac{\Pr(D|M_1)}{\Pr(D|M_2)} = \frac{\int \Pr(\theta_1|M_1)\Pr(D|\theta_1, M_1)\, d\theta_1}{\int \Pr(\theta_2|M_2)\Pr(D|\theta_2, M_2)\, d\theta_2}.$$

Hence, the Bayes factor quantifies the relative evidence, after seeing the data, in favor of one model or the other. The Bayes factor K theoretically runs from 0 to infinity: if $K < 1$, the Bayes factor provides evidence in favor of M_2. If $K > 1$, it provides evidence in favor of M_1, and Bayes factors of $K > 10$ are often considered strong evidence: in this case M_2 would be rejected and M_1 would be accepted. Thus in Bayesian hypothesis testing the selected hypothesis, i.e., the hypothesis encoded in M_1 or M_2, depends on the computed value of K.

The discussion above is a bit abstract but it can easily be clarified using a simple example. Suppose we have a binary random variable and we want to compare a model, M_1, in which the probability of success is $p = \frac{1}{2}$, and another model M_2 where p is unknown and we take a prior distribution for p that is uniform on $[0, 1]$. Thus, we are effectively trying to test whether our data provides evidence that $p = \frac{1}{2}$ compared to p being anywhere between 0 and 1. Suppose we obtain a sample of size $n = 200$ and find 115 successes and 85 failures. The likelihood can now be calculated according to the binomial PDF:

$$\binom{200}{115} p^{115}(1 - p)^{85}$$

Now, since for model M_1 we assume $p = \frac{1}{2}$ we find:

$$\Pr(X = 115 \mid M_1) = \binom{200}{115}\left(\frac{1}{2}\right)^{200} = 0.005956....$$

while for M_2 we assume we do not know p, and hence we obtain

$$\Pr(X = 115 \mid M_2) = \int_0^1 \binom{200}{115} p^{115}(1 - p)^{85}\, dp = \frac{1}{201} = 0.004975....$$

for model M_2.[3] The ratio of these two, and thus the Bayes factor K, is then 1.197, which provides some—but very little—evidence in favor of model M_1. Effectively a Bayesian would conclude that both models are pretty much equally likely given the observed data.

Note that a Frequentist hypothesis test of M_1 (considered as a null hypothesis) would have produced a very different result. Such a test says that M_1 should be rejected at the 5% significance level, since the probability of getting 115 or more successes from a sample of 200 if $p = \frac{1}{2}$ is 0.0400. Hence, while the Frequentist would reject the null hypothesis $H_0 : p = 0.5$ in this case, the Bayesian would not decide that the alternative model M_2 is convincingly more plausible than M_1.[4]

The Bayes factor can be hard to compute for many types of tests, and it is still an active area of research. For a number of standard tests, such as one-sample and two-sample tests for means and proportions, simple to use R packages are available (see for example: https://richarddmorey.github.io/BayesFactor/). For more complicated models computing K can be challenging; however, the flexibility of the Bayes factor in specifying the models means that Bayes factors can often be computed for hypotheses that are hard (if not impossible) to test in a Frequentist framework.

8.5 Challenges Involved in the Bayesian Approach

Above we tried to provide an idea of the Bayesian school of thought in statistics analysis: given a prior distribution function for the parameters of interest, and a sampling model (i.e., the likelihood), we can compute the posterior PDF of the parameters $f(\theta|D)$. This posterior PDF effectively captures all that we can learn from the data, and hence the remainder of a Bayesian data analysis in effect simply considers computing different summaries of $f(\theta|D)$.

While the Bayesian school of thought provides a simple framework for doing data analysis that is extremely flexible and, according to some, philosophically more appealing than the Frequentist school of thought (a discussion we will touch upon in the next section), (applied) Bayesian data analysis is not without difficulties. Although Bayesian thinking has been around for centuries (Thomas Bayes's famous Bayes' Theorem was published in 1763, shortly after his death), a number of issues have hindered its uptake. These issues have often been both philosophical and practical. We will focus on the latter challenges in this section, where we first discuss the issue of prior choice, and subsequently discuss the computational difficulties that are often involved in applied Bayesian analysis using models that are slightly more complex than the models we have discussed up till now.

---

[3] Note that this is the prior times the likelihood and subsequently the parameter p is integrated out.
[4] For more on Bayes factors see Kass and Raftery (1995).

8.5.1 Choosing a Prior

One major challenge to Bayesian thinking has often been the choice of prior distribution: how do we obtain a sensible and valid specification for $f(\theta)$? As was clear from our analysis of both Bernoulli and normal populations, the prior can play a large role in the resulting posterior $f(\theta|D)$ and thus it is likely to also play a large role in the final conclusions, i.e., the reported summaries of $f(\theta|D)$. How then should we choose a good prior?

While some discard choosing a prior altogether on philosophical grounds, here we simply discuss a number of attempts that have been conceived in the large literature on prior choice to tackle the problem of selecting a prior. We have already encountered one class of priors, namely conjugate priors, which are appealing for their computational ease. However, even when settling for a conjugate prior—although one might argue that computational ease should not be the main argument when setting up a data analysis—the values of the parameters in the Bayesian context often called *hyperparameters* still need to be chosen, and hence opting for a conjugate prior does not "solve" the issue of prior choice.

Two approaches to prior choice are often distinguished, effectively highlighting two streams of thought regarding the meaning of the prior. One approach focusses on constructing *uninformative* prior distribution functions. In some sense, uninformative priors seek to formalize the idea that the choice of prior should affect, as little as possible, the resulting posterior. The main challenge for this approach is formalizing what one means exactly by the term "uninformative". The other approach to prior construction aims to construct priors that are *informative*: the prior should reflect the knowledge we already have around the parameter. In this approach the main challenge consists of moving from our current knowledge regarding some data analysis problem—which is often not explicitly quantitative—to a, sometimes high dimensional, probability distribution function that summarizes this knowledge.

8.5.1.1 Uninformative Priors

While the intuition behind uninformative priors is simple enough, the practice turns out to be very challenging. Properly defining *uninformative* has been a challenge, and we will discuss a few common approaches.

One approach that might seem feasible is that of choosing *uniform* priors. A uniform prior for a Bernoulli p—as we analyzed above—can be specified by choosing $f(\theta) = 1, 0 \le \theta \le 1$. It is not hard to extend this to $f(\theta) = [b - a]^{-1}, a \le \theta \le b$ to cover more cases. And, somewhat surprisingly perhaps, extending even further to $f(\theta) = c, -\infty \le \theta \le \infty$, while not a valid probability distribution and therefore often called an *improper* prior, might lead to proper posterior distribution functions and thus can be used for parameters whose ranges are not bounded.

Although uniform priors seem appealing and are very often used, they can often hardly be considered uninformative as uniform priors are often not invariant to transformations: i.e., if we change the parametrization of the model, a uniform prior for one specific parametrization often leads to a (very) informative—in the sense that a lot of prior mass is on specific regions of the parameter space—prior in the new parameter space. Such is for example true for the uniform distribution on p we discussed above: if we change from the probability metric to analyzing the odds-ratio than the previously thought uninformative uniform prior actually places a lot of probability mass on small values of the odds-ratio. Furthermore, uniform priors are often not the "highest variance priors", although often a high variance prior can be motivated to be less informative. For example, when considering the beta prior for the analysis of a Bernoulli p, choosing $\alpha = \beta = 0.5$ would provide less information on the posterior than choosing the uniform $\alpha = \beta = 1$. Actually, in this setting, one could motivate that the highest variance prior $\alpha = \beta \to 0$ is actually the least informative (see below).

In an attempt to solve some of the issues arising with uniform priors, Jeffreys (1946) suggested constructing priors that are invariant to (certain types of) transformations. For a single parameter, Jeffreys' prior is produced by taking the square root of the negative expected value of the second derivative of the likelihood function (i.e., the Fisher information matrix discussed in Chap. 5):

$$f(\theta) = \left[-\mathbb{E}\left(\frac{d^2}{d\theta^2} \log f(x|\theta) \right) \right]^{\frac{1}{2}}. \tag{8.7}$$

While the above expression might look challenging, Jeffreys' prior is often straightforward to calculate in many applied cases.

Another class of priors that is often deemed uninformative, next to Jeffreys' priors and uniform priors, are called *reference priors*: reference priors are constructed in such a way that the likelihood dominates the posterior, i.e., the resulting posterior is as little influenced by the prior as possible. This is formally done by maximizing the Kullback–Leibner (1951) divergence of the posterior distribution to the prior distribution. Reference priors are often easier to compute than Jeffreys' priors for multivariate cases (i.e., multiple parameters) while for the univariate case (i.e., single parameter) both produce the same result.

To make the above discussion of uninformative priors a bit more tangible it is useful to reconsider the results we found for the posterior distribution for a Bernoulli population; we found that the posterior $f(\theta|D)$ was given by a beta PDF with parameters:

- $\alpha' = \alpha + \sum_{i=1}^{n} x_i$,
- $\beta' = \beta + n - \sum_{i=1}^{n} x_i$.

In the specification of the posterior it is clear that the parameters of the prior, i.e., α and β, are combined with the number of successes $\sum_{i=1}^{n} x_i$ and the number of failures $n - \sum_{i=1}^{n} x_i$ in the data. Thus, in this specific case, the prior can be thought of as providing a specific number of prior observations: a prior using $\alpha = \beta = 1$ as parameters effectively encodes the information of two observations, one success and one failure. As we noted before, a beta(1, 1) is uniform on [0, 1]. Clearly, in this case the uniform is not uninformative in the sense that it does effectively provide these two additional data points that carry over to the posterior. Jeffreys' prior (and the reference prior), for this problem is $\alpha = \beta = \frac{1}{2}$ which is far from uniform, but it can be considered as only adding a single observation to the posterior as opposed to two in the uniform case. In this specific case it is also possible to choose $\alpha = \beta \to \infty$; this prior is not proper, but it results in a posterior beta with parameters

- $\alpha' = \sum_{i=1}^{n} x_i,$
- $\beta' = n - \sum_{i=1}^{n} x_i$

which—if at least one failure and one success are observed—can be considered as the least informative. The above discussion mainly goes to show that although choosing uninformative priors might sound appealing, it can be tricky to do so convincingly. The problems are even stronger when larger models, with many more parameters, are considered.

8.5.1.2 Informative Priors

Next to uninformative priors there is a class of priors that is considered informative; these are predominantly separated from the non-informative priors by the fact that they are designed to explicitly contain the information regarding the problem at hand that is already available. Prior studies for example might provide results that one would like to include in a subsequent estimation problem: theoretically, Bayesian analysis methods are extremely well suited to incorporate such prior results by encoding them into the prior distribution that is used in the analysis. We stress the word "theoretically" here, as summarizing the results of prior studies into probability distributions is not at all easy. A large literature on *elicitation* priors exists; these are priors that are explicitly designed to capture knowledge that experts in a domain might have regarding the plausible values of a parameter.

For example, when estimating the average height of a sample of college students (and assuming these to come from a Normal population) in centimeters, one might be realistically able to provide bounds for the mean parameter μ (i.e., $100 < \mu < 200$) and upper bound σ^2 (i.e., $\sigma^2 < 1,000$. However, even with this reasonably defined input—in many cases we know much less about the problem at hand—theoretically still an infinite number of bivariate prior distribution functions for (μ, σ^2) could be conceived that encode this information. Thus, while eliciting knowledge from experts often helps in creating informative priors, there is often still a lot of uncertainty left, even in domains in which much knowledge is available.

Next to elicitation priors, there are also many attempts to include quantitative data from earlier studies into priors for new studies. In a sense this should be easy: if the study examines the exact same question, the posterior of the previous study could be used as the prior for the next (providing an example of the theoretical ease with which Bayesian analysis can combine the results of many studies). However, often studies are not exactly the same, and thus one does not wish to use the posterior exactly; in these cases *power* priors are popular: power priors effectively use a part of the likelihood of a previous study to construct the posterior. Using x_0 as the data from a prior study we can compute

$$f(\theta|x_0, \alpha_0) = f(\theta)[f(x_0|\theta)]^{\alpha_0} \qquad (8.8)$$

where $\alpha_0 \in [0, 1]$ and subsequently use $f(\theta|x_0, \alpha_0)$ as the prior for our next study.

As the use of prior distributions is one of the main criticisms of the Bayesian approach to statistical data analysis, many different ways of constructing priors have emerged in the literature. We refer the interested student to Robert (2007) for a much more extensive discussion of the issue. For now we would like to make two additional remarks: first, conjugate priors can be considered uninformative (i.e., when they are uniform), but can also be extremely informative; conjugacy is orthogonal to informativeness. Second, priors can often be thought of as adding bias to an estimator; in many cases a point estimate provided by a Bayesian analysis is biased (as introduced in Chap. 2). While bias is often considered a drawback in the Frequentist literature, the improvements in MSE that often arise from (informative) priors are considered appealing in the literature on Bayesian data analysis.

8.5.2 Bayesian Computation

Next to prior choice, the complexity of the computations involved in many Bayesian analyses has posed a challenge for the Bayesian paradigm as a whole. For a long time, effectively until we got access to fast computers, applied Bayesian analysis was restricted to a (very) small number of parameters and/or conjugate priors: only in these simple—and often too simplistic—cases was it practically possible to conduct a Bayesian analysis. It is simple to see why Bayesian computations might become challenging quickly; recall Bayes' Theorem:

$$f(\theta|D) = \frac{l(\theta)f(\theta)}{\int l(\theta)f(\theta)d\theta} \qquad (8.9)$$

where the denominator is the integral over the prior times the likelihood (i.e., the marginal likelihood). This integral, especially when the distribution functions considered are highly multivariate, might be very tricky to solve, which means that it becomes hard (or impossible) to work with the posterior distribution.

Luckily, over the last few decades, we have largely solved these issues by resorting to Markov Chain Monte Carlo (MCMC) methods. We have already encountered this approach in Chap. 4: although it might be hard to work with a PMF or PDF analytically, as long as we are able to generate samples from the desired distribution function using a computer we can effectively compute the point estimates and interval estimates of interest. A good and thorough overview of MCMC methods as they are used in Bayesian analysis is provided in Gill (2014); here we highlight the basic ideas of two very popular numerical methods that are often used in applied Bayesian analysis: the *Gibbs sampler* and the *Metropolis–Hastings sampler*.

The basic idea behind Gibbs sampling is relatively simple; suppose we can break up our analysis problem in such a way that, while we might not have analytical expressions or a way to sample from the joint distribution function of the parameters of interest, we can express the parameters of interest conditionally on the other parameters. Furthermore, assume that we are able to generate draws from these posterior distributions. Thus, we might have PDFs $f(x|y)$ and $g(y|x)$ and we would like to obtain samples of their joint distribution function. In this case, Gibbs sampling can be done by, after picking a starting point (x_0, y_0), drawing random values from the conditionals as follows:

$$x_1 \sim f(x|y_0) \quad y_1 \sim f(y|x_1) \tag{8.10}$$

$$x_2 \sim f(x|y_1) \quad y_2 \sim f(y|x_2) \tag{8.11}$$

$$x_3 \sim f(x|y_2) \quad y_3 \sim f(y|x_3) \tag{8.12}$$

$$\sim \dots \quad \dots \tag{8.13}$$

$$x_m \sim f(x|y_{m-1}) \quad y_m \sim f(y|x_m). \tag{8.14}$$

Under relatively mild conditions the generated draws (x_j, y_j) approximate draws from the joint PDF $f(x, y)$ very well (see Gill 2014 for more information). Note that the sequence of samples $x_1, y_1, \dots, x_m, y_m$ is often called a(n) (MCMC) chain.

While Gibbs sampling is often extremely useful, it does rely on the ability to sample from the conditional distribution functions of interest. The Metropolis–Hastings sampler does not have this drawback and effectively allows sampling from any high-dimensional target PDF $f(x_1, \dots, x_n)$. Here we conceptually describe the classical *Metropolis sampler* (the Metropolis–Hastings sampler extends the idea presented here to non-symmetric proposal distributions) for the bivariate case. Suppose we have a PDF $f(x, y)$ that we wish to obtain samples from but we are unable to generate these samples directly. In this case, given a proposal PDF $g(x', y'|x, y)$ that we can sample from, and a starting point (x, y), we can

1. Sample (x', y') from $g(x', y'|x, y)$,
2. Sample a value u from a uniform $[0, 1]$,
3. If $\frac{f(x',y')}{f(x,y)} > u$ then accept (x', y') as the new destination,
4. Or else keep (x, y) as the new destination.

The scheme above will approximately provide (correlated) multivariate draws of $f(x, y)$ for a suitable choice of $g(x', y'|x, y)$ (in this case one that is symmetrical in its arguments, i.e., $g(x', y'|x, y) = g(x, y|x', y')$, although less restrictive versions of the algorithm exist). The intuition behind the method is relatively simple: we construct a sequence (or chain) of draws that spends more time (i.e., less often changes value) in high-density regions of $f(x, y)$ than in low-density ones and thus a histogram over the obtained samples—which will contain more values that are more probable—will converge to the sought-after distribution.

Computational methods have, over the last few decades, become a major ingredient of applied Bayesians data analysis. The Gibbs sampler and the Metropolis–Hastings sampler have become invaluable tools, making it possible to address complex statistical analysis problems using Bayesian methods. However, luckily, for many applied Bayesian analyses it is no longer necessary to derive and implement these samplers yourself; many general purpose software packages exist that allow you to use—out of the box—the samplers described (and more advanced versions thereof) without tedious analytical or engineering work. We will cover an example of this software in the next section.

8.6 Software for Bayesian Analysis

In the previous section we discussed both prior choice and computation as two challenges that have hindered the (practical) adoption of Bayesian data analysis methods. In the next section we will revisit the role of the prior—and some of the philosophical arguments in favor and against the use of priors. In this section, however, we will demonstrate that, for a very large number of problems, the computational challenges have by and large been solved. In recent years a number of software packages have been created which, when provided with a sampling model (i.e., likelihood) and a prior will allow you to obtain samples from the posterior distribution function without the necessity of carrying out analytical derivations or coding up MCMC samplers like the Gibbs sampler or the Metropolis–Hastings sampler we introduced above. Effectively, these software packages—also referred to as probabilistic programming languages—automatically carry out analytical simplifications of known forms, and choose from a variety of MCMC methods the one that is most suited for the problem at hand. For many models, even fairly complex high-dimensional models, these software packages will be able to generate draws from the posterior $f(\theta|D)$ without tedious (manual) computation. As we have highlighted before, once these draws are available it is often simple to provide summaries of the posterior (i.e., expectations, variances), or interval estimates of parameters.

We will provide an example of working with the probabilistic programming language Stan (which is available for download at https://mc-stan.org). Stan is appealing for its simple syntax, and the fact that it interfaces very well with R: it is easy to use R for data processing and cleaning, call Stan from R to do inference, and subsequently use R again to create summaries and plots of the posterior draws provided by Stan.

After installing Stan following the instructions on the website for your specific device, you can install the rstan package from cran (using the usual install.packages() call) to start interfacing between R and Stan. Many users will only interact with Stan via R and thus they solely use the rstan interface. In a session in which you wish to use Stan, you simple instantiate the rstan package:

```
# Including the rstan package
library(rstan)
```

8.6.1   A Simple Bernoulli Model Using Stan

To explain the basic idea behind Stan we will provide a simple example for the
Bayesian analysis of data originating from a Bernoulli population. We will start with
creating some data:

```
# Generating 10 observations from a fair coin:
set.seed(12345)
n <- 10
y <- rbinom(n, 1, 0.5)
y
```

Using this seed we end up with seven successes and three failures.

Although we have already seen how to analyse this data using a conjugate beta
prior leading to a beta$(8, 4)$ posterior for p when using an uninformative prior, we
will focus on using Stan to analyze this same data. To do so, we follow the following
steps:

1. We write the model specification—both the sampling model and the priors—using
   Stan.
2. We fit the model using a call to the `stan()` function specifying our model and
   the data involved.
3. Finally, we extract the posterior draws from the fitted model and create summaries.

8.6.1.1   Specifying the Model

The following code specifies the Beta-Bernoulli model that we discussed earlier in
Stan:

```
write("// Stan model Bernoulli P data { // The input data
  int <lower = 1> N; // sample size
  int y[N]; // vector with data
}

parameters { // The parameter(s) of the model
  real<lower=0,upper=1> p;
}

model { // The model specification
  y ~ bernoulli(p);   // Sampling model
  p ~ beta(1,1);   // Prior
}",
"model1.stan")
```

The call to `write()` effectively writes the string that follows to the file
`model1.stan`. The code is relatively self-explanatory: all Stan needs to oper-
ate is a description of the data involved, a description of the parameters one wishes
to obtain posterior draws for, and the probabilistic models (i.e., the sample model
and the prior) involved in the analysis. Note that in Stan, comments are prefixed by

two slashes ("//"). From here, Stan will be able to do all the work, even if the models are fairly complex (and not conjugate).

8.6.1.2 Fitting the Model

After specifying the model, we can make a call to Stan, supplying both the model and the data:

```
fit <- stan(file = "model1.stan", data = list(N=n, y=y))
```

In the call to Stan we provide a string pointing to the file containing the model description, and we provide the data in a list that contains the names N and y that we used when specifying the model. Note that this is the absolute minimum we should provide to generate inferences: in any serious use of Stan one would like to control the behavior of the samplers involved and set (e.g.,) the number of posterior draws one would like to obtain. See the reference manual for rstan at https://cran. r-project.org/web/packages/rstan/rstan.pdf for more information.

After calling Stan using the default configuration we can simply print the fit object to get some initial results:

```
Chain 4:
> fit
Inference for Stan model: model1.
4 chains, each with iter=2000; warmup=1000; thin=1;
post-warmup draws per chain=1000, total post-warmup draws=4000.

      mean se_mean   sd   2.5%   25%   50%    75% 97.5% n_eff Rhat
p     0.66   0.00  0.13  0.38  0.57  0.67  0.76  0.89  1630    1
lp__ -8.17   0.02  0.74 -10.35 -8.35 -7.88 -7.69 -7.64  1906  1

Samples were drawn using NUTS(diag_e) at Tue Aug 18 11:45:30
    2020.
For each parameter, n_eff is a crude measure of effective sample
    size,
and Rhat is the potential scale reduction factor on split chains
    (at
convergence, Rhat=1).
```

The above shows that we ran four different MCMC chains, each generating 2,000 posterior draws (of which 1,000 are discarded due to the warmup=1,000 argument). The mean of the resulting 4,000 posterior draws is 0.66, which, given the seven successes and three failures (i.e., an MLE estimate of p of $\frac{7}{10}$ and a uniform prior seems reasonable.

8.6.1.3 Summarizing the Posterior

As we have seen above, it is custom in a Bayesian analysis to provide a summary of the posterior. The $\mathtt{fit}$ object we created above contains a total of $m = 4{,}000$ draws from the posterior $f(p|y)$. Using $\mathtt{rstan}$ these can easily be extracted and inspected:

```
> # Extract posterior
> posterior <- extract(fit)
>
> # Dimensions:
> dim(posterior$p)
[1] 4000
>
> # summarize:
> summary(posterior$p)
   Min. 1st Qu. Median   Mean 3rd Qu.   Max.
 0.2399  0.5714 0.6704 0.6587  0.7574 0.9643
```

A call to $\mathtt{plot(density(posterior\$p))}$ provides Fig. 8.3

The example above provided a first glance into the use of Stan for conducting Bayesian Data analysis. Although we demonstrated a very simple one-parameter case, Stan readily extends to much more complicated models: as long as the data, the model parameters, and the sampling model and priors are properly specified Stan will allow you to generate samples from the (often multidimensional) posterior. These draws can subsequently be summarized and/or plotted using R to compute the Bayesian analysis. Thus, the emergence of probabilistic computing languages such as Stan has by and large solved the computational challenges involved in Bayesian data analysis.

Fig. 8.3 Posterior PDF of p generated using Stan

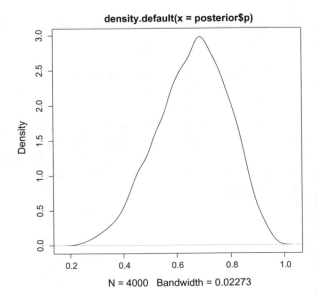

8.7   Bayesian and Frequentist Thinking Compared

In this last section we would like to highlight some of the differences between Bayesian and Frequentists methods. Many text books on Bayesian methods will start with such a discussion, often fiercely arguing that Bayesian methods are superior to Frequentist methods (see, for example, Gill 2014). This is not what we aim to do here; we simply aim to identify differences, and content that both methods have their merits and that students benefit from being familiar with both (and additionally, both methods equally utilize the theoretical concepts, such as random variables and distribution functions, that we introduced in earlier chapters).

A first striking difference between Bayesian and Frequentist thinking is of philosophical nature: Frequentists effectively assume that population values are—albeit unknown—fixed. Frequentists use probability theory, random variables, and distribution functions merely to deal with the randomness that is introduced by the (random) sampling process that selects units from the population. On the contrary, Bayesians use probability theory much more broadly; in a Bayesian framework probability is often considered simply an extension of Boolean logic (see, e.g., Jaynes 2003) to be able to deal with events that are not strictly true or false. This extension can be used to quantify the degree of belief one has regarding all kinds of events; values of population parameters being simply specific events in their own right. More formally, for a Frequentist $f(\theta|D)$ is philosophically meaningless: θ is not a random variable but a fixed quantity. For a Bayesian $f(\theta|D)$ is a natural quantity; it quantifies one's belief regarding θ after observing the data.

For a Frequentist, accepting θ as an random variable is not just philosophically problematic, it also introduces a need for the prior $f(\theta)$, which to a devoted Frequentist is often deemed subjective: we should not be adding "additional" information to our analysis. Data analysis should rely *solely* on the data, and there is thus no proper justification possible for the usage of priors. Conversely, Bayesians argue that any analysis is subjective, as even the choice of the sampling model—which plays a large role in both Bayesian and Frequentist analysis—is subjective. Thus, a statistical analysis never depends solely on the data and the analyst always makes a number of consequential choices. Making these choices and the assumptions involved in an analysis explicit, to a Bayesian, seems better than hiding them in the likelihood function and the estimation method used. Additionally, one can try to choose priors that are objective (see Sect. 8.5.1), thereby attempting to ensure that only the data affects the outcomes of the analysis. Bayesians often support their argument by demonstrating that for any Frequentist estimate a Bayesian prior and loss function can be conceived that leads to the exact same estimate.

We do not think the philosophical disputes raised above will easily be solved. However, they do not need to be resolved to utilize the applied benefits of both methods when conducting a data analysis. Taking a more applied view, a number of differences (and in some ways pros and cons) of Frequentist and Bayesian methods can be identified. First of all, effectively due to the use of priors, the Bayesian analysis framework makes a number of analyses possible that are hard (or sometimes even impossible) in a Frequentist framework. One example thereof includes dealing with relatively extreme datasets. Consider for example the following five observations from a Bernoulli population $(0, 0, 0, 0, 0)$. To a Frequentist the (MLE) estimate of the population p would be 0. This however seems too extreme: can we really conclude that a success is impossible based on five failures? A Bayesian analysis of this same data will—depending on the choice of prior and the loss function—lead to a point estimate that is small, but larger than 0, quantifying the intuition that successes are, albeit perhaps unlikely, still possible. Another example of an extreme dataset would be a single observation from a normal population: in a Frequentist framework it would be impossible to obtain MLE estimates for the mean and the variance in such a case as this would result in determining two unknowns based on a single data point, and no unique solution exists. Depending on the choice of prior, a Bayesian could work with a single data point and update the (joint) posterior for the mean and the variance.

At this point it is useful to note that Bayesian methods often, by virtue of the prior, add bias to an estimator (Gelman et al. 2014). Thus, in many cases, reporting the expected value (or the MAP) of the posterior will not provide an unbiased estimate of the population value. However, often, the increase in bias, especially when the prior is reasonable, leads to a decrease in MSE (Gelman et al. 2014).

Bayesian methods, however, are not always practically more feasible. We already highlighted the computational complexities that arise in many Bayesian analyses; sometimes a Frequentist view is easier to carry out than a Bayesian one (but sometimes this can be reversed). Additionally, one of the main benefits of Frequentist estimators are their known operating characteristics: for example, in a Frequentist framework—if the assumptions are valid—the type I and type II error rates of many hypothesis tests are known. This is due to the fact that error rates over repeated sampling are exactly what is of interest to a Frequentist: the sampling is what drives the uncertainty in the estimate to begin with. Thus, if your interest is in properties of estimators over repeated sampling, the Frequentist method readily provides answers. To find the same answers using Bayesian methods we often have to resort to elaborate simulations (Berry et al. 2010).

8.8 Conclusion

In this chapter we have introduced the Bayesian approach to statistical inference: For a Bayesian the uncertainty regarding parameter estimates does not stem from uncertainty over repeated sampling, but rather quantifies her/his degree of belief regarding the parameter. After specifying a prior belief, we can use Bayes' rule to compute the posterior belief and we can use this posterior distribution function of the parameters for subsequent decision-making. In this chapter we have tried to highlight the general approach, and we have provided a simple analytical example using conjugate priors. Next, we have highlighted how a Bayesian analysis allows us to make decisions: we can either use the posterior distribution function of the parameters given the data directly to make decisions, or we can use more formal approaches such as Bayesian decision theory or the Bayes factor. In both of these latter cases we only scratched the surface: we want to provide students with a look ahead to see what is coming in a more advance (Bayesian) statistics course, without providing all the details. We hope that the current chapter at least provides a good conceptual start for students who wish to study Bayesian methods in more detail.

Problems

8.1 In this assignment we want to explore Bayesian inference for the mean of a Normal population. In this case we assume the data to be Normally distributed, and we are interested in obtaining a posterior PDF for the mean μ after seeing the data. More specifically, we consider the data y to be i.i.d. $\mathcal{N}(\mu, \sigma^2)$. Thus our likelihood is $\Pr(y|\theta) = p(y|\mu, \sigma^2)$. A full Bayesian analysis would require a prior for μ and σ^2; however, to simplify the problem, we are going to assume that σ^2 is known. In this case a Normal prior on μ, that is, $\mu \sim \mathcal{N}(\mu_0, \tau_0^2)$, is conjugate and the posterior for μ is given by

$$p(\mu|y) = \mathcal{N}(\mu|\mu_1, \tau_1^2),$$

where we use μ_1 and τ_1^2 respectively to denote the posterior mean and variance. These are given by

$$\mu_1 = \frac{\frac{\mu_0}{\tau_0^2} + \frac{n\bar{y}}{\sigma^2}}{\frac{1}{\tau_0^2} + \frac{n}{\sigma^2}}$$

and

$$\tau_1^2 = \left(\frac{1}{\tau_0^2} + \frac{n}{\sigma^2}\right)^{-1},$$

where n is the number of observations in y and $\bar{y} = \frac{1}{n}\sum_{i=1}^{n} y_i$ is the sample mean. Given this information, write the following R code:

1. Generate data y containing $n = 50$ observations from a Normal distribution with $\mu = 10$ and $\sigma^2 = 2$.
2. Write a function that takes as arguments the data y and the prior mean and variance μ_0, τ_0^2, and returns the posterior mean and variance μ_1, τ_1^2.
3. Plot the density of the prior distribution function.
4. Add to the previous plot the density of the posterior distribution function in red. Furthermore, add a vertical grey line at $\bar{y}$.
5. Play around with different choices of μ_0, τ_0^2, and n, and see how these affect the posterior PDF $\Pr(\mu|y)$.
6. Consider the relative likelihood $\tilde{L}(\mu) = L(\mu)/L(\hat{\mu})$, where $L(\mu)$ is the regular likelihood as a function of μ and $L(\hat{\mu})$ is the likelihood evaluated at the maximum

likelihood estimate $\hat{\mu}$. Add to the plot of the prior and posterior distribution the relative likelihood $\tilde{L}(\mu)$ in grey.

8.2 Further examine, by writing the R code to simulate it, the example used in this chapter of (a) estimating a Bernoulli θ, and (b) using $\mathbb{E}(\theta|D)$ as a decision to obtain a point estimate of the parameter in more detail. Here are the steps:

1. Write a function to simulate the data. You want to be able to generate $s = \sum_{i=1}^{n} x_i$ for n i.i.d. Bernoulli(θ) observations.
2. Write a function that takes as arguments the parameters α and β for the beta() prior and the (summarized) data s and n and returns the posterior parameters. Set as default values for the prior $\alpha = 1$ and $\beta = 1$.
3. Write a loss function that takes as arguments the decision d and the true value of θ and computes the squared error loss.
4. Write a function that takes the two arguments of the Beta distribution (α and β) and returns its expected value.
5. Choose a "true" θ (for example 0.2) and set $n = 100$. Now repeat the following steps $m = 1000$ times:

    - Simulate $n = 100$ observations for the true θ you chose before.
    - Compute for 100 values in the range $d \in [0, 1]$ the loss. Note that you will have to store all these values in a $100 \times 1,000$ matrix (100 values for each of the $m = 1,000$ repetitions).
    - Compute the posterior mean given your current dataset. Store this in a vector of length $m = 1,000$.

6. Create a figure depicting the average loss over the $m = 1,000$ simulation runs as a function of d, and add a line depicting the average value of the posterior means over the 1,000 simulations. What do you see?

References

S.M. Berry, B.P. Carlin, J.J. Lee, P. Muller, *Bayesian Adaptive Methods for Clinical Trials* (CRC Press, Boca Raton, 2010)

A. Gelman, J.B. Carlin, H.S. Stern, D.B. Dunson, A. Vehtari, D.B. Rubin, *Bayesian Data Analysis*, vol. 2 (CRC Press, Boca Raton, 2014)

J. Gill, *Bayesian Methods: A Social and Behavioral Sciences Approach*, vol. 20 (CRC Press, Boca Raton, 2014)

X. Gu, J. Mulder, M. Deković, H. Hoijtink, Bayesian evaluation of inequality constrained hypotheses. Psychol. Methods **19**(4), 511 (2014)

L. Hespanhol, C.S. Vallio, L.M. Costa, B.T. Saragiotto, Understanding and interpreting confidence and credible intervals around effect estimates. Braz. J. Phys. Ther. **23**(4), 290–301 (2019)

E.T. Jaynes, *Probability Theory: The Logic of Science* (Cambridge University Press, Cambridge, 2003)

H. Jeffreys, An invariant form for the prior probability in estimation problems. Proc. R. Soc. London. Ser. A. Math. Phys. Sci. **186**(1007), 453–461 (1946)

R.E. Kass, A.E. Raftery, Bayes factors. J. Am. Stat. Assoc. **90**(430), 773–795 (1995)

S. Kullback, R.A. Leibler, On information and sufficiency. Ann. Math. Stat. **22**(1), 79–86 (1951)

C. Robert, *The Bayesian Choice: From Decision-Theoretic Foundations to Computational Implementation* (Springer Science & Business Media, 2007)

E.-J. Wagenmakers, J. Verhagen, A. Ly, D. Matzke, H. Steingroever, J.N. Rouder, R.D. Morey, The need for bayesian hypothesis testing in psychological science. *Psychological Science Under Scrutiny: Recent Challenges and Proposed Solutions* (2017), pp. 123–138

Correction to: Statistics for Data Scientists

Correction to:
M. Kaptein and E. van den Heuvel, *Statistics for Data Scientists*, **Undergraduate Topics in Computer Science,**
https://doi.org/10.1007/978-3-030-10531-0

The original version of the book was inadvertently published without incorporating the final corrections in the front matter, which have now been updated. The book and the chapter have been updated with the changes.

The updated version of the book can be found at
https://doi.org/10.1007/978-3-030-10531-0